# MULTIDIMENSIONAL MINIMIZING SPLINES
## *Theory and Applications*

*Grenoble Sciences* was created ten years ago by the Joseph Fourier University of Grenoble, France to evaluate and publish original scientific projects. These are selected with the help of referees whose reports remain confidential. Thereafter, a Reading Committee interacts with the authors as long as necessary to improve the quality of the manuscript. The purpose is to confer the *Grenoble Sciences* label to the best projects, from reference books to text-books.

Editorial Director, Sciences : Professor Jean BORNAREL

Contact : Nicole SAUVAL, tel : (33) 4 76 51 46 95
Grenoble.Sciences@ujf-grenoble.fr

The *Multidimensional Minimizing Splines - Theory and Applications* Reading Committee included the following members :

- Serge DUBUC, Professor, University of Montreal, Canada
- Marc ATTÉIA, Professor, Paul Sabatier University of Toulouse, France
- Jean GACHES, Assistant Professor, Paul Sabatier University of Toulouse, France
- Pierre-Jean LAURENT, Professor, Joseph Fourier University of Grenoble, France
- Marie-Laurence MAZURE, Professor, Joseph Fourier University of Grenoble, France

*Multidimensional Minimizing Splines* is a book for post-graduates. It follows a graduate level title : *Approximation hilbertienne - Splines, ondelettes, fractales* by M. ATTÉIA and J. GACHES, published in the Grenoble Sciences' French series (Collection Grenoble Sciences, EDP Sciences, 1999, 158 p.).

*Grenoble Sciences* is supported by the French Ministry of Higher Education, the Ministry of Research and the « Région Rhône-Alpes ».

# MULTIDIMENSIONAL MINIMIZING SPLINES
## *Theory and Applications*

*by*

**Rémi Arcangéli**

**María Cruz López de Silanes**

**Juan José Torrens**

**SPRINGER SCIENCE+BUSINESS MEDIA, LLC**

**Library of Congress Cataloging-in-Publication**

Arcangéli, Rémi / Cruz López de Silanes, María / José Torrens, Juan
***Multidimensional Minimizing Splines: Theory and Applications***
ISBN 978-1-4757-8843-3 ISBN 978-1-4020-7787-6 (eBook)
DOI 10.1007/978-1-4020-7787-6

---


Originally published by Kluwer Academic Publishers in 2004
Softcover reprint of the hardcover 1st edition 2004

*Printed on acid-free paper.*

# Contents

# *PREFACE*

Generally, we understand *spline functions* as

(a) regular functions whose definition is based on a criterion of minimization in a Hilbert space, or

(b) regular piecewise polynomial or rational functions.

The general definition of splines of type (a) has been formulated by M. Attéia [22, 23], while splines of type (b) appear mostly in American literature.

In dimension $n = 1$, splines have the properties (a) and (b) at the same time, but, when we want to generalize to the case $n > 1$, these two properties cannot be preserved simultaneously.

If we choose to keep the polynomial character of splines, we are led to the introduction of *B-splines* (cf. L. L. Schumaker [129], C. de Boor and K. Höllig [34], W. Dahmen and C. A. Michelli [50]). Related with this topic, we also cite the development of *box splines* (cf. C. de Boor, K. Höllig and S. Riemenschneider [35]) and *radial functions* (cf., for example, N. Dyn [58, 59]).

On the other hand, if a quadratic functional involving derivatives of order $m$ is minimized in a Hilbert space $E$, we obtain *minimizing splines*, in particular, the *$D^m$-splines.* So,

- when $E = \tilde{H}^s$ (a space close to the Sobolev space $H^s(\mathbb{R}^n)$), where $s$ denotes a real number such that $-m + n/2 < s < n/2$, we get the Duchon *$(m, s)$-splines*, also called *polyharmonic splines*, which will be studied in Part A of this book (Chapters I–IV); in particular, the $(m, 0)$-splines are the *$D^m$-splines over* $\mathbb{R}^n$ (or, for $n = 2$ and $m = 2$, *thin plate splines*);

- when $E = L^2(\Omega)$ and $\Omega$ is a bounded domain of $\mathbb{R}^n$, we derive the *$D^m$-splines over* $\Omega$, and we introduce the *discrete $D^m$-splines*, which are approximants of finite element type of the aforesaid $D^m$-splines over $\Omega$. Both are the subject of Part B (Chapters V–VII).

In dimension $n = 1$, the $D^m$-splines over $\Omega = (a, b)$ and the restrictions to $\Omega$ of the $D^m$-splines over $\mathbb{R}$ coincide and are just the classical polynomial splines.

The $B$-splines (concurrently with Bézier surfaces) appear well adapted to problems of geometric design, while the $D^m$-splines constitute a tool of approximation (interpolation or smoothing). Their roles have a different nature, although they may sometimes interfere.

Part C (Chapters VIII–XII) discusses several applications in Geophysics and Geology, arising in oil research, whose source was the French company Société Nationale Elf-Aquitaine. In each case, the problem is to construct a regular surface (i.e. of class

$C^1$ or $C^2$) by interpolating or smoothing a given data set, which can be formed by a finite number of scattered points, or distributed along curves or on an open set. The surface can eventually present discontinuities on a subset of its domain of definition (the fault problem in Geophysics). For every application, we propose a method of approximation based on the theory of discrete $D^m$-splines. Likewise, we establish a convergence result, which validates the method from the theoretic point of view. We also show several numerical examples, where the data are derived from known surfaces. Due to their academic nature, these examples allow an easy comparison between the surface obtained by approximation and the exact surface (the use of terrain data, which is often covered by industrial secret, leads to results which engineers appreciate according to diverse criteria, notably empiric). We point out that, in all the problems considered here, the theory of $D^m$-splines seems to provide a satisfying solution.

An important part of the book is devoted to convergence matters. Indeed, with only one exception, we establish a convergence result for every approximation method presented in these pages. All these results, which share a similar scheme, prove that the corresponding approximants converge to a given function assuming that the parameters on which the approximants depend behave in some suitable way. Owing to their asymptotic nature, neither these results, nor the error estimates stated in Parts A and B, can yield realistic estimates of the approximation error. Hence, our theoretical results do not allow us to deduce numerical values of the parameters in order to compute the approximant. However, they show that this computation makes sense. As pointed out by P. G. Ciarlet about the analysis of convergence questions, "*at least there is a 'negative' aspect that few people contest: presumably, a method should not be used in practice if it were impossible to mathematically prove its convergence...*" (cf. [45, p. 106]). So, the parameters involved in the definition of the approximant must be fixed at best, according to the data of the problem. Only the smoothing parameter $\varepsilon$, introduced in each smoothing method, can be chosen in terms of statistical considerations, for example, by means of the generalized cross validation method (cf. P. Craven and G. Wahba [49], C. Gu [76], G. Wahba [149]). We proceed in this way to deal with the numerical examples in Part C. We think that there is no contradiction between first using a deterministic method, which yields theoretical results, and then resorting to a statistical method, which permits us to get good estimates of the "optimal" value of $\varepsilon$ (in particular when data are noisy), provided the two following conditions are verified: on the one hand, the formulation of the discrete problem to be solved is justified by a convergence result, and, on the other hand, the discrete problem depends only on $\varepsilon$.

* * *

We would like to dedicate this book to the memory of the French mathematicians Noël Gastinel (1925–1984) and Charles Goulaouic (1938–1983). The first author of this book is deeply grateful to them for helping him during a difficult time in his career. The second and third authors did not have the good fortune to meet them, but they know the importance of the contribution to the fields of Mathematics of Computation by N. Gastinel, and Functional Analysis and Partial Differential Equations by C. Goulaouic.

We are indebted to Professor P. G. Ciarlet, who took an interest in some aspects of this work, and Professor P.-J. Laurent, whose support was decisive for its publication. We are additionally indebted to the Director, M. J. Bornarel, of Editions Grenoble Sciences and his staff, specially Mme. N. Sauval, for their aid and encouragement. We thank Ms. M. Michel for helping us translate our manuscript into English.

And for personal reasons, we wish to thank our families for their support, patience and understanding during the writing of this book.

* * *

This work has been partially supported by the Ministry of Science and Technology (Spain), Project I+D, BFM2000-1058.

* * *

The various surface fitting methods developed in Part C have been implemented using *Mathematica*[†] 4.1. The corresponding *Mathematica* notebooks and packages have not been reproduced in this book. However, they can be obtained free of charge by e-mailing the third author at the following address: `jjtorrens@unavarra.es`.

[†] *Mathematica* is a registered trademark of Wolfram Research, Inc.

# PRELIMINARIES

For any $a, b \in \mathbb{R}$, with $a \leq b$, we write $(a,b) = \{x \in \mathbb{R} \mid a < x < b\}$, $[a,b] = \{x \in \mathbb{R} \mid a \leq x \leq b\}$, $[a,b) = \{x \in \mathbb{R} \mid a \leq x < b\}$ and $(a,b] = \{x \in \mathbb{R} \mid a < x \leq b\}$. Similarly, we use the notations $[a,+\infty)$, $(a,+\infty)$, $(-\infty,b]$ and $(-\infty,b)$.

Let $n$ be an integer $\geq 1$. For any $x, y \in \mathbb{R}^n$, we denote by $x \cdot y$ the Euclidean scalar product of $x$ and $y$. Likewise, we denote by $|\cdot|$ (resp. by $\delta(\,\cdot\,,\,\cdot\,)$) the Euclidean norm (resp. the Euclidean distance) in $\mathbb{R}^n$. The closure, the interior and the boundary of any subset $B$ of $\mathbb{R}^n$ will be denoted by $\overline{B}$, $\overset{\circ}{B}$ and $\partial B$, respectively.

For any $\alpha = (\alpha_1, \ldots, \alpha_n) \in \mathbb{N}^n$, we write $|\alpha| = \alpha_1 + \cdots + \alpha_n$, $\alpha! = \alpha_1! \cdots \alpha_n!$, and $\partial^\alpha = \frac{\partial^{|\alpha|}}{\partial x_1^{\alpha_1} \cdots \partial x_n^{\alpha_n}}$. For any $x = (x_1, \ldots, x_n) \in \mathbb{R}^n$, we also write $x^\alpha = x_1^{\alpha_1} \cdots x_n^{\alpha_n}$.

We denote by $\mathcal{D} = \mathcal{D}(\mathbb{R}^n)$, $\mathcal{D}' = \mathcal{D}'(\mathbb{R}^n)$, $\mathcal{S} = \mathcal{S}(\mathbb{R}^n)$ and $\mathcal{S}' = \mathcal{S}'(\mathbb{R}^n)$ (resp., for any nonempty open set $\Omega$ in $\mathbb{R}^n$, by $\mathcal{D}(\Omega)$ and $\mathcal{D}'(\Omega)$) the classical spaces of functions or distributions over $\mathbb{R}^n$ (resp. over $\Omega$) (cf. L. Schwartz [131], F. Trèves [144]). Every time that this has a meaning, we denote by $T.\varphi$ the value of the distribution $T$ at the test function $\varphi$, and by $S * T$, the convolution product of the distributions $S$ and $T$. We denote by $\mathcal{F}v$ or $\hat{v}$ the Fourier transform of the element $v \in \mathcal{S}'$, defined on $\mathcal{S}$ by

$$\hat{v}(\xi) = \int_{\mathbb{R}^n} \exp(-2i\pi x \cdot \xi) v(x)\, dx.$$

Let $\Omega$ be a nonempty open set in $\mathbb{R}^n$. For any $l \in \mathbb{N}$, we denote by $H^l(\Omega)$ the Sobolev space of integer order $l$, endowed with the norm

$$\|v\|_{l,\Omega} = \left( \sum_{|\alpha| \leq l} \int_\Omega |\partial^\alpha v(x)|^2\, dx \right)^{1/2},$$

where the integral, like any other in this book, is taken in the Lebesgue sense. We shall also use the scalar semi-products

$$\forall j = 0, \ldots, l,\ (u,v)_{j,\Omega} = \sum_{|\alpha| = j} \int_\Omega \partial^\alpha u(x) \overline{\partial^\alpha v(x)}\, dx,$$

and the associated semi-norms

$$\forall j = 0, \ldots, l,\ |v|_{j,\Omega} = \left((v,v)_{j,\Omega}\right)^{1/2}.$$

If $r \in (0,+\infty) \setminus \mathbb{N}$, we write $r = l + \theta$, with $l \in \mathbb{N}$ and $\theta \in (0,1)$, and we denote by $H^r(\Omega)$ the Sobolev space of noninteger order $r$, endowed with the norm

$$\|v\|_{r,\Omega} = \left( \sum_{j=0}^{l} |v|_{j,\Omega}^2 + |v|_{r,\Omega}^2 \right)^{1/2},$$

where

$$|v|_{r,\Omega} = \left( \sum_{|\alpha|=l} \int_{\Omega\times\Omega} \frac{|\partial^\alpha v(x) - \partial^\alpha v(y)|^2}{|x-y|^{n+2\theta}}\, dx\, dy \right)^{1/2} .$$

For any $r \in \mathbb{R}$, the space $H^r(\mathbb{R}^n)$ can also be defined by the Fourier transform: $H^r(\mathbb{R}^n)$ is the space of distributions $v \in \mathcal{S}'$ whose Fourier transform $\hat{v}$ is a locally integrable function such that the function $\xi \mapsto (1+|\xi|^2)^{r/2}\,\hat{v}(\xi)$ belongs to $L^2(\mathbb{R}^n)$. For any nonempty open set $\Omega$ in $\mathbb{R}^n$, the space $H^r(\Omega)$ could then be defined from $H^r(\mathbb{R}^n)$ by restriction to $\Omega$. We clarify that, for any compact set $K \subset \mathbb{R}^n$ whose interior $\mathring{K}$ is nonempty and for any $r \geq 0$, for the sake of simplicity we shall write $H^r(K)$ instead of $H^r(\mathring{K})$, and $|\cdot|_{j,K}$ and $|\cdot|_{r,K}$, instead of $|\cdot|_{j,\mathring{K}}$ and $|\cdot|_{r,\mathring{K}}$. For the study of Sobolev spaces, we refer the reader to R. A. Adams [1], H. Brézis [40], C. Goulaouic [71], J. L. Lions and E. Magenes [94], and J. Nečas [109]. The main results appear in P. Grisvard [75].

Let $\Omega$ be a nonempty open subset of $\mathbb{R}^n$. We denote by $L^1_{\text{loc}}(\Omega)$ the space of locally integrable functions over $\Omega$. For any $r \in \mathbb{R}$, we denote by $H^r_{\text{loc}}(\Omega)$ the space of the distributions $v$ over $\Omega$ such that, for any $\varphi \in \mathcal{D}(\Omega)$, $\varphi v \in H^r(\Omega)$. For any $\mu \in \mathbb{N}$, we write $C^\mu(\Omega)$ for the Fréchet space of functions of class $C^\mu$ over $\Omega$. When $\Omega$ is bounded, we denote (cf. R. A. Adams [1]) by $C^\mu(\overline{\Omega})$ (resp. by $C^{\mu,\lambda}(\overline{\Omega})$, with $0 \leq \lambda \leq 1$) the Banach space of functions of class $C^\mu$ (resp. $C^{\mu,\lambda}$) over $\overline{\Omega}$, endowed with the usual norm

$$\|v\|_{C^\mu(\overline{\Omega})} = \max_{|\alpha|\leq\mu} \sup_{x\in\Omega} |\partial^\alpha v(x)|$$

$$\left( \text{resp. } \|v\|_{C^{\mu,\lambda}(\overline{\Omega})} = \|v\|_{C^\mu(\overline{\Omega})} + \max_{|\alpha|\leq\mu} \sup_{\substack{x,y\in\Omega \\ x\neq y}} \frac{|\partial^\alpha v(x) - \partial^\alpha v(y)|}{|x-y|^\lambda} \right).$$

By *an open set in* $\mathbb{R}^n$ *with a Lipschitz-continuous boundary* we shall understand a bounded, connected, nonempty, open subset of $\mathbb{R}^n$ with a Lipschitz-continuous boundary in the J. Nečas [109] sense. When the open set $\Omega$ has a Lipschitz-continuous boundary, the space $H^r(\Omega)$ verifies the following properties (cf. P. Grisvard [75]), where $\hookrightarrow$ and $\overset{c}{\subset}$ denote, respectively, the continuous and the compact injection:

$$\forall r > 0,\ \forall \mu \in \mathbb{N},\ r > n/2 + \mu,\ \exists \lambda \in (0,1],\ H^r(\Omega) \hookrightarrow C^{\mu,\lambda}(\overline{\Omega}) \tag{1}$$

(Sobolev's Hölder Imbedding Theorem);

$$\forall r, r' \in \mathbb{R},\ r > r',\ H^r(\Omega) \overset{c}{\subset} H^{r'}(\Omega) \tag{2}$$

$$\forall r > 0,\ \forall \mu \in \mathbb{N},\ r > n/2 + \mu,\ H^r(\Omega) \overset{c}{\subset} C^\mu(\overline{\Omega}) \tag{3}$$

(Rellich-Kondrašov Compact Imbedding Theorems);

$$\forall r > 0,\ \forall \mu \in \mathbb{N},\ r > n/2 + \mu,\ H^r(\Omega) \hookrightarrow C^\mu(\overline{\Omega}) \tag{4}$$

(Sobolev's Continuous Imbedding Theorem, which results from (1) or (3));

$$\left|\begin{array}{l}\text{for any } r > 0, \text{ there exists a linear continuous operator } P \text{ from } H^r(\Omega) \\ \text{into } H^r(\mathbb{R}^n) \text{ such that, for any } v \in H^r(\Omega),\ Pv|_\Omega = v\end{array}\right. \tag{5}$$

(Existence Theorem of an Extension Operator).

The *Local Sequential Weak Compactness Theorem* (cf., for example, K. Yosida [151, p. 126]) holds for reflexive Banach spaces: if $X$ is a reflexive Banach space and $(x_n)_{n\in\mathbb{N}}$ is any bounded sequence in $X$, then there exists a subsequence $(x_{n_p})_{p\in\mathbb{N}}$ extracted from $(x_n)$ and an element $x^* \in X$ such that, as $p \to +\infty$,

$$x_{n_p} \to x^*, \quad \text{weakly in } X.$$

In the sequel, we shall use the following derived result many times.

***Corollary* 1** – *Let $X$ be a reflexive Banach space, $E$, a subset of a metric space which admits at least one accumulation point, and $(x_i)_{i\in E}$, any bounded family in $X$. Then, for any accumulation point of $E$, there exists a sequence $(x_{i_l})_{l\in\mathbb{N}}$ in $X$, extracted from $(x_i)_{i\in E}$, and an element $x^* \in X$ such that, as $l \to +\infty$,*

$$x_{i_l} \to x^*, \quad \textit{weakly in } X.$$

**Proof** – Let $\bar{e}$ be an accumulation point of $E$. Hence, there exists a sequence $(i_l)_{l\in\mathbb{N}}$ in $E$ such that $i_l \to \bar{e}$ in the metric space in which $E$ is contained. Then, it suffices to apply the theorem to the sequence $(x_{i_l})_{l\in\mathbb{N}}$. □

Corollary 1 will be used in the following situation: the reflexive Banach space $X$ is a Hilbert space, the metric space is the space $\mathbb{R}^p$, with $1 \leq p \leq 4$, and the accumulation point $\bar{e}$ to be considered is a point, as the case may be, such as $0$, $(0,\varepsilon)$, $(0,0)$, $(0,\varepsilon,0)$, $(0,0,\varepsilon,0)$, $(0,0,0,0)$, where $\varepsilon$ denotes some positive number.

For any $l \in \mathbb{N}$, we denote by $P_l = P_l(\mathbb{R}^n)$ the space of polynomial functions defined over $\mathbb{R}^n$ of degree $\leq l$ with respect to the set of variables and, for any $l \in \mathbb{N}$ and for any nonempty connected open subset $\Omega$ in $\mathbb{R}^n$, by $P_l(\Omega)$, the space of restrictions to $\Omega$ of the functions in $P_l$.

We shall denote by $\mathfrak{M}$ the dimension of $P_{m-1}$, where $m$ is given by the condition $-m + n/2 < s < n/2$ (with $s \in \mathbb{R}$) in Chapters I–IV and by the condition $m > n/2$ in Chapters V–VII. Likewise, in Chapters II and III, the dimension of the space $P_k$, with $k$ defined by the relation (II–6.1), will be denoted by $\mathfrak{K}$.

For any $l \in \mathbb{N}$, we shall say (cf. P. G. Ciarlet [45]) that a set $A = \{a_1, \ldots, a_N\}$ of $N$ points of $\mathbb{R}^n$ is $P_l$-*unisolvent* if

$$\forall \{\alpha_1, \ldots, \alpha_N\} \subset \mathbb{R},\ \exists! \psi \in P_l,\ \forall i = 1, \ldots, N,\ \psi(a_i) = \alpha_i.$$

In this case, we shall also say that the $N$-tuple $a = (a_1, \ldots, a_N) \in (\mathbb{R}^n)^N$ is $P_l$-unisolvent. It is clear that a necessary condition for the set $A$ (or for the $N$-tuple $a$) to be $P_l$-unisolvent is that $\dim P_l = N$.

Finally, with the same letter $C$ we shall denote various strictly positive constants.

# PART A

## $(m, s)$-*SPLINES*

# INTRODUCTION

The theory of $(m, s)$-splines was introduced and developed by J. Duchon [53, 54, 55, 56, 57]. The study is situated in the framework of certain functional spaces of the "Sobolev type", the spaces $X^{m,s}$. These spaces are semi-Hilbert spaces, i.e. vector spaces endowed with a scalar semi-product and the associated semi-norm, and which are complete for this semi-norm.

The method elaborated by J. Duchon follows M. Attéia's [22, 23] ideas about abstract spline functions: it appeals to the notion of *reproducing kernel* of a semi-Hilbert space (cf. N. Aronszajn [21], L. Schwartz [130]). This method provides an explicit characterization of the $(m, s)$-splines from the knowledge of a reproducing kernel of the space $X^{m,s}$. A part of the results can be deduced directly from P.-J. Laurent [88, Chapter 4].

The point of view adopted in our work is different: we do not use reproducing kernels, but we treat the spaces $X^{m,s}$ as Hilbert spaces, by equipping them with suitable norms. This method allows us to obtain all the results in a relatively simple way and, on the other hand, it appears well adapted to establish error estimates.

Chapter I is devoted to the study of properties of the spaces $X^{m,s}$. In particular, we define (following J. Nečas [109]) a norm which makes $X^{m,s}$ a Hilbert space. We point out that the spaces introduced in this part are spaces of complex valued functions, due to the use of the Fourier transform.

Chapter II discusses interpolating splines, essentially for the model problem of Lagrange interpolation. By endowing $X^{m,s}$ with a norm associated with the interpolation conditions, we establish with no difficulty the existence, the uniqueness and two characterizations of the interpolating $(m, s)$-splines. To get them explicitly, we make use of the Fourier transform of functions of the Euclidean distance: the corresponding result is obtained by solving a problem of division in $\mathcal{S}'$. We then give some examples: thin plate splines, pseudo-polynomial splines, and splines defined by local mean values. We obtain the linear system which determines the interpolating spline and we verify that the matrix of the system is regular. Using a convenient norm for the space $X^{m,s}$, we show the convergence of the interpolating $(m, s)$-splines. Finally, by means of several technical results, we can establish estimates of the interpolation error for a function belonging to the Sobolev space $H^{m+s}(\Omega)$ in terms of the Hausdorff distance $d$ between $\overline{\Omega}$ and the set of data points.

Chapter III reconsiders for smoothing splines the questions which are treated in Chapter II for the interpolating splines: existence, uniqueness, characterization and computation. Likewise, we prove the convergence of the smoothing splines to an interpolating spline when the smoothing parameter $\varepsilon$ tends to 0. We then prove the convergence of the smoothing splines and we establish error estimates. Finally, we study the problem of convergence for noisy data.

Chapter IV gives a résumé of the $(m, l, s)$-splines, which are a generalization of the $(m, s)$-splines.

# *CHAPTER I*

# *THE SPACES $X^{m,s}$*

## *1. DEFINITION*

Let $n \in \mathbb{N}^*$. For any real number $s$, we write (cf. J. Peetre [116])

$$\widetilde{H}^s = \left\{ v \in \mathcal{S}' \,\middle|\, \hat{v} \in L^1_{\text{loc}}(\mathbb{R}^n),\ \int_{\mathbb{R}^n} |\xi|^{2s} |\hat{v}(\xi)|^2 \, d\xi < +\infty \right\}.$$

We have the following result.

***Proposition* 1.1** – *Suppose that $s < n/2$. Then, the space $\widetilde{H}^s$, endowed with the norm*

$$|v|_{0,s} = \left( \int_{\mathbb{R}^n} |\xi|^{2s} |\hat{v}(\xi)|^2 \, d\xi \right)^{1/2},$$

*is a Hilbert space, contained in $\mathcal{S}'$ with continuous injection.*

**Proof** – (J. Duchon). For any $f \in L^2(\mathbb{R}^n)$, the function $\xi \mapsto |\xi|^{-s} f(\xi)$ is locally integrable over $\mathbb{R}^n$, given that, for any compact $K \subset \mathbb{R}^n$,

$$\int_K |\xi|^{-s} |f(\xi)| \, d\xi \leq \left( \int_K |\xi|^{-2s} \, d\xi \right)^{1/2} \left( \int_K |f(\xi)|^2 \, d\xi \right)^{1/2}$$

and that $\xi \mapsto |\xi|^\lambda$ is locally integrable for $\lambda > -n$. Thus, this function defines a distribution, in fact, a tempered distribution, since

$$\forall \varphi \in \mathcal{S}, \left| \int_{\mathbb{R}^n} |\xi|^{-s} f(\xi) \varphi(\xi) \, d\xi \right| \leq \left( \int_{\mathbb{R}^n} |f(\xi)|^2 \, d\xi \right)^{1/2} \left( \int_{\mathbb{R}^n} |\xi|^{-2s} |\varphi(\xi)|^2 \, d\xi \right)^{1/2}$$

and

$$(\varphi \to 0 \text{ in } \mathcal{S}) \Longrightarrow \left( \int_{\mathbb{R}^n} |\xi|^{-2s} |\varphi(\xi)|^2 \, d\xi \to 0 \right).$$

As $\mathcal{F}$ is an automorphism of $\mathcal{S}'$, we deduce that the mapping $f \mapsto \mathcal{F}^{-1}(|\xi|^{-s} f)$ is an isometry from $L^2(\mathbb{R}^n)$ on $\widetilde{H}^s$, so $\widetilde{H}^s$ is a Hilbert space. On the other hand, if $f_j \to 0$ in $L^2(\mathbb{R}^n)$,

$$\forall \varphi \in \mathcal{S}, \int_{\mathbb{R}^n} |\xi|^{-s} f_j(\xi) \varphi(\xi) \, d\xi \to 0.$$

Hence, $|\xi|^{-s} f_j \to 0$ in $\mathcal{S}'$. It has thus been proved that $\widetilde{H}^s \hookrightarrow \mathcal{S}'$. □

For any $m \in \mathbb{N}^*$ and for any $s \in \mathbb{R}$, we shall write

$$X^{m,s} = \{ v \in \mathcal{D}' \mid \forall \alpha \in \mathbb{N}^n,\ |\alpha| = m,\ \partial^\alpha v \in \widetilde{H}^s \},$$

$$\forall u, v \in X^{m,s},\ (u,v)_{m,s} = \sum_{|\alpha|=m} \frac{m!}{\alpha!} \int_{\mathbb{R}^n} |\xi|^{2s} \mathcal{F}\partial^\alpha u(\xi) \overline{\mathcal{F}\partial^\alpha v(\xi)}\, d\xi,$$

and

$$\forall v \in X^{m,s},\ |v|_{m,s} = \left((v,v)_{m,s}\right)^{1/2}.$$

If $u \in X^{m,s}$ and $v \in X^{m,s}$ are real valued, then the scalar semi-product $(u,v)_{m,s}$ is also real and we have $|u+iv|^2_{m,s} = |u|^2_{m,s} + |v|^2_{m,s}$.

***Proposition* 1.2** – *Suppose that $s < n/2$. Then, the space $X^{m,s}$, endowed with the semi-norm $|\cdot|_{m,s}$, is a semi-Hilbert space (i.e. complete for this semi-norm), contained in $\mathcal{S}'$.*

**Proof** – (J. Duchon). Let $(u_j)_{j\in\mathbb{N}}$ be a Cauchy sequence in $(X^{m,s}, |\cdot|_{m,s})$. For any $\alpha \in \mathbb{N}^n$, with $|\alpha| = m$, $(\partial^\alpha u_j)$ is a Cauchy sequence in $\widetilde{H}^s$. Then, $(\partial^\alpha u_j)$ converges to $v_\alpha$ in $\widetilde{H}^s$ and in $\mathcal{S}'$. But, due to the continuity of the derivation in $\mathcal{S}'$,

$$\forall \alpha, \beta \in \mathbb{N}^n,\ |\alpha| = |\beta| = m,\ \partial^\beta v_\alpha = \partial^\alpha v_\beta.$$

In consequence (cf. L. Schwartz [131]), there exists $u \in \mathcal{D}'$ such that

$$\forall \alpha \in \mathbb{N}^n,\ |\alpha| = m,\ v_\alpha = \partial^\alpha u.$$

So there exists $u \in X^{m,s}$ such that $\lim_{j\to\infty} |u - u_j|_{m,s} = 0$ ($u$ is not unique, since $u_j \to u + P_{m-1}$). This implies that $(X^{m,s}, |\cdot|_{m,s})$ is complete.

All that remains is to show that $X^{m,s} \subset \mathcal{S}'$. Now, the distributions belonging to $X^{m,s}$ are tempered, due to the fact that every distribution $T$ whose first derivatives are tempered is also tempered. In fact (cf. L. Schwartz [131]), it is necessary and sufficient that the regularizations $t \mapsto (\varphi * T)(t)$, with $\varphi \in \mathcal{D}$, are slowly increasing (i.e. they are bounded by polynomials when $|t| \to +\infty$). By the Mean Value Theorem ($\varphi * T$ is of class $C^\infty$), we get

$$\forall t \in \mathbb{R}^n,\ |(\varphi * T)(t)| \le |(\varphi * T)(0)| + |t| \sup_{\theta\in[0,1]} \left( \sum_{j=1}^n \left| \frac{\partial(\varphi * T)}{\partial x_j}(\theta t) \right|^2 \right)^{1/2}.$$

But, for $j = 1, \dots, n$, $\dfrac{\partial(\varphi * T)}{\partial x_j} = \varphi * \dfrac{\partial T}{\partial x_j}$ is a slowly increasing function. Hence,

$$\exists C > 0,\ \exists k \in \mathbb{N},\ \forall t \in \mathbb{R}^n,\ |(\varphi * T)(t)| \le |(\varphi * T)(0)| + C(1 + |t|^2)^{k+1},$$

and the Proposition follows. □

The spaces $X^{m,s}$ are Beppo Levi spaces (cf. J. Deny and J. L. Lions [51] and, for $s = 0$, J. Nečas [109]). Since $\widetilde{H}^0 = L^2(\mathbb{R}^n)$, the space $X^{m,0}$, also denoted by $V_2^{(m)}(\mathbb{R}^n)$ or $D^{-m}L^2(\mathbb{R}^n)$, is just the space $\{ v \in \mathcal{D}' \mid \forall \alpha \in \mathbb{N}^n,\ |\alpha| = m,\ \partial^\alpha v \in L^2(\mathbb{R}^n) \}$.

## 2. *IMBEDDINGS AND NORMS*

***Proposition* 2.1** – *Suppose that $-m - n/2 < s$. Then $\mathcal{S} \subset X^{m,s}$.*

**Proof** – Let $v \in \mathcal{S}$. It suffices to verify that

$$\forall \alpha \in \mathbb{N}^n,\ |\alpha| = m,\ \int_{\mathbb{R}^n} |\xi|^{2s} |\mathcal{F}\partial^\alpha v(\xi)|^2 \, d\xi < +\infty.$$

But this is a simple consequence of the following inequality:

$$|\xi|^{2s} |\mathcal{F}\partial^\alpha v(\xi)|^2 \le (2\pi)^{2m} |\xi|^{2m+2s} |\hat{v}(\xi)|^2,$$

taking into account that the right-hand member is integrable, since $2m + 2s > -n$ and $\hat{v} \in \mathcal{S}$. □

For any nonempty open set $\Omega$ in $\mathbb{R}^n$, for any $m \in \mathbb{N}^*$ and for any $s \in \mathbb{R}$, we write $X^{m,s}_\Omega$ (resp. $\widetilde{H}^s_\Omega$, resp. $H^s_\Omega$) for the space of restrictions to $\Omega$ of the distributions belonging to $X^{m,s}$ (resp. to $\widetilde{H}^s$, resp. to $H^s(\mathbb{R}^n)$).

***Theorem* 2.1** – *Suppose that $-m - n/2 < s$. Then, for any nonempty bounded open set $\Omega$ in $\mathbb{R}^n$, we have $X^{m,s}_\Omega = H^{m+s}_\Omega$.*

**Proof** – (J. Duchon).

1) Let us prove that

$$\widetilde{H}^s_\Omega \subset H^s_\Omega. \tag{2.1}$$

Firstly, suppose that $s > 0$. Let us remark that, for any $s > 0$, there exists a ball $B \in \mathbb{R}^n$ such that

$$\forall \xi \in \mathbb{R}^n \setminus B,\ (1 + |\xi|^2)^s \le 2|\xi|^{2s}.$$

Let $u \in \widetilde{H}^s$. The function $\hat{u}$ can be written as $\hat{u} = v_1 + v_2$, with

$$v_1 = \begin{cases} 0, & \text{on } B, \\ \hat{u}, & \text{on } \mathbb{R}^n \setminus B. \end{cases}$$

The function $v_1$ verifies

$$\int_{\mathbb{R}^n} (1 + |\xi|^2)^s |v_1(\xi)|^2 \, d\xi = \int_{\mathbb{R}^n \setminus B} (1 + |\xi|^2)^s |\hat{u}(\xi)|^2 \, d\xi$$
$$\le 2 \int_{\mathbb{R}^n} |\xi|^{2s} |\hat{u}(\xi)|^2 \, d\xi < +\infty.$$

Hence, $\mathcal{F}^{-1} v_1 \in H^s(\mathbb{R}^n)$. As the distribution $v_2$ has a compact support, $\mathcal{F}^{-1} v_2$ is a function of class $C^\infty$ (cf. L. Schwartz [131]) and then, as $s > 0$, $\mathcal{F}^{-1} v_2 \in H^s_{\text{loc}}(\mathbb{R}^n)$. Thus, $u$ can be written as the sum of an element of $H^s(\mathbb{R}^n)$ and an element of $H^s_{\text{loc}}(\mathbb{R}^n)$. We deduce that $\widetilde{H}^s_\Omega \subset H^s_\Omega$ if $s > 0$.

When $s \le 0$, we have $\widetilde{H}^s \subset H^s(\mathbb{R}^n)$, since, for any $\xi \in \mathbb{R}^n$, $(1+|\xi|^2)^s \le |\xi|^{2s}$. Therefore, (2.1) holds.

2) Let $u \in H_\Omega^{m+s}$. By definition, $u$ is the restriction to $\Omega$ of an element $\tilde{u} \in H^{m+s}(\mathbb{R}^n)$ that we can always suppose to have a compact support. Then $\mathcal{F}\tilde{u}$ is a function of class $C^\infty$. On the other hand, for $2m+2s > -n$, the function $\xi \mapsto |\xi|^{2m+2s}$ is locally integrable and there exists a ball $B \subset \mathbb{R}^n$ such that

$$\forall \xi \in \mathbb{R}^n \setminus B,\ |\xi|^{2m+2s} \le 2(1+|\xi|^2)^{m+s}.$$

So, we deduce that, for any $\alpha \in \mathbb{N}^n$, with $|\alpha| = m$,

$$\int_{\mathbb{R}^n} |\xi|^{2s} |\mathcal{F}\partial^\alpha \tilde{u}(\xi)|^2 \, d\xi \le (2\pi)^{2m} \int_B |\xi|^{2m+2s} |\mathcal{F}\tilde{u}(\xi)|^2 \, d\xi$$
$$+ 2(2\pi)^{2m} \int_{\mathbb{R}^n} (1+|\xi|^2)^{m+s} |\mathcal{F}\tilde{u}(\xi)|^2 \, d\xi < +\infty,$$

and, in consequence, $u \in X_\Omega^{m,s}$. Hence, $H_\Omega^{m+s} \subset X_\Omega^{m,s}$.

Conversely, let $u \in X^{m,s}$ and let $\Omega'$ be a bounded open set in $\mathbb{R}^n$ which contains $\overline{\Omega}$. For any $\alpha \in \mathbb{N}^n$, with $|\alpha| = m$, $\partial^\alpha u|_{\Omega'}$ belongs to $\widetilde{H}_{\Omega'}^s$ and then, by (2.1), it belongs to $H_{\Omega'}^s$. Therefore, for any $\alpha \in \mathbb{N}^n$, with $|\alpha| = m+1$, $\partial^\alpha u|_{\Omega'}$ belongs to $H_{\Omega'}^{s-1}$. Let us write

$$P = \Delta^l,$$

where $\Delta$ is the Laplacian operator, and $l \in \mathbb{N}^*$ is such that $m = 2l$ if $m$ is even and $m = 2l-1$ if $m$ is odd. We have

$$(Pu)|_{\Omega'} \in \begin{cases} H_{\Omega'}^s, & m = 2l, \\ H_{\Omega'}^{s-1}, & m = 2l-1. \end{cases}$$

Now, $P$ is an elliptic operator of order $2l$ with constant coefficients. It follows from Friedrichs' Theorem (cf. J. L. Lions and E. Magenes [94, Theorem 3.2, p. 138]) that, in both cases, $u|_{\Omega'} \in H_{\mathrm{loc}}^{m+s}(\Omega')$. We conclude that $u|_\Omega \in H_\Omega^{m+s}$.

Therefore, $X_\Omega^{m,s} \subset H_\Omega^{m+s}$ and the Theorem follows. □

Hereafter, we assume that $m \in \mathbb{N}^*$ and $s \in \mathbb{R}$ satisfy the hypothesis

$$-m + \frac{n}{2} < s < \frac{n}{2}, \tag{2.2}$$

(which obviously implies the condition $-m - \frac{n}{2} < s < \frac{n}{2}$).

The following result will allow us to handle the spaces $X^{m,s}$ as Hilbert spaces.

***Theorem 2.2*** – *Suppose that (2.2) holds. Let $\Omega^*$ be a bounded, connected, nonempty, open subset of $\mathbb{R}^n$. Then, the space $X^{m,s}$, endowed with the norm*

$$\|v\|_{m,s}^{\Omega^*} = \left( \int_{\Omega^*} |v(x)|^2 \, dx + |v|_{m,s}^2 \right)^{1/2},$$

*is a Hilbert space, whose topology is independent of $\Omega^*$.*

**Proof** –

1) By Theorem 2.1 and (2.2), $X^{m,s}_{\Omega^*}$ is contained in $L^2(\Omega^*)$, so the mapping $\|\cdot\|^{\Omega^*}_{m,s}$ is well defined. It is clear that $\|\cdot\|^{\Omega^*}_{m,s}$ is a norm on $X^{m,s}$. We shall prove that $X^{m,s}$ is complete for this norm.

Let $(u_j)_{j\in\mathbb{N}}$ be a Cauchy sequence in $(X^{m,s}, \|\cdot\|^{\Omega^*}_{m,s})$. From the definition of the norm it follows that there exists $\tilde{u} \in X^{m,s}$ such that $\lim_{j\to+\infty}|u_j - \tilde{u}|_{m,s} = 0$ (since $X^{m,s}$ is complete for the semi-norm $|\cdot|_{m,s}$). As $\widetilde{H}^s$ is contained in $\mathcal{S}'$ (cf. Proposition 1.1), we have

$$\forall\alpha \in \mathbb{N}^n,\ |\alpha| = m,\ \partial^\alpha(u_j|_{\Omega^*}) \to \partial^\alpha(\tilde{u}|_{\Omega^*}) \text{ in } \mathcal{D}'(\Omega^*).$$

We also deduce that there exists $u^* \in L^2(\Omega^*)$ such that

$$\forall\alpha \in \mathbb{N}^n,\ |\alpha| = m,\ \partial^\alpha(u_j|_{\Omega^*}) \to \partial^\alpha(u^*) \text{ in } \mathcal{D}'(\Omega^*).$$

Hence, there exists $\psi \in P_{m-1}$ such that $u^* = (\tilde{u} + \psi)|_{\Omega^*}$ and, in consequence, $\lim_{j\to+\infty} u_j = \tilde{u} + \psi$. Thus, the sequence $(u_j)_{j\in\mathbb{N}}$ is convergent and, therefore, the space $(X^{m,s}, \|\cdot\|^{\Omega^*}_{m,s})$ is complete.

2) Let $\Omega^*$ and $\Omega^{**}$ be two bounded, connected, nonempty, open subsets of $\mathbb{R}^n$, and let $\|\cdot\|^{\Omega^*}_{m,s}$ and $\|\cdot\|^{\Omega^{**}}_{m,s}$ be the corresponding norms. There exists a bounded, connected, open set $\Omega$ which contains $\Omega^*$ and $\Omega^{**}$. As $\Omega^* \subset \Omega$, the canonical injection from $(X^{m,s}, \|\cdot\|^{\Omega}_{m,s})$ to $(X^{m,s}, \|\cdot\|^{\Omega^*}_{m,s})$ is continuous and then, by the Banach Isomorphism Theorem (cf. H. Brézis [40, Corollary II.6]), it is bicontinuous. Thus, the norms $\|\cdot\|^{\Omega}_{m,s}$ and $\|\cdot\|^{\Omega^*}_{m,s}$ are equivalent. Likewise, the norms $\|\cdot\|^{\Omega}_{m,s}$ and $\|\cdot\|^{\Omega^{**}}_{m,s}$ are equivalent. Therefore, the topology in $X^{m,s}$ is independent of $\Omega^*$. □

From now on, if nothing else is specified, we shall assume that $X^{m,s}$ is endowed with a norm $\|\cdot\|^{\Omega^*}_{m,s}$, which we shall simply write as $\|\cdot\|_{m,s}$, without making any reference to a particular open set $\Omega^*$ (since the norm chosen over $X^{m,s}$ does not play a role by itself in the theory of $(m,s)$-splines, it can be replaced by an equivalent norm).

***Corollary* 2.1** – *Suppose that* (2.2) *holds. Let $\Omega$ be an open subset of $\mathbb{R}^n$ with a Lipschitz-continuous boundary (cf. Preliminaries). Then, the operator $R_\Omega$ of restriction to $\Omega$ is linear and continuous from $X^{m,s}$ onto $H^{m+s}(\Omega)$.*

**Proof** – Taking into account that $H^{m+s}_\Omega = H^{m+s}(\Omega)$ when $\Omega$ has a Lipschitz-continuous boundary, it follows from Theorem 2.1 that $R_\Omega X^{m,s} = H^{m+s}(\Omega)$.

Let $(u_j)_{j\in\mathbb{N}}$ be a sequence in $X^{m,s}$ such that

$$\exists u \in X^{m,s},\ u_j \to u \text{ in } X^{m,s}, \tag{2.3}$$

$$\exists v \in H^{m+s}(\Omega),\ u_j|_\Omega \to v \text{ in } H^{m+s}(\Omega). \tag{2.4}$$

By (2.3) and Theorem 2.2, we have

$$u_j|_\Omega \to u|_\Omega \text{ in } L^2(\Omega).$$

On the other hand, from (2.4) and (2.2), we deduce that

$$u_j|_\Omega \to v \text{ in } L^2(\Omega).$$

Then, the Closed Graph Theorem (cf. H. Brézis [40, Theorem II.7]) proves that $R_\Omega$ is continuous. □

**Remark 2.1** – If the hypothesis (2.2) is replaced by the condition $-m \leq s < \frac{n}{2}$, Theorem 2.2 and Corollary 2.1 remain true. □

***Corollary* 2.2** – *Suppose that (2.2) holds. Then, $X^{m,s}$ is contained in $C^0(\mathbb{R}^n)$ with continuous injection.*

**Proof** – For any $R > 0$, let $\Omega_R$ be the open ball of centre 0 and radius $R$ in $\mathbb{R}^n$. Taking into account that, as $m + s > \frac{n}{2}$, $H^{m+s}(\Omega_R)$ is continuously imbedded into $C^0(\overline{\Omega}_R)$ (cf. (4) in Preliminaries), it follows from Corollary 2.1 that every function $v \in X^{m,s}$ is continuous over $\mathbb{R}^n$. On the other hand, for any compact $K$ of $\mathbb{R}^n$, there exists $R = R(K) > 0$ such that $\Omega_R \supset K$. Then, there exists $C(K) > 0$ such that

$$\forall v \in X^{m,s}, \ \sup_{x \in K} |v(x)| \leq C(K) \|v\|_{m+s,\Omega_R}.$$

Therefore, we deduce that $X^{m,s}$ is contained in $C^0(\mathbb{R}^n)$ with continuous injection. □

We finish this section with two important results for the study of spline functions. Let us denote by $\mathfrak{M}$ the dimension of $P_{m-1}$, with $m$ given by (2.2).

***Proposition* 2.2** – *Suppose that (2.2) holds. Let $E$ be a finite subset of $\mathbb{R}^n$ and let $[\![ \cdot ]\!]_{E,m,s}$ be the mapping defined by*

$$\forall v \in X^{m,s}, \ [\![v]\!]_{E,m,s} = \left( \sum_{a \in E} |v(a)|^2 + |v|_{m,s}^2 \right)^{1/2}. \tag{2.5}$$

*Then, if $E$ contains a $P_{m-1}$-unisolvent subset, $[\![ \cdot ]\!]_{E,m,s}$ is a Hilbertian norm on $X^{m,s}$ equivalent to the norm $\| \cdot \|_{m,s}$. Conversely, if $[\![ \cdot ]\!]_{E,m,s}$ is a norm on $X^{m,s}$, the set $E$ contains a $P_{m-1}$-unisolvent subset and $[\![ \cdot ]\!]_{E,m,s}$ is equivalent to $\| \cdot \|_{m,s}$.*

**Proof** – Let $N = \operatorname{card} E$.

1) Suppose that $E$ contains a $P_{m-1}$-unisolvent subset.

a) It is clear that $[\![ \cdot ]\!]_{E,m,s}$ is a norm on $X^{m,s}$ associated with a scalar product. On the other hand, it is easily seen, taking Corollary 2.2 into account, that it is sufficient to prove the result when $E$ is a $P_{m-1}$-unisolvent set. This is what we shall suppose in the following points b) and c).

b) From Corollary 2.2, it follows that there exists a constant $C(E)$ such that

$$\forall v \in X^{m,s}, \ \forall a \in E, \ |v(a)| \leq C(E) \|v\|_{m,s},$$

and hence

$$\forall v \in X^{m,s}, \ [\![v]\!]_{E,m,s} \leq \left(1 + \mathfrak{M}(C(E))^2\right)^{1/2} \|v\|_{m,s}.$$

c) Let $(u_j)_{j\in\mathbb{N}}$ be a Cauchy sequence in $(X^{m,s}, [\![\cdot]\!]_{E,m,s})$. Then

$$\exists \tilde{u} \in X^{m,s},\ \lim_{j\to+\infty} |u_j - \tilde{u}|_{m,s} = 0,$$

and

$$\forall a \in E,\ \exists \eta_a \in \mathbb{C},\ \lim_{j\to+\infty} u_j(a) = \eta_a.$$

As $E$ is $P_{m-1}$-unisolvent, there exists $\psi \in P_{m-1}$ such that

$$\forall a \in E,\ (\tilde{u} + \psi)(a) = \eta_a.$$

In consequence, $\lim_{j\to+\infty} u_j = \tilde{u} + \psi$. Thus, the sequence $(u_j)_{j\in\mathbb{N}}$ is convergent and, therefore, the space $(X^{m,s}, [\![\cdot]\!]_{E,m,s})$ is complete.

d) The equivalence of norms is then a consequence of points b) and c) and Banach's Isomorphism Theorem (cf. H. Brézis [40, Corollary II.6]).

2) Suppose that $[\![\cdot]\!]_{E,m,s}$ is a norm on $X^{m,s}$. Let $\psi$ be any element of $P_{m-1}$ such that

$$\forall a \in E,\ \psi(a) = 0.$$

We have $[\![\psi]\!]_{E,m,s} = 0$, and hence $\psi = 0$. This proves that the linear mapping $\Phi : \psi \in P_{m-1} \mapsto (\psi(a))_{a\in E} \in \mathbb{R}^N$ is an injection. Then, $N \geq \mathfrak{M}$ and $\operatorname{rank}\Phi = \mathfrak{M}$. We deduce that $E$ contains a $P_{m-1}$-unisolvent subset. The equivalence of norms results from point 1). □

**Proposition 2.3** – *Suppose that* (2.2) *holds. Let $\Omega$ be an open subset of $\mathbb{R}^n$ with a Lipschitz-continuous boundary (cf. Preliminaries) and let $B_0 = \{b_{01}, \ldots, b_{0\mathfrak{M}}\}$ be a $P_{m-1}$-unisolvent subset of $\overline{\Omega}$. For any $r > 0$, we denote by $\mathcal{B}_r$ the family of all subsets $B = \{b_1, \ldots, b_{\mathfrak{M}}\}$ of $\overline{\Omega}$ which satisfy the following condition*

$$\forall j = 1, \ldots, \mathfrak{M},\ |b_j - b_{0j}| \leq r. \tag{2.6}$$

*Then, there exists $r_0 > 0$ such that the family $\mathcal{B}_{r_0}$ is formed by $P_{m-1}$-unisolvent subsets and the mapping $[\![\cdot]\!]_{B,m,s}$, defined for any subset $B = \{b_1, \ldots, b_{\mathfrak{M}}\}$ of $\overline{\Omega}$ by*

$$\forall v \in X^{m,s},\ [\![v]\!]_{B,m,s} = \left(\sum_{j=1}^{\mathfrak{M}} |v(b_j)|^2 + |v|_{m,s}^2\right)^{1/2},$$

*is, for every $B \in \mathcal{B}_{r_0}$, a norm on $X^{m,s}$, uniformly equivalent over $\mathcal{B}_{r_0}$ to the norm $\|\cdot\|_{m,s}$.*

**Proof** –

1) From Corollary 2.2, it follows that

$$\exists C > 0,\ \forall r > 0,\ \forall B \in \mathcal{B}_r,\ \forall v \in X^{m,s},\ [\![v]\!]_{B,m,s} \leq C\|v\|_{m,s}.$$

2) For any $v \in X^{m,s}$ and for any $B \in \mathcal{B}_r$, we obviously have

$$\frac{1}{2}\sum_{j=1}^{\mathfrak{M}} |v(b_{0j})|^2 \leq \sum_{j=1}^{\mathfrak{M}} |v(b_{0j}) - v(b_j)|^2 + \sum_{j=1}^{\mathfrak{M}} |v(b_j)|^2.$$

From Sobolev's Hölder Imbedding Theorem for the space $H^{m+s}(\Omega)$ (cf. (1) in Preliminaries), (2.6) and Corollary 2.1, we deduce that

$$\forall \gamma > 0,\ \exists r > 0,\ \forall B \in \mathcal{B}_r,\ \forall v \in X^{m,s},\ \sum_{j=1}^{\mathfrak{M}} |v(b_{0j}) - v(b_j)|^2 \leq \gamma^2 \|v\|_{m,s}^2.$$

Hence,

$$\begin{aligned}&\forall \gamma > 0,\ \exists r > 0,\ \forall B \in \mathcal{B}_r,\ \forall v \in X^{m,s},\\ &\qquad \frac{1}{2}\sum_{j=1}^{\mathfrak{M}} |v(b_{0j})|^2 + |v|_{m,s}^2 - \gamma^2 \|v\|_{m,s}^2 \leq [\![v]\!]_{B,m,s}^2.\end{aligned}$$

Now, by Proposition 2.2, the mapping $v \mapsto \left(\frac{1}{2}\sum_{j=1}^{\mathfrak{M}} |v(b_{0j})|^2 + |v|_{m,s}^2\right)^{1/2}$ is a norm on $X^{m,s}$ which is equivalent to the norm $\|\cdot\|_{m,s}$. We deduce that there exists $C' > 0$ such that

$$\forall \gamma > 0,\ \exists r = r(\gamma) > 0,\ \forall B \in \mathcal{B}_r,\ \forall v \in X^{m,s},\ (C'^2 - \gamma^2)\|v\|_{m,s}^2 \leq [\![v]\!]_{B,m,s}^2.$$

We choose any $\gamma_0 \in (0, C')$ and then we set $r_0 = r(\gamma_0)$. It follows that, for any $B \in \mathcal{B}_{r_0}$, $[\![\cdot]\!]_{B,m,s}$ is a norm on $X^{m,s}$ and that $[\![\cdot]\!]_{B,m,s}$ is uniformly equivalent over $\mathcal{B}_{r_0}$ to the norm $\|\cdot\|_{m,s}$.

3) Let $B \in \mathcal{B}_{r_0}$. For any $\psi \in P_{m-1}$ such that

$$\forall a \in B,\ \psi(a) = 0,$$

we have $[\![\psi]\!]_{B,m,s} = 0$, and hence $\psi = 0$. We conclude that any $B \in \mathcal{B}_{r_0}$ is $P_{m-1}$-unisolvent. □

# CHAPTER II

# INTERPOLATING $(m, s)$-SPLINES

We restrict ourselves to the study of the model problem of *Lagrange interpolation*. The case of *Hermite interpolation* can be treated in the same way. The most general case where the interpolation data are the values of distributions with compact support is studied in Section 3 (with the example of the local mean value conditions).

Throughout this chapter, except in Section 3, we shall suppose, without mentioning it, that $m$, $n$ and $s$ are, respectively, two positive integers and a real number which satisfy (I–2.2). Hence, $s < n/2 < m+s$. Let us observe, in particular, that $m+s > 0$.

## 1. DEFINITION AND FIRST PROPERTIES

Suppose we are given an ordered set $A$ of $N$ distinct points of $\mathbb{R}^n$ which contains a $P_{m-1}$-unisolvent subset. The set $A$ constitutes the set of points of Lagrange interpolation in the space $X^{m,s}$ (it is known, from Corollary I–2.2, that the respective interpolation conditions make sense).

We denote by $\rho \in \mathcal{L}(X^{m,s}, \mathbb{C}^N)$ the operator defined by

$$\rho v = \big(v(a)\big)_{a \in A},$$

whose continuity is a consequence of Corollary I–2.2.

Let $\beta \in \mathbb{C}^N$. We consider the affine linear variety

$$\mathcal{K} = \{\, v \in X^{m,s} \mid \rho v = \beta \,\}$$

and also the associated vector subspace

$$\mathcal{K}_0 = \{\, v \in X^{m,s} \mid \rho v = 0 \,\}.$$

Then we call *interpolating* $(m, s)$*-spline relative to* $A$ *and* $\beta$ any solution, if any exists, of the problem: find $\sigma$ solution of

$$\begin{cases} \sigma \in \mathcal{K}, \\ \forall v \in \mathcal{K},\ |\sigma|_{m,s} \le |v|_{m,s}. \end{cases} \tag{1.1}$$

***Theorem* 1.1** – *Problem* (1.1) *has a unique solution* $\sigma$.

**Proof** – By a corollary of Urysohn's Theorem (cf. L. Hörmander [80, Theorem 1.2.2, p. 4]), one can find real functions $\varphi_a \in \mathcal{D}$ such that

$$\forall a, b \in A,\ \varphi_a(b) = \begin{cases} 1, & b = a, \\ 0, & b \ne a. \end{cases}$$

Hence, for any $\beta = (\beta_a)_{a\in A} \in \mathbb{C}^N$, the function $\sum_{a\in A} \beta_a \varphi_a$ belongs to $\mathcal{K}$. It is then clear that $\mathcal{K}$ is a nonempty, closed, convex subset of $X^{m,s}$.

Now, using (I–2.5) with $E = A$, we derive from Proposition I–2.2 that (1.1) is equivalent to

$$\begin{cases} \sigma \in \mathcal{K}, \\ \forall v \in \mathcal{K},\ [\![\sigma]\!]_{A,m,s} \le [\![v]\!]_{A,m,s}. \end{cases}$$

We deduce that (1.1) admits a unique solution, namely the element of minimal norm $[\![\cdot]\!]_{A,m,s}$ in $\mathcal{K}$. □

When $\beta \in \mathbb{R}^N$, $\sigma$ is *real valued.* To see this, let us observe that, if $\sigma = \sigma_1 + i\,\sigma_2$, then $\sigma_1 \in \mathcal{K}$ and $|\sigma_1|^2_{m,s} \le |\sigma_1|^2_{m,s} + |\sigma_2|^2_{m,s} = |\sigma|^2_{m,s}$. Hence $\sigma = \sigma_1$ and $\sigma_2 = 0$.

***Proposition* 1.1** – *The solution $\sigma$ of problem* (1.1) *is characterized by*

$$\begin{cases} \sigma \in \mathcal{K}, \\ \forall w \in \mathcal{K}_0,\ (\sigma, w)_{m,s} = 0. \end{cases} \tag{1.2}$$

**Proof** – The element $\sigma$ is the projection (for the distance associated with the norm $[\![\cdot]\!]_{A,m,s}$) of the origin over the complete, convex and nonempty subset $\mathcal{K}$ of the pre-Hilbert space $X^{m,s}$. Then, $\sigma$ is characterized by

$$\forall v \in \mathcal{K},\ \Re[\![0 - \sigma, v - \sigma]\!]_{A,m,s} \le 0,$$

where the symbol $\Re$ denotes the real part and $[\![\cdot\,,\,\cdot]\!]_{A,m,s}$ is the scalar product associated with the norm $[\![\cdot]\!]_{A,m,s}$. The Proposition is then a simple consequence. □

***Proposition* 1.2** – *There exists one and only one pair $(\sigma, \lambda) \in X^{m,s} \times \mathbb{C}^N$, with $\lambda = (\lambda_a)_{a\in A}$, such that*

$$\begin{cases} \sigma \in \mathcal{K}, \\ \forall v \in X^{m,s},\ (\sigma, v)_{m,s} = \sum\limits_{a\in A} \lambda_a \overline{v(a)}, \end{cases} \tag{1.3}$$

*where $\sigma$ is just the solution of problem* (1.1).

**Proof** – If $(\sigma, \lambda)$ is a solution of (1.3), then $\sigma \in \mathcal{K}$ and, for any $w \in \mathcal{K}_0$, $(\sigma, w)_{m,s} = 0$. Hence, $\sigma$ is the solution of (1.1) and $\sigma$ is unique. Now, if $(\sigma, \lambda')$ and $(\sigma, \lambda'')$ are two solutions of (1.3), with $\lambda' = (\lambda'_a)_{a\in A}$ and $\lambda'' = (\lambda''_a)_{a\in A}$, we deduce that, for any $v \in X^{m,s}$, $\sum_{a\in A}(\lambda'_a - \lambda''_a)\overline{v(a)} = 0$, which implies that $\lambda' = \lambda''$. Therefore, there exists, at most, one solution of (1.3).

On the other hand, for any $v \in X^{m,s}$, the function $w = v - \sum_{a\in A} v(a)\varphi_a$, where $\varphi_a$ are the functions introduced in the proof of Theorem 1.1, belongs to $\mathcal{K}_0$. Then, (1.3) results from (1.2) if we take, for any $a \in A$, $\lambda_a = (\sigma, \varphi_a)_{m,s}$ (we note that, if $\sigma$ is real valued, then any $\lambda_a$ is real). Hence, $(\sigma, \lambda)$, with $\lambda = (\lambda_a)_{a\in A}$, is a solution of (1.3). This completes the proof. □

The vector $-2\lambda$, where $\lambda = (\lambda_a)_{a\in A}$ is the vector introduced in (1.3), is just the *Lagrange multiplier* of problem (1.1).

## 2. THE SPACE OF $(m,s)$-SPLINE FUNCTIONS. EXAMPLES

We write

$$S = \{ u \in X^{m,s} \mid \forall w \in \mathcal{K}_0,\ (u,w)_{m,s} = 0 \}. \tag{2.1}$$

***Proposition* 2.1** – *The set $S$ is a subspace of dimension $N$ of $X^{m,s}$. Moreover, the restriction $\rho_\perp$ of $\rho$ to $S$ is an isomorphism from $S$ onto $\mathbb{C}^N$ and, for any $\beta \in \mathbb{C}^N$, $\rho_\perp^{-1}(\beta)$ is just the interpolating $(m,s)$-spline relative to $A$ and $\beta$.*

**Proof** – The set $S$ is evidently a subspace (the orthogonal complement of $\mathcal{K}_0$ in $(X^{m,s}, [\![ \cdot ]\!]_{A,m,s})$). By Proposition 1.1, we have

$$\forall \beta \in \mathbb{C}^N,\ \exists ! \sigma \in S,\ \rho\sigma = \beta.$$

Thus, we deduce that $\rho_\perp$ is a (linear) bijection from $S$ to $\mathbb{C}^N$ and hence $\dim S = N$. The Proposition then follows. □

For any $\beta \in \mathbb{C}^N$, the subspace $S$ contains the interpolating $(m,s)$-spline relative to $A$ and $\beta$. We shall show in Section III–1 that, for any $\beta \in \mathbb{C}^N$ and for any $\varepsilon > 0$, $S$ also contains the *smoothing $(m,s)$-spline relative to $A$, $\beta$ and $\varepsilon$*. We say that $S$ is the *space of the $(m,s)$-spline functions relative to $A$.*

For any $\lambda > 0$, we introduce the function $K_\lambda$, defined by

$$\forall x \in \mathbb{R}^n,\ K_\lambda(x) = \begin{cases} |x|^\lambda, & \lambda \notin 2\mathbb{N}^*, \\ |x|^\lambda \log|x|, & \lambda \in 2\mathbb{N}^*, \end{cases} \tag{2.2}$$

where log denotes the Neperian logarithm.

***Proposition* 2.2** – *Let $\nu = m+s-n/2$. The function $K_{2\nu}$ satisfies the following properties:*

*(i)* $\mathcal{F}^{-1}(\mathrm{Pf}|\xi|^{-2m-2s})(x) = \begin{cases} C_1 K_{2\nu}(x), & \nu \notin \mathbb{N}^*, \\ C_2 K_{2\nu}(x) + C_3|x|^{2\nu}, & \nu \in \mathbb{N}^*, \end{cases}$

*where $\mathrm{Pf}|\xi|^{-2m-2s}$ is the pseudo-function distribution given by*

$$\mathrm{Pf}|\xi|^{-2m-2s}.\varphi = \text{finite part of} \int_{\mathbb{R}^n} |\xi|^{-2m-2s}\varphi(\xi)\, d\xi$$

*and $C_1$, $C_2$ and $C_3$ are real constants.*

*(ii) For any distribution $\mu \in H^{-m-s}(\mathbb{R}^n)$ with compact support such that*

$$\forall q \in P_{m-1},\ \mu.q = 0 \tag{2.3}$$

*(where $\mu.q$ replaces $\mu(q)$, the value of the distribution $\mu$ at $q$), the convolution product $\mu * K_{2\nu}$ belongs to $X^{m,s}$.*

**Proof** – (J. Duchon).

1) The relation (i) is a classical result concerning the Fourier transforms of functions of the Euclidean distance. See L. Schwartz [131, p. 257].

2) Let $\mu \in H^{-m-s}(\mathbb{R}^n)$ with compact support and let $\alpha \in \mathbb{N}^n$ such that $|\alpha| = m$. We have

$$\mathcal{F}(\partial^\alpha(\mu * K_{2\nu}))(\xi) = (2i\pi\xi)^\alpha \hat{\mu}(\xi)\widehat{K}_{2\nu}(\xi).$$

Now, it is clear that $\hat{\mu}$ has all its derivatives of order $\leq m-1$ null at the origin. To prove this, let us observe that, for any distribution $U$ with compact support, $\widehat{U}$ is a function of class $C^\infty$, since

$$\forall \xi \in \mathbb{R}^n,\ \widehat{U}(\xi) = U_x.e^{-2i\pi x.\xi}$$

and the successive derivatives of $\widehat{U}$ are obtained by derivation "under the integration sign". Thus, for the distribution $\mu$, it follows from (2.3) that

$$\forall \beta \in \mathbb{N}^n,\ |\beta| \leq m-1,\ \partial^\beta \hat{\mu}(0) = \hat{\mu}.(-2i\pi x)^\beta = 0,$$

as stated above. Using Leibniz's formula, we derive from this result that the function $\xi \mapsto \xi^\alpha \hat{\mu}(\xi)$ has all its derivatives of order $\leq 2m-1$ null at the origin.

Now, taking into account that (cf. L. Schwartz [131])

$$\forall j \in \mathbb{N},\ \mathcal{F}(|x|^{2j}) = \left(\frac{-1}{4\pi^2}\right)^j \Delta^j \delta,$$

$\Delta$ being the Laplace operator, we obtain, according to (i),

$$\widehat{K}_{2\nu} = \begin{cases} \frac{1}{C_1}\operatorname{Pf}|\xi|^{-2m-2s}, & \nu \notin \mathbb{N}^*, \\ \frac{1}{C_2}\operatorname{Pf}|\xi|^{-2m-2s} + b\Delta^\nu\delta, & \nu \in \mathbb{N}^*, \end{cases}$$

where $b$ is a non-null constant.

Let us consider the product distribution $g$ of the distribution $\operatorname{Pf}|\xi|^{-2m-2s}$ and the function $\xi \mapsto \xi^\alpha \hat{\mu}(\xi)$ (which belongs to $C^\infty(\mathbb{R}^n)$). We have

$$\forall \varphi \in \mathcal{D},\ g.\varphi = \operatorname{Pf}\int_{\mathbb{R}^n} |\xi|^{-2m-2s}\xi^\alpha \hat{\mu}(\xi)\varphi(\xi)\,d\xi. \tag{2.4}$$

Since $\hat{\mu}$ has all its derivatives of order $\leq m-1$ null at the origin, then, in a neighbourhood of the origin,

$$\exists C > 0,\ |\hat{\mu}(\xi)| \leq C|\xi|^m,$$

from which we have

$$\exists C > 0,\ |\xi|^{-2m-2s}|\xi^\alpha \hat{\mu}(\xi)| \leq C|\xi|^{-2s}.$$

As $s < n/2$, the integral given in (2.4) is convergent. Hence, the distribution $g$ is, in fact, a locally integrable function, and there exists a constant $C$ such that

$$|\xi|^{2s}|g(\xi)|^2 \leq C|\xi|^{-2s}$$

in a neighbourhood of the origin and

$$|\xi|^{2s}|g(\xi)|^2 \leq C(1+|\xi|^2)^{-m-s}|\hat{\mu}(\xi)|^2$$

outside such a neighbourhood (since $m+s > 0$ and hence $|\xi|^{-2m-2s} \leq C(1+|\xi|^2)^{-m-s}$ if $|\xi|$ is large enough). Taking into account that $\mu \in H^{-m-s}(\mathbb{R}^n)$, we deduce that $\mathcal{F}^{-1}g \in \widetilde{H}^s$.

Finally, let us consider the product distribution $h$ of the distribution $\Delta^\nu\delta$, with $\nu \in \mathbb{N}^*$, and the function $\xi \mapsto \xi^\alpha\hat{\mu}(\xi)$. We have

$$\forall \varphi \in \mathcal{D},\ h.\varphi = \Delta^\nu\big(\xi \mapsto \xi^\alpha\hat{\mu}(\xi)\varphi(\xi)\big)(0)$$

Now, all the derivatives of order $2\nu$ of the function $\xi \mapsto \xi^\alpha\hat{\mu}(\xi)\varphi(\xi)$ are null at the origin, since, by (I–2.2), $2\nu \leq 2m-1$ and the derivatives of order $\leq 2m-1$ of the function $\xi \mapsto \xi^\alpha\hat{\mu}(\xi)$ have null values at the origin. Thus $h = 0$.

The Proposition then follows. □

**Remark 2.1** – The constants $C_1$, $C_2$ and $C_3$, introduced in the preceding result, can be explicitly computed from L. Schwartz [131, p. 257]. In particular, one has

$$C_1 = \frac{\pi^{2m+2s-n/2}\,\Gamma(n/2-m-s)}{\Gamma(m+s)} \quad \text{and} \quad C_2 = 2\frac{(-1)^{m+s+1-n/2}\,\pi^{2m+2s-n/2}}{\Gamma(m+s)\,(m+s-n/2)!}.$$

Notice that, in the expression of $C_1$, $\Gamma(n/2-m-s)$ denotes the value at the point $n/2-m-s$ of the *extension* to $(-\infty,0)\setminus\mathbb{Z}$ of the elementary Eulerian function $\Gamma$. The constant $C_3$ will not appear again. The exact values of $C_1$ and $C_2$ are of purely theoretical importance, because they are neither needed for the computation of interpolating splines nor, in practice, for that of smoothing $(m,s)$-splines (see Section III–2). □

***Theorem* 2.1** – *Every element $u \in S$ can be written in a unique way in the form*

$$u(x) = \sum_{a\in A} \lambda_a^* K_{2m+2s-n}(x-a) + p(x), \tag{2.5}$$

*where $K_{2m+2s-n}$ is the function defined in (2.2), $p \in P_{m-1}$ and*

$$\forall q \in P_{m-1},\ \sum_{a\in A} \lambda_a^* q(a) = 0. \tag{2.6}$$

**Proof** –

1) Reasoning as in the proof of Proposition 1.2, it follows from the definition of $S$ that an element $u$ of $X^{m,s}$ belongs to $S$ if and only if there exist unique constants $\lambda_a = \lambda_a(u)$, with $a \in A$, which verify the relation

$$\forall v \in X^{m,s},\ (u,v)_{m,s} = \sum_{a\in A} \lambda_a \overline{v(a)}. \tag{2.7}$$

Let $\mu = \sum_{a \in A} \lambda_a \delta_a$, where $\lambda_a$ are the constants introduced in (2.7). Then, the distribution $\mu$ belongs to $H^{-m-s}(\mathbb{R}^n)$, since $\delta \in H^{-n/2-\theta}$ for any $\theta > 0$, and $\mu$ also satisfies the condition (2.3), since, taking $v = \overline{q}$ in (2.7), with $q \in P_{m-1}$, we get

$$\forall q \in P_{m-1}, \sum_{a \in A} \lambda_a q(a) = 0.$$

Let $u \in S$. By Proposition I–2.1, the relation (2.7) is necessarily verified for any $v \in \mathcal{S}$. Let us note that the product of the function $\xi \mapsto \xi^\alpha$, with $|\alpha| = m$, and the locally integrable function $\xi \mapsto \xi^\alpha \hat{u}(\xi)$, associated with the product distribution of the distribution $\hat{u}$ and the function $\xi \mapsto \xi^\alpha$, is just the locally integrable function $\xi \mapsto \xi^{2\alpha}\hat{u}(\xi)$, associated with the product distribution of the distribution $\hat{u}$ and the function $\xi \mapsto \xi^{2\alpha}$. On the other hand,

$$\sum_{|\alpha|=m} \frac{m!}{\alpha!} \xi^{2\alpha} = |\xi|^{2m}.$$

Using the definition of the scalar semi-product $(\,\cdot\,,\,\cdot\,)_{m,s}$, we obtain the relation

$$\forall \varphi \in \mathcal{S},\ (u, v)_{m,s} = (2\pi)^{2m} \int_{\mathbb{R}^n} |\xi|^{2s}\big(|\xi|^{2m}\hat{u}(\xi)\big)\overline{\hat{\varphi}(\xi)}\, d\xi,$$

where $\xi \mapsto |\xi|^{2s}\big(|\xi|^{2m}\hat{u}(\xi)\big)$ represents the product of the function $\xi \mapsto |\xi|^{2s}$ and the locally integrable function $\xi \mapsto |\xi|^{2m}\hat{u}(\xi)$. The function $\xi \mapsto |\xi|^{2s}\big(|\xi|^{2m}\hat{u}(\xi)\big)$ is locally integrable, since it is a linear combination of functions $\xi \mapsto |\xi|^{2s}\big(\xi^{2\alpha}\hat{u}(\xi)\big)$, with $|\alpha| = m$, each one of which is written as a product of the function $\xi \mapsto |\xi|^s \xi^\alpha$ (a locally square integrable function, since $m+s > 0$) and the function $\xi \mapsto |\xi|^s\big(\xi^\alpha \hat{u}(\xi)\big)$ (which is square integrable, since $u \in X^{m,s}$). The function $\xi \mapsto |\xi|^{2s}\big(|\xi|^{2m}\hat{u}(\xi)\big)$ then defines a distribution (tempered, since $\hat{u}$ is tempered) by means of an integral over $\mathbb{R}^n$. We deduce that

$$\forall \varphi \in \mathcal{S},\ (u, \varphi)_{m,s} = (2\pi)^{2m}\big(|\xi|^{2s}(|\xi|^{2m}\hat{u})\big).\overline{\hat{\varphi}}.$$

Remarking that

$$\forall \varphi \in \mathcal{S},\ \overline{\varphi} = \overline{\mathcal{F}^{-1}(\hat{\varphi})} = \mathcal{F}\overline{\hat{\varphi}}$$

and using the definition of the Fourier transform in $\mathcal{S}'$, we obtain the relation

$$\forall \varphi \in \mathcal{S}, \sum_{a \in A} \lambda_a \overline{\varphi(a)} = \mu.\mathcal{F}\overline{\hat{\varphi}} = \hat{\mu}.\overline{\hat{\varphi}}.$$

Thus, every $u \in S$ is a solution of

$$\begin{cases} u \in X^{m,s}, \\ |\xi|^{2s}\big(|\xi|^{2m}\hat{u}\big) = (2\pi)^{-2m}\hat{\mu}, \end{cases} \tag{2.8}$$

with $\mu = \sum_{a \in A} \lambda_a \delta_a$, where $\lambda_a$, for $a \in A$, are the constants introduced in (2.7). Therefore, the determination of the elements of $S$ leads to a problem of division in $\mathcal{S}'$.

2) Let $v$ and $v_0$ be two solutions of (2.8). Then, $w = v - v_0$ verifies that, in the sense of functions, $|\xi|^{2s}(|\xi|^{2m}\hat{w}) = 0$. Hence, $|\xi|^{2m}\hat{w} = 0$ in the sense of distributions. Since $\hat{w}$ has $\{0\}$ as support, $\hat{w}$ is a finite linear combination of derivatives of the Dirac measure (cf. L. Schwartz [131]) and, in consequence, $w$ is a polynomial function. Now, since $w \in X^{m,s}$, $w$ belongs to $P_{m-1}$. We deduce that the set of solutions of (2.8) is the set $v_0 + P_{m-1}$, where $v_0$ is a particular solution of (2.8).

Let

$$T = (2\pi)^{-2m}\hat{\mu}\,\mathrm{Pf}|\xi|^{-2m-2s},$$

where $\mu$ is the distribution introduced in 1). Taking into account that $\hat{\mu} \in C^\infty(\mathbb{R}^n)$ and $s < n/2$, it follows from the properties of the pseudo-functions distributions that

$$|\xi|^{2m}T = (2\pi)^{-2m}\,\mathrm{Pf}(\hat{\mu}|\xi|^{-2s}) = (2\pi)^{-2m}\hat{\mu}|\xi|^{-2s}.$$

Thus we can take $\hat{v}_0 = T$. In consequence, by Proposition 2.2, the solutions of (2.8) can be written in the form

$$v = \mu * \left(aK_{2m+2s-n} + b|x|^{2m+2s-n}\right) + \psi,$$

where $K_{2m+2s-n}$ is the function defined in (2.2), $a$ and $b$ are constants, with $b = 0$ if $m + s - n/2 \notin \mathbb{N}^*$, and $\psi \in P_{m-1}$.

Now, $\mu$ verifies (2.3). We deduce that, if $\nu = m + s - n/2 \in \mathbb{N}^*$, then

$$\mu * |x|^{2\nu} \in P_{m-1}.$$

To see this, it suffices to observe that, by (I–2.2), $\nu \leq m - 1$ and hence

$$\forall \alpha \in \mathbb{N}^n,\ |\alpha| = m,\ \partial^\alpha(\mu * |x|^{2\nu}) = \mu * \partial^\alpha |x|^{2\nu} = 0.$$

Therefore, the solutions $v$ of (2.8) verify (2.5) and (2.6) with

$$\lambda_a^* = \begin{cases} (2\pi)^{-2m}C_1\lambda_a, & 2m + 2s - n \notin 2\mathbb{N}^*, \\ (2\pi)^{-2m}C_2\lambda_a, & 2m + 2s - n \in 2\mathbb{N}^*, \end{cases} \tag{2.9}$$

where $C_1$ and $C_2$ are the constants introduced in Proposition 2.2. Then, every $u \in S$ verifies (2.5) and (2.6) with the coefficients $\lambda_a^* = \lambda_a^*(u)$ defined by (2.9), proved to be unique, and a polynomial $p = p(u) \in P_{m-1}$, also unique, since $A$ contains a $P_{m-1}$-unisolvent subset. This completes the proof. □

**Remark 2.2** – In fact, Theorem 2.1 admits a reciprocal result: every element $u$ of the form (2.5), where $K_{2m+2s-n}$ denotes the function introduced in (2.2), $p \in P_{m-1}$ and $(\lambda_a^*)_{a\in A}$ verifies (2.6), belongs to $S$. This result can be derived from Theorem 4.1 (see Section 4). □

**Remark 2.3** – When $s = 0$, the solutions of the homogeneous equation associated with (2.8), i.e. the equation

$$\mathcal{F}(\Delta^m u) = 0,$$

are the *polyharmonic polynomials* (cf. L. Schwartz [131]). By analogy with this definition, the $(m, s)$-splines are often called *polyharmonic splines*. □

As a consequence of Theorem 2.1, the solution $\sigma$ of problem (1.1) can be written in a unique way in the form

$$\sigma(x) = \sum_{a\in A} \lambda_a^* K_{2m+2s-n}(x-a) + p(x),$$

with $p = p(\sigma) \in P_{m-1}$ and $\lambda_a^* = \lambda_a^*(\sigma) \in \mathbb{C}$, for $a \in A$, verifying (2.6). Let us observe that, if $\sigma$ is real valued (i.e. if $\beta \in \mathbb{R}^N$), then the coefficients $\lambda_a^*$ and the polynomial $p$ are real.

By Corollary I–2.1 and Sobolev's Continuous Imbedding Theorem (cf. (4) in Preliminaries), we have

$$X^{m,s} \hookrightarrow C^r(\mathbb{R}^n)$$

for any integer $r$ such that $m+s > r+n/2$. The following result shows that the $(m,s)$-spline functions relative to $A$ are *more regular* than what could be supposed due to their belonging to the space $X^{m,s}$. This is the property of *super-regularity* of the spline functions.

***Corollary* 2.1** – *The space $S$ of the $(m,s)$-spline functions relative to $A$ verifies the inclusion $S \subset C^\eta(\mathbb{R}^n)$, where*

$$\eta = \begin{cases} \textit{integer part of } 2m+2s-n, & 2m+2s-n \notin \mathbb{N}^*, \\ 2m+2s-n-1, & \textit{otherwise.} \end{cases}$$

**Proof** – To prove this result, it suffices to show that all the derivatives of order $\le \eta$ of the function $K_{2m+2s-n}$ are null at the origin. This fact is easily verified if $2m+2s-n \in 2\mathbb{N}^*$, while, for the case $2m+2s-n \notin 2\mathbb{N}^*$, it is a consequence of the following relation:

$$\forall\alpha \in \mathbb{N}^n,\ |\alpha| \le \eta,\ \partial^\alpha K_{2m+2s-n}(|x|) = |x|^{2m+2s-n-|\alpha|} \sum_{\beta\le\alpha} C_{\alpha,\beta} \left(\frac{x}{|x|}\right)^\beta,$$

where, given $\beta = (\beta_1, \dots, \beta_n) \in \mathbb{N}^n$, the notation $\beta \le \alpha$ means that, for $j = 1, \dots, n$, $\beta_j \le \alpha_j$, and $C_{\alpha,\beta}$ are constants. □

To finish this section, we give some examples of $(m,s)$-splines (cf. J. Duchon [55]). Let $\{p_1, \dots, p_{\mathfrak{M}}\}$, where $\mathfrak{M}$ denotes the dimension of $P_{m-1}$, be a basis of $P_{m-1}$. We recall that $\mathfrak{M} = \binom{n+m-1}{m-1}$.

**Example 2.1 (Thin plate splines)** – For any integer $m > n/2$, the $(m,0)$-splines are called *$D^m$-splines over $\mathbb{R}^n$*. In particular, taking $n=2$ and $m=2$, we obtain the *$D^2$-splines over $\mathbb{R}^2$* or *thin plate splines*, which can be written in the form

$$u(x) = \sum_{a\in A} \lambda_a^* |x-a|^2 \log|x-a| + \sum_{j=1}^{3} c_j p_j(x),$$

where

$$\forall j = 1,2,3,\ \sum_{a\in A} \lambda_a^* p_j(a) = 0.$$

By Corollary 2.1, $u$ is of class $C^1$. We shall see in Section 4 that the coefficients $\lambda_a^*$, with $a \in A$, $c_1$, $c_2$ and $c_3$ of the thin plate interpolating spline $\sigma$ relative to $A$ and $\beta = (\beta_b)_{b\in A} \in \mathbb{C}^N$ are determined in a unique way by solving the linear system

$$\begin{cases} \sum_{a\in A} \lambda_a^* |b-a|^2 \log|b-a| + \sum_{j=1}^{3} c_j p_j(b) = \beta_b, \quad b \in A, \\ \sum_{a\in A} \lambda_a^* p_j(a) = 0, \quad j = 1,2,3 \end{cases} \tag{2.10}$$

(with $|b-a|^2 \log|b-a| = 0$ when $b = a$).

For $n = 1$, if we denote by $a_1, \ldots, a_N$ the points of $A$, with $a_j < a_{j+1}$ for $j = 1, \ldots, N-1$, we shall show later (see Remark VII–1.2) that the space $S$ of $D^m$-splines over $\mathbb{R}$ relative to $A$ is the space of functions $u \in C^{2m-2}(\mathbb{R})$ such that $u|_{(-\infty,a_1]}$, $u|_{[a_N,+\infty)}$ and, for $j = 1, \ldots, N-1$, $u|_{[a_j,a_{j+1}]}$ are polynomial functions of degree less than or equal to $m-1$, $m-1$ and $2m-1$, respectively. □

**Example 2.2 (Pseudo-polynomial splines)** – Taking $s = \frac{n-1}{2}$, we get the *pseudo-polynomial splines*, in particular,

- for $m = 1$, the *multiquadric functions* (cf. R. L. Hardy [78, 79])

$$u(x) = \sum_{a\in A} \lambda_a^* |x-a| + c_1,$$

with

$$\sum_{a\in A} \lambda_a^* = 0;$$

- for $m = 2$, the *pseudo-cubic splines*

$$u(x) = \sum_{a\in A} \lambda_a^* |x-a|^3 + \sum_{j=1}^{n+1} c_j p_j(x),$$

with

$$\forall j = 1, \ldots, n+1, \ \sum_{a\in A} \lambda_a^* p_j(a) = 0.$$

The multiquadrics and the pseudo-cubic splines are functions of classes $C^0$ and $C^2$, respectively. The *pseudo-quintic splines*, obtained for $m = 3$, are of class $C^4$.

It can be proved (see the general result given in Section 4) that the coefficients $\lambda_a^*$, with $a \in A$, and $c_1, \ldots, c_{\mathfrak{M}}$ of any interpolating $(m, \frac{n-1}{2})$-spline relative to $A$ and $\beta \in \mathbb{C}^N$ are determined in a unique way by solving a linear system of order $N + \mathfrak{M}$ analogous to (2.10).

For $n = 1$, any pseudo-polynomial spline is also a $D^m$-spline over $\mathbb{R}$. □

# 3. OTHER PROBLEMS. SPLINES DEFINED BY LOCAL MEAN VALUES

Instead of the case of Lagrange interpolation, we can consider the more general situation in which the interpolating data are the values of continuous linear functionals $\mu_1, \dots, \mu_N$ over $C^r(\mathbb{R}^n)$, that is, the values of distributions with compact support of order $\leq r$, with $r \in \mathbb{N}$ such that $m + s > r + n/2$. In this way, we can treat, in particular, the problem of Hermite interpolation.

For any $\beta = (\beta_j)_{1 \leq j \leq N} \in \mathbb{C}^N$, let

$$\mathcal{K} = \{ v \in X^{m,s} \mid \forall j = 1, \dots, N, \ \mu_j(v) = \beta_j \}. \tag{3.1}$$

Under the hypothesis

$$\forall p \in P_{m-1}, \ \big(\mu_1(p) = \cdots = \mu_N(p) = 0\big) \Rightarrow \big(p = 0\big),$$

the results of previous sections can be adapted to prove that there exists a unique solution $\sigma$ of problem (1.1), with $\mathcal{K}$ defined by (3.1), and that $\sigma$ can be written in a unique way in the form

$$\sigma = \sum_{j=1}^{N} \lambda_j^* \mu_j * K_{2m+2s-n} + p,$$

with $p \in P_{m-1}$ and, for $j = 1, \dots, N$, $\lambda_j^* \in \mathbb{C}$ verifying the relation

$$\forall q \in P_{m-1}, \ \sum_{j=1}^{N} \lambda_j^* \mu_j(q) = 0.$$

One can also envisage (cf. J. Duchon [55]) the case of an *infinity* of continuous linear functionals $\mu$ which have a compact support contained in the closure of a bounded open subset $\Omega$ of $\mathbb{R}^n$, belonging to a closed subspace $\mathcal{M}$ of $H^{-m-s}(\mathbb{R}^n)$, with $m+s > 0$. Then, for any given function $f$ in $H^{m+s}(\Omega)$, one proves, under the hypothesis

$$\forall p \in P_{m-1}, \ \big(\forall \mu \in \mathcal{M}, \ \mu(p) = 0\big) \Rightarrow \big(p = 0\big),$$

the existence and the uniqueness of a function $\sigma \in X^{m,s}$ with minimal semi-norm $|\cdot|_{m,s}$ satisfying the relation

$$\forall \mu \in \mathcal{M}, \ \mu(\sigma) = \mu(f).$$

However, in the general case, the problem of expressing $\sigma$ explicitly remains open.

The example of *splines defined by local mean values* corresponds to the following situation. We first introduce an interpolating operator $\rho$ of the form

$$\rho v = \left( \frac{1}{\operatorname{meas} \omega_j} \int_{\omega_j} v(x)\, dx \right)_{1 \leq j \leq N}, \tag{3.2}$$

where $\{\omega_j \mid 1 \le j \le N\}$ denotes a set of $N$ nonempty bounded open subsets of $\mathbb{R}^n$, pairwise disjoint. Then, we consider the Beppo-Levi space $X^{1,0}$ (denoted at times by $D^{-1}L^2(\mathbb{R}^n)$ or even $V_2^{(1)}(\mathbb{R}^n)$). Although (except for $n = 1$) the condition (I–2.2) is not verified, it is known, as pointed out in Remark I–2.1, that Theorem I–2.2 and Corollary I–2.1 remain valid for $m = 1$ and $s = 0$. We deduce that $\rho$ is a linear continuous operator from $L^2_{\text{loc}}(\mathbb{R}^n)$ into $\mathbb{C}^N$, hence from $X^{1,0}$ into $\mathbb{C}^N$.

Taking up the study of the preceding section, if we replace the Lagrange interpolating operator by the operator defined in (3.2), we can prove without difficulty the existence and the uniqueness of an interpolating (1,0)-spline $\sigma$ relative to $\rho$ and $\beta = (\beta_j)_{1\le j\le N} \in \mathbb{C}^N$, i.e. the solution of problem (1.1) with $\rho$ defined by (3.2), $m = 1$ and $s = 0$. We can verify that Propositions 1.1, 1.2 and 2.1 remain valid in this case.

Now, let

$$\mu = \sum_{j=1}^{N} \frac{\lambda_j}{\operatorname{meas} \omega_j} \chi_{\omega_j},$$

where, for $j = 1, \ldots, N$, $\chi_{\omega_j}$ denotes the characteristic function of $\omega_j$ and $\lambda_1, \ldots, \lambda_N$ are constants verifying the relation

$$\sum_{j=1}^{N} \lambda_j = 0.$$

Likewise, let $H$ be the elemental solution of the operator $\Delta$, defined by

$$H(x) = \begin{cases} \frac{1}{2}|x|, & n = 1, \\ \frac{1}{2\pi}\log|x|, & n = 2, \\ \frac{1}{(2-n)S_n}|x|^{2-n}, & n > 2, \end{cases}$$

where $S_n$ stands for the area of the unit sphere in $\mathbb{R}^n$. By rewriting the proof of Proposition 2.2, simpler in this case, we can prove that

$$H * \mu \in X^{1,0}.$$

Using (2.8) with $u = \sigma$, or the equivalent equation $\Delta\sigma = \mu$, we then show that the interpolating spline $\sigma$ defined by local mean values is written in a unique way in the form

$$\sigma(x) = \sum_{k=1}^{N} \frac{\lambda_k}{\operatorname{meas} \omega_k} \int_{\omega_k} H(x-y)\, dy + C,$$

with $C \in \mathbb{C}$ and $\lambda_1, \ldots, \lambda_N \in \mathbb{C}$ verifying the relation

$$\sum_{k=1}^{N} \lambda_k = 0.$$

A direct reasoning on the expression obtained for $\sigma$ shows that $\sigma \in X^{2,0} \cap C^1(\mathbb{R}^n)$.

The coefficients $\lambda_k$ and $C$ are obtained by solving the linear system

$$\begin{cases} \displaystyle\sum_{k=1}^{N} \alpha_{jk}\lambda_k + C = \beta_j, & 1 \le j \le N, \\ \displaystyle\sum_{k=1}^{N} \lambda_k = 0, & \end{cases}$$

where, for $j, k = 1, \dots, N$,

$$\alpha_{jk} = \frac{1}{\operatorname{meas}\omega_j \operatorname{meas}\omega_k} \int_{\omega_j}\int_{\omega_k} H(x-y)\,dx\,dy.$$

**Remark 3.1** – The computation of the coefficients $\alpha_{jk}$ poses problems which cannot be solved in the general case. But when the sets $\omega_j$ are Euclidean balls of centre $a_j$ and radius $r_j$, we have, for $n \ge 2$, the following formulae (given by M. N. Benbourhim [26, 27]):

$$\alpha_{jk} = \begin{cases} H(a_j - a_k), & j \ne k, \\ H(r_j) - \dfrac{r_j^{2-n}}{(n+2)S_n}, & j = k, \end{cases}$$

where

$$H(r_j) = \begin{cases} \frac{1}{2\pi}\log r_j, & n = 2, \\ \frac{1}{(2-n)S_n} r_j^{2-n}, & n > 2. \end{cases}$$

To establish these formulae, we proceed in two steps.

1) For $k \in \{1, \dots, N\}$, we consider the function $\psi_k$ defined over $\mathbb{R}^n$ by

$$\psi_k(x) = \frac{1}{\operatorname{meas}\omega_k}\int_{\omega_k} H(x-y)\,dy = \left(\frac{\chi_{\omega_k}}{\operatorname{meas}\omega_k} * H\right)(x).$$

We have

$$\Delta\psi_k = \frac{\chi_{\omega_k}}{\operatorname{meas}\omega_k},$$

and hence, by Friedrichs' Theorem (cf. J. L. Lions and E. Magenes [94, Theorem 3.2, p. 138]), $\psi_k \in H^2_{\text{loc}}(\mathbb{R}^n)$. We deduce that $\psi_k$ admits traces (both equal) on each side of the boundary $\partial\omega_k$ of $\omega_k$.

For any $x$ such that $|x - a_k| \ge r_k$, the function $y \mapsto H(x-y)$ is harmonic on $\omega_k$. Therefore, by the mean value formula, $\psi_k(x) = H(x - a_k)$ if $|x - a_k| \ge r_k$.

On the other hand, let us consider the function $u$ defined over $\mathbb{R}^n$ by

$$u(x) = \frac{1}{2n}\left(|x - a_k|^2 - r_k^2\right) + (\operatorname{meas}\omega_k)H(r_k).$$

We have

$$\begin{cases} \Delta u(x) = 1, & x \in \mathbb{R}^n, \\ u(x) = (\operatorname{meas}\omega_k)H(r_k), & x \in \partial\omega_k. \end{cases}$$

Then, we get

$$\begin{cases} \Delta\Big(\psi_k - \dfrac{u}{\operatorname{meas}\omega_k}\Big) = 0, & \text{on } \omega_k, \\ \psi_k - \dfrac{u}{\operatorname{meas}\omega_k} = 0, & \text{on } \partial\omega_k. \end{cases}$$

We deduce (solution of the homogeneous Dirichlet problem for the operator $-\Delta$) that, if $|x - a_k| \leq r_k$,

$$\psi_k(x) = \frac{1}{2n \operatorname{meas}\omega_k}\big(|x - a_k|^2 - r_k^2\big) + H(r_k).$$

2) For $j, k = 1, \dots, N$, we next compute

$$\alpha_{jk} = \frac{1}{\operatorname{meas}\omega_j}\int_{\omega_j} \psi_k(x)\, dx,$$

and the result follows, taking into account that, for $j = 1, \dots, N$,

$$\int_{\omega_j} |x - a_j|^2\, dx = S_n \frac{r_j^{n+2}}{n+2}. \quad \square$$

## 4. *COMPUTATION OF INTERPOLATING SPLINES*

Let us return to the problem of Lagrange interpolation introduced in Section 1. According to Theorem 2.1, the interpolating spline $\sigma$ relative to $A$ and $\beta$, solution of problem (1.1), is written in a unique way in the form (2.5)–(2.6). Denoting by $\{p_1, \dots, p_{\mathfrak{M}}\}$ a basis of $P_{m-1}$ and by $c_j$, for $j = 1, \dots, \mathfrak{M}$, a constant, we have

$$\sigma(x) = \sum_{a\in A} \lambda_a^* K_{2m+2s-n}(x - a) + \sum_{j=1}^{\mathfrak{M}} c_j p_j(x),$$

with

$$\forall j = 1, \dots, \mathfrak{M}, \ \sum_{a\in A} \lambda_a^* p_j(a) = 0.$$

Therefore, to determine $\sigma$ is equivalent to computing the $N + \mathfrak{M}$ unknown coefficients $\lambda_a^*$, for $a \in A$, and $c_j$, for $j = 1, \dots, \mathfrak{M}$, by solving the linear system of order $N + \mathfrak{M}$

$$\begin{cases} \displaystyle\sum_{a\in A} \lambda_a^* K_{2m+2s-n}(b - a) + \sum_{j=1}^{\mathfrak{M}} c_j p_j(b) = \beta_b, \quad b \in A, \\ \displaystyle\sum_{a\in A} \lambda_a^* p_j(a) = 0, \quad 1 \leq j \leq \mathfrak{M}. \end{cases} \tag{4.1}$$

Let us denote by $G$ the matrix of this linear system. We can immediately verify that $G$ is *symmetric* and *almost full* (only the block of the $\mathfrak{M}$ last rows and $\mathfrak{M}$ last columns is null). Likewise, $G$ is not positive definite, since $X^T G X = 0$ for any vector $X = \binom{0}{X_2} \in \mathbb{R}^{N+\mathfrak{M}}$, with $X_2 \in \mathbb{R}^{\mathfrak{M}}$ (here, $X^T$ denotes the transpose of $X$).

However, it is possible to reduce the resolution of (4.1) to that of a linear system with positive definite matrix (cf. L. Paihua [111], where diverse algorithms for computing $(m, s)$-splines are presented; see also N. Dyn, D. Levin and S. Rippa [60]).

Finally, the matrix $G$ is *regular*. This fact is not evident a priori: the uniqueness of $\sigma$ does not imply that (4.1) cannot have solutions other than the coefficients of $\sigma$. We shall need the following result.

***Theorem* 4.1** – *The space $\mathcal{S} + P_{m-1}$ is dense in $X^{m,s}$.*

**Proof** –

1) It suffices to prove that $\mathcal{S}$ is dense in the semi-Hilbert space $(X^{m,s}, |\cdot|_{m,s})$. To see this, let us observe that, if $A_0$ denotes a $P_{m-1}$-unisolvent subset of $\overline{\Omega}$, then, by Proposition I–2.2, the norm $[\![\cdot]\!]_{A_0,m,s}$, defined in (I–2.5) with $E = A_0$, is equivalent on $X^{m,s}$ to the norm $\|\cdot\|_{m,s}$. Given that

$$\forall v \in X^{m,s},\ \exists \psi \in P_{m-1},\ \forall a \in A_0,\ (v+\psi)(a) = 0,$$

we deduce that there exists $C > 0$ such that

$$\forall v \in X^{m,s},\ \exists \psi \in P_{m-1},\ \|v+\psi\|_{m,s} \le C|v|_{m,s}.$$

Hence, the density of $\mathcal{S}$ in $(X^{m,s}, |\cdot|_{m,s})$ implies that of $\mathcal{S}+P_{m-1}$ in $(X^{m,s}, \|\cdot\|_{m,s})$.

2) Let us suppose that $u \in X^{m,s}$, with

$$\forall \varphi \in \mathcal{S},\ (u, \varphi)_{m,s} = 0.$$

Taking up again the reasoning of the proof of Theorem 2.1, we obtain the relation, analogous to (2.8),

$$|\xi|^{2s}(|\xi|^{2m}\hat{u}) = 0.$$

We deduce that the support of $\hat{u}$ is the set $\{0\}$. Thus, $\hat{u}$ is a finite linear combination of derivatives of the Dirac measure. Consequently, $u$ is a polynomial and hence, since $u \in X^{m,s}$, $u \in P_{m-1}$. We conclude that $\mathcal{S}$ is dense in $(X^{m,s}, |\cdot|_{m,s})$. □

***Theorem* 4.2** – *The matrix $G$ is regular.*

**Proof** – Let us prove that the homogeneous system associated with (4.1) admits only the null solution. For this, let us suppose that the coefficients $\lambda_a^*$, for $a \in A$, and $c_1, \ldots, c_{\mathfrak{M}}$ are a solution of the system (4.1) for $\beta = 0$. Now, let

$$u = \mu^* * K_{2m+2s-n} + \sum_{j=1}^{\mathfrak{M}} c_j p_j,$$

with $\mu^* = \sum_{a\in A} \lambda_a^* \delta_a$. By hypothesis, $\mu^*$ belongs to $H^{-m-s}(\mathbb{R}^n)$ and verifies (2.3). Hence, by Proposition 2.2, $u \in X^{m,s}$. On the other hand (cf. the proof of Proposition 2.2), for any $\alpha \in \mathbb{N}^n$, with $|\alpha| = m$,

$$\mathcal{F}(\partial^\alpha(\mu^* * K_{2m+2s-n}))(\xi) = C(2i\pi\xi)^\alpha \widehat{\mu^*}(\xi)\,\mathrm{Pf}|\xi|^{-2m-2s},$$

where $C = 1/C_1$ if $2m+2s-n \notin 2\mathbb{N}^*$, or $C = 1/C_2$ otherwise ($C_1$ and $C_2$ denote the constants introduced in Proposition 2.2). We deduce that

$$\forall \varphi \in \mathcal{S},\ (u,\varphi)_{m,s} = C(2\pi)^{2m} \int_{\mathbb{R}^n} |\xi|^{2s} \left( |\xi|^{2m} \widehat{\mu^*}(\xi)\, \mathrm{Pf}|\xi|^{-2m-2s} \right) \overline{\widehat{\varphi}}(\xi)\, d\xi$$
$$= C(2\pi)^{2m} \int_{\mathbb{R}^n} \widehat{\mu^*}(\xi) \overline{\widehat{\varphi}}(\xi)\, d\xi = C(2\pi)^{2m} \widehat{\mu^*}.\overline{\widehat{\varphi}}.$$

Therefore (see point 1) of the proof of Theorem 2.1),

$$\forall \varphi \in \mathcal{S},\ (u,\varphi)_{m,s} = C(2\pi)^{2m} \sum_{a\in A} \lambda_a^* \overline{\varphi(a)}.$$

Since $\mathcal{S} + P_{m-1}$ is dense in $X^{m,s}$ (cf. Theorem 4.1), taking into account that, for $j = 1, \ldots, \mathfrak{M}$, $\sum_{a\in A} \lambda_a^* p_j(a) = 0$, and applying Corollary I–2.2, we obtain

$$\forall v \in X^{m,s},\ (u,v)_{m,s} = C(2\pi)^{2m} \sum_{a\in A} \lambda_a^* \overline{v(a)}. \tag{4.2}$$

We deduce that $u$ is a solution of the problem

$$\begin{cases} u \in \mathcal{K}_0, \\ \forall v \in \mathcal{K}_0,\ (u,v)_{m,s} = 0, \end{cases}$$

which is just (1.2) with $\beta = 0$. Hence $u = 0$. Thus, (4.2) implies that $\lambda_a^* = 0$, for any $a \in A$. Finally, since $A$ contains a $P_{m-1}$-unisolvent subset, $c_j = 0$, for $j = 1, \ldots, \mathfrak{M}$. The Theorem follows. □

## 5. A CONVERGENCE RESULT

Let $\Omega$ be a bounded open subset of $\mathbb{R}^n$ and let $f \in H^{m+s}(\Omega)$. We shall prove that, under suitable hypotheses, the interpolating $(m, s)$-spline relative to $A$ and $\big(f(a)\big)_{a\in A}$ converges to $f$ in $H^{m+s}(\Omega)$ as $N \to +\infty$. To see this, we first need to introduce new notations and hypotheses, modifying those of Section 1.

Suppose we are given

- an open subset $\Omega$ of $\mathbb{R}^n$ with a Lipschitz-continuous boundary (cf. Preliminaries),
- a subset $\mathbb{D}$ of $(0, +\infty)$ such that $0 \in \overline{\mathbb{D}}$,
- for any $d \in \mathbb{D}$, an ordered set $A^d$ of $N = N(d)$ distinct points in $\overline{\Omega}$.

Later, we shall suppose that, for any $d \in \mathbb{D}$, the set $A^d$ verifies the condition

$$\sup_{x\in\Omega} \delta(x, A^d) = d, \tag{5.1}$$

where $\delta$ denotes the Euclidean distance in $\mathbb{R}^n$. Let us observe that the left-hand side of (5.1) is just the Hausdorff distance between $A^d$ and $\overline{\Omega}$. Consequently, (5.1) implies that $\mathbb{D}$ is bounded and that this distance tends to 0 as $d$ does (classical condition in the study of the convergence of spline functions). Likewise, let us remark the ambiguity in the meaning of $d$, defined first as an index and next, independently, in (5.1). This situation is analogous to that found in the Finite Element theory. We finally point out that hypotheses $0 \in \overline{\mathbb{D}}$ and (5.1) imply the weaker condition

$$\lim_{d\to 0} \sup_{x\in\Omega} \delta(x, A^d) = 0. \tag{5.2}$$

***Proposition* 5.1** – *Suppose that* (5.2) *holds. Then, there exists $\eta > 0$ and, for any $d \in \mathbb{D}$, a $P_{m-1}$-unisolvent subset $A_0^d$ of $A^d$ such that the mapping $[\![ \cdot ]\!]_{A_0^d,m,s}$, defined, for any $d \in \mathbb{D}$, by*

$$\forall v \in X^{m,s},\ [\![v]\!]_{A_0^d,m,s} = \left( \sum_{a\in A_0^d} |v(a)|^2 + |v|_{m,s}^2 \right)^{1/2}, \tag{5.3}$$

*is, for any $d \in \mathbb{D} \cap (0, \eta]$, a norm on $X^{m,s}$, uniformly equivalent over $\mathbb{D} \cap (0, \eta]$ to the norm $\| \cdot \|_{m,s}$.*

**Proof** – This result is a corollary of Proposition I–2.3. Let $B_0 = \{b_{01}, \ldots, b_{0\mathfrak{M}}\}$ be any $P_{m-1}$-unisolvent subset of $\overline{\Omega}$. By hypotheses $0 \in \overline{\mathbb{D}}$ and (5.2), we have

$$\forall j = 1, \ldots, \mathfrak{M},\ \exists (a_{0j}^d)_{d\in\mathbb{D}},\ (\forall d \in \mathbb{D},\ a_{0j}^d \in A^d) \text{ and } (b_{0j} = \lim_{d\to 0} a_{0j}^d).$$

For any $d \in \mathbb{D}$, let $A_0^d$ be the set $\{a_{01}^d, \ldots, a_{0\mathfrak{M}}^d\}$. Then, Proposition I–2.3 shows that there exists a positive constant $\eta$ such that, for any $d \le \eta$, the set $A_0^d$ is $P_{m-1}$-unisolvent and the mapping $[\![ \cdot ]\!]_{A_0^d,m,s}$ is a norm on $X^{m,s}$, uniformly equivalent over the set $\mathbb{D} \cap (0, \eta]$ (which is nonempty, since $0 \in \overline{\mathbb{D}}$) to the norm $\| \cdot \|_{m,s}$. □

**Remark 5.1** – Since $\mathbb{D}$ is bounded, we may assume, for simplicity, that the constant $\eta$ introduced in the preceding result is an upper bound of $\mathbb{D}$. Therefore, without any loss of generality in the study of the convergence, we suppose that, for any $d \in \mathbb{D}$, $A^d$ contains a $P_{m-1}$-unisolvent subset. □

For any $d \in \mathbb{D}$, we now denote by $\rho^d \in \mathcal{L}(X^{m,s}, \mathbb{C}^N)$ the operator defined by

$$\rho^d v = \big(v(a)\big)_{a\in A^d}, \tag{5.4}$$

and we write

$$\mathcal{K}_0^d = \{ v \in X^{m,s} \mid \rho^d v = 0 \}.$$

Let $f$ be a given function in $H^{m+s}(\Omega)$. For any $d \in \mathbb{D}$, denoting by $\beta^d$ the element $\big(f(a)\big)_{a\in A^d}$ of $\mathbb{C}^N$, we write

$$\mathcal{K}^d = \{ v \in X^{m,s} \mid \rho^d v = \beta^d \}.$$

For any $d \in \mathbb{D}$, we denote by $\sigma^d$ the interpolating $(m, s)$-spline relative to $A^d$ and $\beta^d$, i.e. the solution of the problem

$$\begin{cases} \sigma^d \in \mathcal{K}^d, \\ \forall v \in \mathcal{K}^d, \ |\sigma^d|_{m,s} \leq |v|_{m,s}. \end{cases}$$

Let us remark that, for any $d \in \mathbb{D}$, $\sigma^d$ exists and is unique, since $A^d$ contains a $P_{m-1}$-unisolvent subset.

The following theorem, obtained by J. Duchon, is a convergence result of the interpolating spline $\sigma^d$.

***Theorem* 5.1** – *Suppose that* (5.2) *holds. Then,*

$$\lim_{d \to 0} \|\sigma^d - f^\Omega\|_{m,s} = 0,$$

*where $f^\Omega$ denotes the unique element of minimal semi-norm $|\cdot|_{m,s}$ in the set $\{ v \in X^{m,s} \mid v|_\Omega = f \}$.*

**Proof** –

1) The set defined above is convex and, by virtue of Corollary I–2.1, nonempty and closed in $X^{m,s}$. Hence, there exists a unique element $f^\Omega$ of minimal norm $\|\cdot\|^\Omega_{m,s}$ (cf. Theorem I–2.2 with $\Omega^* = \Omega$), and it is clear that $f^\Omega$ is also the unique element of minimal semi-norm $|\cdot|_{m,s}$ in that set.

2) For any $d \in \mathbb{D}$, let $A_0^d$ be the $P_{m-1}$-unisolvent subset of $A^d$ given by Proposition 5.1. By definition of $\sigma^d$ and the norm (5.3), we have

$$\forall d \in \mathbb{D}, \ [\![\sigma^d]\!]_{A_0^d,m,s} \leq [\![f^\Omega]\!]_{A_0^d,m,s},$$

from which we deduce, according to Proposition 5.1, that the family $(\sigma^d)_{d\in\mathbb{D}}$ is bounded in $X^{m,s}$. Thus (cf. Corollary 1 in Preliminaries), there exists a sequence $(\sigma^{d_l})_{l\in\mathbb{N}}$, with $\lim_{l\to+\infty} d_l = 0$, extracted from the family $(\sigma^d)_{d\in\mathbb{D}}$, and an element $f^* \in X^{m,s}$ such that, as $l \to +\infty$,

$$\sigma^{d_l} \to f^*, \quad \text{weakly in } X^{m,s}.$$

3) We shall prove that $f^*|_\Omega = f$. Let $x$ be any point in $\Omega$. Hypotheses $0 \in \overline{\mathbb{D}}$ and (5.2) imply that

$$\exists (x^d)_{d\in\mathbb{D}}, \ (\forall d \in \mathbb{D}, \ x^d \in A^d) \text{ and } (x = \lim_{d\to 0} x^d).$$

Then, taking into account that

$$\forall l \in \mathbb{N}, \ \rho^{d_l}\sigma^{d_l} = \rho^{d_l} f^\Omega,$$

we have

$$\forall l \in \mathbb{N}, \ f(x) - \sigma^{d_l}(x) = \big(f^\Omega(x) - f^\Omega(x^{d_l})\big) + \big(\sigma^{d_l}(x^{d_l}) - \sigma^{d_l}(x)\big).$$

By Corollary I–2.2, the mapping $v \mapsto v(x)$ is strongly continuous from $X^{m,s}$ into $\mathbb{C}$ and is also weakly continuous from $X^{m,s}$ into $\mathbb{C}$. Then

$$\lim_{l\to+\infty} \sigma^{d_l}(x) = f^*(x).$$

Likewise, the continuity of $f^\Omega$ implies that

$$\lim_{l\to+\infty} f^\Omega(x^{d_l}) = f^\Omega(x).$$

Finally, by (I–2.2) and Sobolev's Hölder Imbedding Theorem for the space $H^{m+s}(\Omega)$ (cf. (1) in Preliminaries), we have

$$\lim_{l\to+\infty} \left(\sigma^{d_l}(x^{d_l}) - \sigma^{d_l}(x)\right) = 0.$$

The last four relations imply that $f^*|_\Omega = f$.

4) By (I–2.2) and (2) in Preliminaries, the injection from $H^{m+s}(\Omega)$ into $L^2(\Omega)$ is compact. It follows from Corollary I–2.1 and point 2) that, as $l \to +\infty$,

$$\sigma^{d_l} \to f^*, \quad \text{strongly in } L^2(\Omega). \tag{5.5}$$

On the other hand, one evidently has

$$|\sigma^{d_l} - f^*|^2_{m,s} = |\sigma^{d_l}|^2_{m,s} + |f^*|^2_{m,s} - 2\Re(\sigma^{d_l}, f^*)_{m,s},$$

where $\Re$ stands for the real part. Now, for any $l \in \mathbb{N}$, $f^*$ belongs to $\mathcal{K}^{d_l}$. Thus,

$$|\sigma^{d_l} - f^*|^2_{m,s} \le 2|f^*|^2_{m,s} - 2\Re(\sigma^{d_l}, f^*)_{m,s},$$

and, therefore,

$$\lim_{l\to+\infty} |\sigma^{d_l} - f^*|_{m,s} = 0.$$

Taking limits in the inequality $|\sigma^{d_l}|_{m,s} \le |f^\Omega|_{m,s}$, we get

$$|f^*|_{m,s} \le |f^\Omega|_{m,s}.$$

This relation shows, using the definition of $f^\Omega$ and the result of point 3), that $f^* = f^\Omega$. From (5.5), it follows that

$$\lim_{l\to+\infty} \|\sigma^{d_l} - f^\Omega\|_{m,s} = 0.$$

5) To conclude the proof, we argue by contradiction. Assume that $\|\sigma^d - f^\Omega\|_{m,s}$ does not tend to 0 when $d$ does. Then, there exists a real number $\alpha > 0$ and a sequence $(d'_l)_{l\in\mathbb{N}} \subset \mathbb{D}$, convergent to 0, such that

$$\forall l \in \mathbb{N}, \ \|\sigma^{d'_l} - f^\Omega\|_{m,s} > \alpha. \tag{5.6}$$

But the sequence $(\sigma^{d'_l})_{l\in\mathbb{N}}$ is bounded in $X^{m,s}$. A similar argument to that of points 2), 3) and 4) shows that there exists a subsequence of $(\sigma^{d'_l})_{l\in\mathbb{N}}$ which converges to $f^\Omega$ in $X^{m,s}$, leading to a contradiction with (5.6). This completes the proof. □

# 6. AUXILIARY RESULTS

In this section, we are going to introduce some results which we shall need later.

For any $t \in \mathbb{R}^n$ and for any $r > 0$, we denote by $\overline{B}(t, r)$ the closed ball of centre $t$ and radius $r$. We also recall that $\chi_E$ stands for the characteristic function of the set $E$.

***Proposition* 6.1** – *Let $\Omega$ be an open set in $\mathbb{R}^n$ with a Lipschitz-continuous boundary (cf. Preliminaries). Then, there exist constants $M > 1$, $M_1 > 1$ and $\lambda_0 > 0$ such that, for any $\lambda \in (0, \lambda_0]$, there exists $T_\lambda \subset \Omega$ verifying*

*(i)* $\forall t \in T_\lambda,\ \overline{B}(t, \lambda) \subset \Omega,$

*(ii)* $\Omega \subset \bigcup_{t \in T_\lambda} \overline{B}(t, M\lambda),$

*(iii)* $\sum_{t \in T_\lambda} \chi_{\overline{B}(t, M\lambda)} \le M_1.$

**Proof** – (J. Duchon).

1) Let $\nu = 2/\sqrt{n}$. It is clear that, for any $x \in \mathbb{R}^n$, $\delta(x, \mathbb{Z}^n) \le 1/\nu$, where $\delta$ stands for the Euclidean distance. Thus, for any $x \in \mathbb{R}^n$ and $\lambda > 0$, $\delta(x, \nu\lambda\mathbb{Z}^n) \le \lambda$. Therefore,

$$\forall x \in \mathbb{R}^n,\ \forall \lambda > 0,\ \exists t \in \nu\lambda\mathbb{Z}^n,\ \overline{B}(t, \lambda) \subset \overline{B}(x, 2\lambda).$$

2) Since the open set $\Omega$ has a Lipschitz-continuous boundary, it also has the cone property (cf. P. Grisvard [75]). Thus, there exists $\theta \in (0, \pi/2]$, $r > 0$ and, for any $x \in \Omega$, a unit vector $\xi(x) \in \mathbb{R}^n$ such that the cone

$$C\big(x, \xi(x), \theta, r\big) = \{\, x + \alpha\eta \mid \eta \in \mathbb{R}^n,\ |\eta| = 1,\ \eta \cdot \xi(x) \ge \cos\theta,\ 0 \le \alpha \le r \,\}$$

is contained in $\Omega$.

3) Let $M = 2(1 + \sin\theta)/\sin\theta$ and $\lambda_0 = r/M$. For any $\lambda \in (0, \lambda_0]$, let

$$T_\lambda = \{\, t \in \nu\lambda\mathbb{Z}^n \mid \overline{B}(t, \lambda) \subset \Omega \,\}.$$

Obviously, point (i) holds by definition of $T_\lambda$. Now, let $\lambda \in (0, \lambda_0]$. For any $x \in \Omega$, it is readily seen that

$$\overline{B}\Big(x + \frac{2\lambda}{\sin\theta}\,\xi(x), 2\lambda\Big) \subset C\big(x, \xi(x), \theta, M\lambda\big).$$

Likewise, since $M\lambda \le r$, by point 2),

$$C\big(x, \xi(x), \theta, M\lambda\big) \subset C\big(x, \xi(x), \theta, r\big) \subset \Omega.$$

But, by point 1), there exists $t \in \nu\lambda\mathbb{Z}^n$ such that

$$\overline{B}(t, \lambda) \subset \overline{B}\Big(x + \frac{2\lambda}{\sin\theta}\,\xi(x), 2\lambda\Big).$$

We deduce that $\overline{B}(t,\lambda) \subset \Omega$ and $t \in C\big(x, \xi(x), \theta, M\lambda\big)$, which implies that $t \in T_\lambda$ and $|t-x| \leq M\lambda$. Hence, $x \in \overline{B}(t, M\lambda)$ for some $t \in T_\lambda$. Therefore, point (ii) is satisfied.

4) Let $M_1 = (M\sqrt{n}+1)^n$. For any $\lambda \in (0, \lambda_0]$ and $x = (x_1, \dots, x_n) \in \mathbb{R}^n$, the closed ball $\overline{B}(x, M\lambda)$ is contained in the closed interval $\prod_{j=1}^n [x_j - M\lambda, x_j + M\lambda]$, which, in turn, contains at most $M_1$ points of $\nu\lambda\mathbb{Z}^n$. This means that $x$ belongs to at most $M_1$ closed balls with centres in $T_\lambda$ and radius $M\lambda$. In consequence, point (iii) holds. □

Let $k$ be the integer defined as follows:

$$k = \begin{cases} \text{integer part of } m+s, & m+s \notin \mathbb{N}^*, \\ m+s-1, & \text{otherwise.} \end{cases} \tag{6.1}$$

***Proposition 6.2*** – *Let $\Omega$ be an open subset of $\mathbb{R}^n$ with a Lipschitz-continuous boundary (cf. Preliminaries). Then, there exists a positive constant $C$ such that*

$$\forall v \in H^{m+s}(\Omega), \quad \min_{\psi \in P_k(\Omega)} \|v - \psi\|_{m+s,\Omega} \leq C|v|_{m+s,\Omega}.$$

**Proof** – This is a classical result when $m+s \in \mathbb{N}^*$ (cf., for example, J. Nečas [109]). For the case $m+s \notin \mathbb{N}^*$, see A. M. Sanchez and R. Arcangéli [127]. □

Let $\Omega$ be now an open subset of $\mathbb{R}^n$ with a Lipschitz-continuous boundary (cf. Preliminaries), $P$, a prolongation operator from $H^{m+s}(\Omega)$ into $H^{m+s}(\mathbb{R}^n)$, $I$, the identity operator in $H^{m+s}(\Omega)$, $\Pi_k$, the projection operator from $H^{m+s}(\Omega)$ onto $P_k(\Omega)$, and $E$, the operator that assigns to any polynomial function over $\Omega$ the same polynomial function over $\mathbb{R}^n$. We write

$$\widetilde{P} = P(I - \Pi_k) + E\Pi_k. \tag{6.2}$$

It is clear that $\widetilde{P}$ is a prolongation operator from $H^{m+s}(\Omega)$ into the semi-normed space $\{\, v \in \mathcal{D}' \mid |v|_{m+s,\mathbb{R}^n} < +\infty \,\}$. Moreover, we have the following result, established by G. Geymonat (cf. G. Strang [136]).

***Proposition 6.3*** – *Let $\Omega$ be an open subset of $\mathbb{R}^n$ with a Lipschitz-continuous boundary (cf. Preliminaries) and let $\widetilde{P}$ be the operator defined in (6.2). Then, there exists a constant $C > 0$ such that*

$$\forall v \in H^{m+s}(\Omega), \ |\widetilde{P}v|_{m+s,\mathbb{R}^n} \leq C|v|_{m+s,\Omega}.$$

**Proof** – Since the semi-norm $|\cdot|_{m+s,\mathbb{R}^n}$ is null over $P_k$,

$$\forall v \in H^{m+s}(\Omega), \ |\widetilde{P}v|_{m+s,\mathbb{R}^n} = |P(I-\Pi_k)v|_{m+s,\mathbb{R}^n} \leq \|P(I-\Pi_k)v\|_{m+s,\mathbb{R}^n}.$$

By definition of $P$, we have

$$\exists C > 0, \ \forall v \in H^{m+s}(\Omega), \ |\widetilde{P}v|_{m+s,\mathbb{R}^n} \leq C\|(I-\Pi_k)v\|_{m+s,\Omega},$$

from which, by definition of $\Pi_k$, we deduce that

$$\exists C > 0, \ \forall v \in H^{m+s}(\Omega), \ \forall \psi \in P_k(\Omega), \ |\widetilde{P}v|_{m+s,\mathbb{R}^n} \leq C\|(I-\Pi_k)(v-\psi)\|_{m+s,\Omega}.$$

Consequently,

$$\exists C > 0,\ \forall v \in H^{m+s}(\Omega),\ \forall \psi \in P_k(\Omega),\ |\widetilde{P}v|_{m+s,\mathbb{R}^n} \leq C\|(v - \psi)\|_{m+s,\Omega}.$$

Proposition 6.2 then gives the result. □

We recall that two open subsets $\Omega$ and $\widehat{\Omega}$ of $\mathbb{R}^n$ are *affine-equivalent* if there exists an invertible affine mapping $F : \hat{x} \mapsto L\hat{x} + c$, with $L \in \mathcal{L}(\mathbb{R}^n, \mathbb{R}^n)$ and $c \in \mathbb{R}^n$, such that $\Omega = F(\widehat{\Omega})$ (cf. P. G. Ciarlet [45]). Any function $\hat{v} : \widehat{\Omega} \to \mathbb{R}$ (respectively, any function $v : \Omega \to \mathbb{R}$) is associated with the function $v = \hat{v} \circ F^{-1}$ (respectively, the function $\hat{v} = v \circ F$). In these conditions, we have

***Proposition* 6.4** – *Let $\Omega$ and $\widehat{\Omega}$ be two affine-equivalent open subsets of $\mathbb{R}^n$ with a Lipschitz-continuous boundary (cf. Preliminaries). Then, for any function $v$ belonging to $H^{m+s}(\Omega)$, the function $\hat{v}$ belongs to $H^{m+s}(\widehat{\Omega})$ and vice versa. Moreover, there exists a constant $C > 0$, which only depends on $n$ and $m + s$, such that*

*(i)* $\forall v \in H^{m+s}(\Omega),\ |\hat{v}|_{m+s,\widehat{\Omega}} \leq C\|L\|^{\mu}\,|\det L|^{-\nu}\,|v|_{m+s,\Omega},$

*(ii)* $\forall \hat{v} \in H^{m+s}(\widehat{\Omega}),\ |v|_{m+s,\Omega} \leq C\|L^{-1}\|^{\mu}\,|\det L|^{\nu}\,|\hat{v}|_{m+s,\widehat{\Omega}},$

*where either $\mu = m + s$ and $\nu = 1/2$, if $m + s \in \mathbb{N}^*$, or $\mu = m + s + n/2$ and $\nu = 1$, otherwise.*

**Proof** – If $m + s \in \mathbb{N}^*$, cf. P. G. Ciarlet [45]. Otherwise, cf. A. M. Sanchez and R. Arcangéli [127]. □

We conclude this section with two results relative to Lagrange interpolation in $\mathbb{R}^n$. Here, as well as in Sections 7 and III–4, we denote by $\mathfrak{K}$ the dimension of the space $P_k$.

***Proposition* 6.5** – *Let $\Omega$ be an open subset of $\mathbb{R}^n$ with a Lipschitz-continuous boundary (cf. Preliminaries). Let $B \subset \Omega^{\mathfrak{K}}$ be a compact set of $P_k$-unisolvent $\mathfrak{K}$-tuples $b = (b_1, \ldots, b_{\mathfrak{K}})$, and, for any $b \in B$, let $\Pi^b$ be the Lagrange $P_k$-interpolating operator, defined, for any $v \in H^{m+s}(\Omega)$, by*

$$\Pi^b v \in P_k(\Omega) \text{ and, for } j = 1, \ldots, \mathfrak{K},\ \Pi^b v(b_j) = v(b_j).$$

*Then, there exists $C > 0$ (dependent on $\Omega$, $B$ and $m + s$) such that*

$$\forall b \in B,\ \forall v \in H^{m+s}(\Omega),\ \|v - \Pi^b v\|_{m+s,\Omega} \leq C|v|_{m+s,\Omega}.$$

**Proof** – (cf. J. Duchon [56]).

1) By (I–2.2) and Sobolev's Continuous Imbedding Theorem (cf. (4) in Preliminaries), for any $b \in B$, $I - \Pi^b$ is a linear continuous operator from $H^{m+s}(\Omega)$ into $H^{m+s}(\Omega)$. We deduce that, for any $b \in B$, there exists a constant $C(b)$ such that

$$\forall v \in H^{m+s}(\Omega),\ \|v - \Pi^b v\|_{m+s,\Omega} \leq C(b)\|v\|_{m+s,\Omega}.$$

2) Let us prove that

$$\sup_{b\in B} C(b) < +\infty. \tag{6.3}$$

For this, it is sufficient to show that, for any $v \in H^{m+s}(\Omega)$, the set $\{\Pi^b v \mid b \in B\}$ is bounded in $H^{m+s}(\Omega)$. The relation (6.3) then follows by applying the Banach-Steinhaus Theorem (cf. H. Brézis [40, Theorem II.1]) to the family of operators $(I - \Pi^b)_{b\in B}$.

Let $p_1, \dots, p_{\mathfrak{K}}$ be a basis of $P_k(\Omega)$ and, for any $b \in B$, let $M(b) = \big(p_i(b_j)\big)_{1\le i,j\le \mathfrak{K}}$. The matrix $M(b)$ is regular, since $b$ is $P_k$-unisolvent. Denoting by $m'_{ij}(b)$ the generic element of the inverse matrix $M(b)^{-1}$, for any $v \in H^{m+s}(\Omega)$, we have

$$\Pi^b v = \sum_{i,j=1}^{\mathfrak{K}} v(b_i) m'_{ij}(b) p_j.$$

But, on the one hand, $v \in H^{m+s}(\Omega)$ is continuous on $\overline{\Omega}$ and, on the other hand, since matrix inversion is a continuous operation, each function $m'_{ij}$ is bounded on the compact set $B$. We deduce that, for any $v \in H^{m+s}(\Omega)$, $\|\Pi^b v\|_{m+s,\Omega}$ remains bounded when $b$ varies in $B$.

3) Since

$$\forall b \in B,\ \forall \psi \in P_k(\Omega),\ \Pi^b \psi = \psi,$$

by the preceding points, there exists a constant $C$ such that

$$\forall b \in B,\ \forall v \in H^{m+s}(\Omega),\ \forall \psi \in P_k(\Omega),\ \|v - \Pi^b v\|_{m+s,\Omega} \le C\|v - \psi\|_{m+s,\Omega}.$$

Then, using Proposition 6.2, the result follows. □

***Proposition* 6.6** – *There exists $R > 1$ (dependent on $n$ and $m+s$) and, for any $M \ge 1$, a constant $C$ (dependent on $M$, $n$ and $m+s$) verifying the following property: for any $d > 0$ and any $t \in \mathbb{R}^n$, the ball $\overline{B}(t, Rd)$ contains $\mathfrak{K}$ closed balls $\mathcal{B}_1, \dots, \mathcal{B}_{\mathfrak{K}}$ of radius $d$ such that, for any $v \in H^{m+s}\big(\overline{B}(t, MRd)\big)$ which is null in at least one point of each of the balls $\mathcal{B}_1, \dots, \mathcal{B}_{\mathfrak{K}}$, we have*

$$\forall l = 0, \dots, k,\ |v|_{l,\overline{B}(t,MRd)} \le C d^{m+s-l} |v|_{m+s,\overline{B}(t,MRd)}.$$

**Proof** – (Cf. J. Duchon [56]).

1) Let $b^0 = (b^0_1, \dots, b^0_{\mathfrak{K}}) \in (\mathbb{R}^n)^{\mathfrak{K}}$ be a $P_k$-unisolvent $\mathfrak{K}$-tuple. Let us see that there exists $r_0 > 0$ such that any $\mathfrak{K}$-tuple $b \in \prod_{j=1}^{\mathfrak{K}} \overline{B}(b^0_j, r_0)$ is $P_k$-unisolvent. For this, we reason as in the proof of Proposition I–2.3.

Let $\Omega$ be an open subset with a Lipschitz-continuous boundary (cf. Preliminaries) such that $b^0 \in \Omega^{\mathfrak{K}}$. Since $k+1 \ge m+s$, then $H^{k+1}(\Omega) \hookrightarrow C^0(\overline{\Omega})$. Using the inequality

$$\frac{1}{2}\sum_{j=1}^{\mathfrak{K}} |v(b^0_j)|^2 \le \sum_{j=1}^{\mathfrak{K}} |v(b^0_j) - v(b_j)|^2 + \sum_{j=1}^{\mathfrak{K}} |v(b_j)|^2$$

and Sobolev's Hölder Imbedding Theorem for the space $H^{k+1}(\Omega)$ (cf. (1) in Preliminaries), it is seen that there exists $r_0 > 0$ such that, for any $b = (b_1, \dots, b_{\mathfrak{K}}) \in \prod_{j=1}^{\mathfrak{K}} \overline{B}(b_j^0, r_0)$, the mapping $[\![ \cdot ]\!]_{b,k+1,\Omega}$, defined by

$$[\![v]\!]_{b,k+1,\Omega} = \left( \sum_{j=1}^{\mathfrak{K}} |v(b_j)|^2 + |v|_{k+1,\Omega}^2 \right)^{1/2},$$

is a norm on $H^{k+1}(\Omega)$.

Now, let $b = (b_1, \dots, b_{\mathfrak{K}}) \in \prod_{j=1}^{\mathfrak{K}} \overline{B}(b_j^0, r_0)$. For any $\psi \in P_k(\Omega)$ such that, for $j = 1, \dots, \mathfrak{K}$, $\psi(b_j) = 0$, we have $[\![\psi]\!]_{b,k+1,\Omega} = 0$ and hence $\psi = 0$. Therefore, $b$ is $P_k$-unisolvent.

2) By a homothecy of reason $1/r_0$, writing $\hat{\alpha}_j = (1/r_0) b_j^0$, we obtain $\mathfrak{K}$ balls $\overline{B}(\hat{\alpha}_j, 1)$ such that the product $\prod_{j=1}^{\mathfrak{K}} \overline{B}(\hat{\alpha}_j, 1)$, by point 1), is a compact subset of $(\mathbb{R}^n)^{\mathfrak{K}}$ formed by $P_k$-unisolvent $\mathfrak{K}$-tuples. The set $\bigcup_{j=1}^{\mathfrak{K}} \overline{B}(\hat{\alpha}_j, 1)$ is bounded and so contained in a ball $\overline{B}(\hat{a}, R)$ whose radius $R > 1$ depends on $n$ and $k$, and hence, on $n$ and $m+s$.

Let $M \geq 1$. We can apply Proposition 6.5 to the open ball $B(\hat{a}, MR)$. Thus, there exists a constant $C$ (depending on $M$, $n$ and $m+s$) such that, for any $\hat{v} \in H^{m+s}(B(\hat{a}, MR))$ and for any $b \in \prod_{j=1}^{\mathfrak{K}} \overline{B}(\alpha_j, 1)$,

$$\|\hat{v} - \Pi^b \hat{v}\|_{m+s,\overline{B}(\hat{a},MR)} \leq C |\hat{v}|_{m+s,\overline{B}(\hat{a},MR)}.$$

In other words, there exists $C$ such that, for any integer $l = 0, \dots, k$ and for any $\hat{v} \in H^{m+s}(\overline{B}(\hat{a}, MR))$ which is null in one point of each of the balls $\overline{B}(\hat{\alpha}_j, 1)$,

$$|\hat{v}|_{l,\overline{B}(\hat{a},MR)} \leq C |\hat{v}|_{m+s,\overline{B}(\hat{a},MR)}.$$

For any $d > 0$ and any $t \in \mathbb{R}^n$, let $F_t^d$ be the invertible affine mapping $x \mapsto t + d(x - \hat{a})$. This mapping transforms the ball $\overline{B}(\hat{a}, MR)$ into the ball $\overline{B}(t, MRd)$, the ball $\overline{B}(\hat{a}, R)$ into the ball $\overline{B}(t, Rd)$ and, for any $j = 1, \dots, \mathfrak{K}$, the ball $\overline{B}(\hat{\alpha}_j, 1)$ into a ball $\mathcal{B}_j$ of radius $d$ contained in the ball $\overline{B}(t, Rd)$.

Writing $v = \hat{v} \circ (F_t^d)^{-1}$ and applying Proposition 6.4 (where, in this case, $\|L\| = d$, $\|L^{-1}\| = 1/d$ and $\det L = d^n$), we see that there exists a constant $C$ such that, for any integer $l = 0, \dots, k$ and for any $v \in H^{m+s}(\overline{B}(t, MRd))$ which is null in one point of each ball $\mathcal{B}_j$,

$$|v|_{l,\overline{B}(t,MRd)} \leq C d^{-l+n/2} |\hat{v}|_{l,\overline{B}(\hat{a},MR)}$$

and

$$|\hat{v}|_{m+s,\overline{B}(\hat{a},MR)} \leq C d^{m+s-n/2} |v|_{m+s,\overline{B}(t,MRd)}.$$

The result then follows. □

## 7. *ESTIMATES OF THE APPROXIMATION ERROR*

Let us consider again the framework introduced in Section 5. We recall that $\Omega$ is an open subset of $\mathbb{R}^n$ with a Lipschitz-continuous boundary (cf. Preliminaries), $f$ is a given function in $H^{m+s}(\Omega)$ and $\mathbb{D}$ is a subset of $(0,+\infty)$ such that $0 \in \overline{\mathbb{D}}$. Likewise, for any $d \in \mathbb{D}$, $A^d$ denotes a finite subset of $\overline{\Omega}$ verifying (5.1), and $\sigma^d$ stands for the interpolating $(m,s)$-spline relative to $A^d$ and $\beta^d = \big(f(a)\big)_{a\in A^d}$.

From now on, for simplicity, we shall write $\sigma^d$ instead of $\sigma^d|_\Omega$.

***Theorem* 7.1** – *Suppose that* (5.1) *holds. Then, there exist positive constants* $R$, $\lambda_0$ *and* $C$ *such that, for any* $d \in \mathbb{D} \cap (0, \lambda_0/R]$, *we have*

$$\forall l = 0,\ldots,k,\ |f-\sigma^d|_{l,\Omega} \le C d^{m+s-l} |f-\sigma^d|_{m+s,\Omega},$$

*where* $k$ *is the integer defined in* (6.1).

**Proof** – (Cf. J. Duchon [56]).

1) Let $R$ be the constant of Proposition 6.6 and let $M > 1$, $\lambda_0 > 0$ and $M_1 > 1$ be the constants introduced in Proposition 6.1. We recall that, for any $\lambda \in (0, \lambda_0]$, there exists a set $T_\lambda \subset \Omega$ verifying

$$\Omega \subset \bigcup_{t\in T_\lambda} \overline{B}(t, M\lambda).$$

For any $d \in \mathbb{D} \cap (0,\lambda_0/R]$, let $u^d = f - \sigma^d$. By the above property of $T_\lambda$, with $\lambda = Rd$, we have

$$\forall d \in \mathbb{D} \cap (0,\lambda_0/R],\ \forall l = 0,\ldots,k,\ |u^d|^2_{l,\Omega} \le |\tilde{u}^d|^2_{l,\bigcup_{t\in T_{Rd}} \overline{B}(t,MRd)},$$

where $\tilde{u}^d = \widetilde{P}u^d$, $\widetilde{P}$ being the operator defined by (6.2). Then,

$$\forall d \in \mathbb{D} \cap (0,\lambda_0/R],\ \forall l = 0,\ldots,k,\ |u^d|^2_{l,\Omega} \le \sum_{t\in T_{Rd}} |\tilde{u}^d|^2_{l,\overline{B}(t,MRd)}.$$

Now, it is clear that, for any $d \in \mathbb{D} \cap (0,\lambda_0/R]$ and any $t \in T_{Rd}$, $\tilde{u}^d$ belongs to $H^{m+s}(\overline{B}(t, MRd))$. Likewise, by (5.1), $\tilde{u}^d$ is null in at least one point of each of the balls $\mathcal{B}_1,\ldots,\mathcal{B}_{\mathfrak{K}}$ of radius $d$ (where $\mathfrak{K}$ denotes the dimension of $P_k$), contained in the ball $\overline{B}(t, Rd)$, associated with $d$ and $t$ by Proposition 6.6. Using this proposition, we deduce that there exists a constant $C$, which depends on $M$, $n$ and $m+s$, such that

$$\forall d \in \mathbb{D} \cap (0,\lambda_0/R],\ \forall t \in T_{Rd},$$
$$\forall l = 0,\ldots,k,\ |\tilde{u}^d|_{l,\overline{B}(t,MRd)} \le C d^{m+s-l} |\tilde{u}^d|_{m+s,\overline{B}(t,MRd)}.$$

Therefore,

$$\forall d \in \mathbb{D} \cap (0,\lambda_0/R],\ \forall l = 0,\ldots,k,\ |u^d|^2_{l,\Omega} \le C^2 d^{2(m+s-l)} \sum_{t\in T_{Rd}} |\tilde{u}^d|^2_{m+s,\overline{B}(t,MRd)}.$$

2) If $m+s \in \mathbb{N}^*$, for any $d \in \mathbb{D} \cap (0,\lambda_0/R]$, we have

$$\sum_{t\in T_{Rd}} |\tilde{u}^d|^2_{m+s,\overline{B}(t,MRd)} = \sum_{t\in T_{Rd}} \int_{\mathbb{R}^n} \chi_{\overline{B}(t,MRd)} \Big( \sum_{|\alpha|=m+s} |\partial^\alpha \tilde{u}^d(x)|^2 \Big)\, dx,$$

and so, using point (iii) of Proposition 6.1, we obtain

$$\forall d \in \mathbb{D} \cap (0, \lambda_0/R], \ \sum_{t \in T_{Rd}} |\tilde{u}^d|^2_{m+s,\overline{B}(t,MRd)} \le M_1 |\tilde{u}^d|^2_{m+s,\mathbb{R}^n}. \qquad (7.1)$$

If $m+s \notin \mathbb{N}^*$, using the same result, we get

$$\forall d \in \mathbb{D} \cap (0, \lambda_0/R], \ \sum_{t \in T_{Rd}} \chi_{Q(t)} \le M_1,$$

where $Q(t) = \overline{B}(t, MRd) \times \overline{B}(t, MRd)$. This relation, together with the definition of the semi-norm $|\cdot|_{m+s,\overline{B}(t,MRd)}$, implies that, in this case, (7.1) also holds.

3) From points 1) and 2), we deduce that there exists $C > 0$ such that

$$\forall d \in \mathbb{D} \cap (0, \lambda_0/R], \ \forall l = 0, \ldots, k, \ |u^d|_{l,\Omega} \le C d^{m+s-l} |\tilde{u}^d|_{m+s,\mathbb{R}^n}. \qquad (7.2)$$

The Theorem is then a simple consequence of (7.2) and Proposition 6.3. □

***Corollary* 7.1** – *Suppose that* (5.1) *holds. Then, we have*

$$\forall l = 0, \ldots, k, \ |f - \sigma^d|_{l,\Omega} = o(d^{m+s-l}), \ d \to 0,$$

*and also evidently*

$$|f - \sigma^d|_{m+s,\Omega} = o(1), \ d \to 0.$$

**Proof** – It is sufficient to apply Theorems 7.1 and 5.1 and Corollary I–2.1. □

**Remark 7.1** – Let $n = 2$. For the thin plate splines (i.e. $(2,0)$-splines), of class $C^1$, we obtain error estimates of the form

$$|f - \sigma^d|_{l,\Omega} = o(d^{2-l}), \ d \to 0, \quad l = 0, 1, 2,$$

whereas for the pseudo-cubic splines (i.e. $(2, 1/2)$-splines), of class $C^2$ and easier to compute, we obtain

$$|f - \sigma^d|_{l,\Omega} = o(d^{5/2-l}), \ d \to 0, \quad l = 0, 1, 2. \ \square$$

# *CHAPTER III*

# *SMOOTHING* $(m, s)$*-SPLINES*

As in Chapter II, hereafter we assume that $m \in \mathbb{N}^*$, $n \in \mathbb{N}^*$ and $s \in \mathbb{R}$ satisfy the hypothesis (I–2.2).

## *1. DEFINITION AND FIRST PROPERTIES*

We keep the definition of $A$, $N$, $\rho$ and $\mathcal{K}_0$ given in Section II–1. Likewise, we denote by $\langle \cdot \rangle$ (respectively by $\langle \cdot , \cdot \rangle$) the Hermitian norm (respectively, the Hermitian scalar product) in $\mathbb{C}^N$. For any $\varepsilon > 0$ and any $\beta \in \mathbb{C}^N$, we write

$$\forall v \in X^{m,s},\ J_\varepsilon(v) = \langle \rho v - \beta \rangle^2 + \varepsilon |v|^2_{m,s}.$$

Then, we call *smoothing* $(m, s)$*-spline relative to* $A$*,* $\beta$ *and* $\varepsilon$ any solution, if any exists, of the problem: find $\sigma_\varepsilon$ such that

$$\begin{cases} \sigma_\varepsilon \in X^{m,s}, \\ \forall v \in X^{m,s},\ J_\varepsilon(\sigma_\varepsilon) \leq J_\varepsilon(v). \end{cases} \tag{1.1}$$

**Theorem 1.1** – *Problem* (1.1) *admits a unique solution* $\sigma_\varepsilon$*, which is also the unique solution of the variational problem: find* $\sigma_\varepsilon$ *such that*

$$\begin{cases} \sigma_\varepsilon \in X^{m,s}, \\ \forall v \in X^{m,s},\ \langle \rho\sigma_\varepsilon, \rho v \rangle + \varepsilon(\sigma_\varepsilon, v)_{m,s} = \langle \beta, \rho v \rangle. \end{cases} \tag{1.2}$$

*In addition, the smoothing* $(m, s)$*-spline* $\sigma_\varepsilon$ *relative to* $A$*,* $\beta$ *and* $\varepsilon$ *belongs to the space* $S$ *of the* $(m, s)$*-spline functions relative to* $A$*, defined in* (II–2.1).

**Proof** – Let us endow $X^{m,s}$ with the norm $[\![ \cdot ]\!]_{A,m,s}$, defined in (I–2.5) with $E = A$, for which, by Theorem I–2.2 and Proposition I–2.2, $X^{m,s}$ is a Hilbert space. On the one hand, taking into account that $\rho \in \mathcal{L}(X^{m,s}, \mathbb{C}^N)$, the mapping

$$(u, v) \mapsto \langle \rho u, \rho v \rangle + \varepsilon(u, v)_{m,s}$$

is sesquilinear with hermitian symmetry, continuous on $X^{m,s} \times X^{m,s}$, since

$$\forall u, v \in X^{m,s},\ |\langle \rho u, \rho v \rangle + \varepsilon(u, v)_{m,s}| \leq \max(1, \varepsilon)[\![u]\!]_{A,m,s}[\![v]\!]_{A,m,s},$$

and $X^{m,s}$-elliptic, since

$$\forall v \in X^{m,s},\ \langle \rho v \rangle^2 + \varepsilon |v|^2_{m,s} \geq \min(1, \varepsilon)[\![v]\!]^2_{A,m,s}.$$

On the other hand, the mapping $v \mapsto \langle \beta, \rho v \rangle$ is continuous antilinear on $X^{m,s}$. The Lax-Milgram Lemma (cf. P. G. Ciarlet [45, Theorem 1.1.3]) then proves that (1.1) and (1.2) admit the same unique solution $\sigma_\varepsilon$. Finally, we derive from (1.2) that, for all $w \in \mathcal{K}_0$, $(\sigma_\varepsilon, w)_{m,s} = 0$, and then $\sigma_\varepsilon \in S$. □

The following result shows that the smoothing $(m, s)$-spline $\sigma_\varepsilon$ relative to $A$, $\beta$ and $\varepsilon$ is an *approximant*, as $\varepsilon \to 0$, of the interpolating $(m, s)$-spline $\sigma$ relative to $A$ and $\beta$.

***Theorem* 1.2** – *As $\varepsilon \to 0$, we have $\|\sigma_\varepsilon - \sigma\|_{m,s} = O(\varepsilon)$.*

**Proof** – Taking into account that $\beta = \rho\sigma$ and the definition of the scalar product $[\![\cdot, \cdot]\!]_{A,m,s}$ associated with the norm $[\![\cdot]\!]_{A,m,s}$, from (1.2) we deduce that

$$\begin{aligned}\forall \varepsilon > 0,\ \forall v \in X^{m,s},\ (1-\varepsilon)\langle \rho(\sigma_\varepsilon - \sigma), \rho v\rangle + \varepsilon[\![\sigma_\varepsilon - \sigma, v]\!]_{A,m,s} \\ = -\varepsilon[\![\sigma, v]\!]_{A,m,s} + \varepsilon\langle \rho\sigma, \rho v\rangle,\end{aligned}$$

from which, denoting by $\rho^* \in \mathcal{L}(\mathbb{C}^N, X^{m,s})$ the Hilbertian adjoint operator of $\rho$ for the scalar product $[\![\cdot, \cdot]\!]_{A,m,s}$, we get

$$\forall \varepsilon > 0,\ (1-\varepsilon)\rho^*\rho(\sigma_\varepsilon - \sigma) + \varepsilon(\sigma_\varepsilon - \sigma) = -\varepsilon\sigma + \varepsilon\rho^*\rho\sigma. \tag{1.3}$$

Now, since $\dim(\operatorname{Im}\rho^*) < +\infty$, $\sigma \in S = (\operatorname{Ker}\rho)^\perp = \overline{\operatorname{Im}\rho^*} = \operatorname{Im}\rho^*$, where $\operatorname{Ker}\rho$ and $\operatorname{Im}\rho^*$ are, respectively, the null space of $\rho$ and the range of $\rho^*$. Hence, there exists $\xi \in \mathbb{C}^N$ such that $-\sigma = \rho^*\xi$. Likewise, by Proposition II–2.1, there exists $y \in S$ such that $\xi = \rho y$. Let $z = \sigma + y$. It follows from (1.3) that

$$\forall \varepsilon > 0,\ (1-\varepsilon)\rho^*\rho\frac{\sigma_\varepsilon - \sigma}{\varepsilon} + \varepsilon\frac{\sigma_\varepsilon - \sigma}{\varepsilon} = \rho^*\rho z,$$

or, equivalently,

$$\forall \varepsilon > 0,\ (1-\varepsilon)\rho^*\rho(v_\varepsilon - z) + \varepsilon(v_\varepsilon - z) = \varepsilon(\rho^*\rho z - z), \tag{1.4}$$

with $v_\varepsilon = \frac{\sigma_\varepsilon - \sigma}{\varepsilon}$. If we form the scalar product $[\![\cdot, \cdot]\!]_{A,m,s}$, of both sides of (1.4) with $v_\varepsilon - z$, we deduce that

$$\forall \varepsilon > 0,\ (1-\varepsilon)\langle\rho(v_\varepsilon - z)\rangle^2 + \varepsilon[\![v_\varepsilon - z]\!]^2_{A,m,s} \le \varepsilon[\![\rho^*\rho z - z]\!]_{A,m,s}[\![v_\varepsilon - z]\!]_{A,m,s}.$$

This implies that

$$\forall \varepsilon \in (0,1),\ [\![v_\varepsilon - z]\!]_{A,m,s} \le [\![\rho^*\rho z - z]\!]_{A,m,s},$$

and, therefore, the Theorem follows. □

## 2. COMPUTATION OF SMOOTHING SPLINES

By Theorem 1.1, the smoothing $(m, s)$-spline $\sigma_\varepsilon$ relative to $A$, $\beta$ and $\varepsilon$ belongs to $S$. Thus, by Theorem II–2.1, $\sigma_\varepsilon$ is written in a unique way in the form

$$\sigma_\varepsilon(x) = \sum_{a \in A} \lambda_a^* K_{2m+2s-n}(x - a) + \sum_{j=1}^{\mathfrak{M}} c_j p_j(x),$$

with

$$\forall j = 1, \ldots, \mathfrak{M}, \ \sum_{a \in A} \lambda_a^* p_j(a) = 0,$$

where $\{p_1, \ldots, p_{\mathfrak{M}}\}$ denotes a basis of $P_{m-1}$, and $\lambda_a^*$, for $a \in A$, and $c_j$, for $j = 1, \ldots, \mathfrak{M}$, are coefficients. Likewise, reasoning as in the proof of Theorem II–2.1 (see the relations (II–2.7) with $u = \sigma_\varepsilon$ and (II–2.9)), we deduce that

$$\forall v \in X^{m,s}, \ (\sigma_\varepsilon, v)_{m,s} = C^* \sum_{a \in A} \lambda_a^* \overline{v(a)},$$

where

$$C^* = \begin{cases} (2\pi)^{2m}/C_1, & 2m + 2s - n \notin 2\mathbb{N}^*, \\ (2\pi)^{2m}/C_2, & 2m + 2s - n \in 2\mathbb{N}^*. \end{cases}$$

Here, $C_1$ and $C_2$ denote the constants introduced in Proposition II–2.2, explicitly given in Remark II–2.1. In particular, for the cases of thin plate and pseudo-cubic splines in $\mathbb{R}^2$ (cf. Examples II–2.1 and II–2.2), the values of $C^*$ are, respectively, $8\pi$ and 9.

For any $b \in A$, let $\varphi_b \subset X^{m,s}$ be a function such that

$$\forall a \in A, \ \varphi_b(a) = \begin{cases} 1, & a = b, \\ 0, & a \neq b. \end{cases}$$

Taking $v = \varphi_b$ in (1.2), we obtain the linear system of order $N + \mathfrak{M}$

$$\begin{cases} \varepsilon C^* \lambda_b^* + \sum_{a \in A} \lambda_a^* K_{2m+2s-n}(b - a) + \sum_{j=1}^{\mathfrak{M}} c_j p_j(b) = \beta_b, & b \in A, \\ \sum_{a \in A} \lambda_a^* p_j(a) = 0, \quad 1 \leq j \leq \mathfrak{M}. \end{cases} \tag{2.1}$$

Thus, it suffices to solve (2.1) in order to determine $\sigma_\varepsilon$. Let us observe that the preceding linear system can be straightforwardly written in terms of the parameter $\varepsilon^* = \varepsilon C^*$. So, one can decide to fix the value of $\varepsilon^*$ instead of that of $\varepsilon$. In real problems, the choice of $\varepsilon^*$ can be made directly, either by *trial and error*, or by statistical methods, such as the *generalized cross validation method* (cf. P. Craven and G. Wahba [49], C. Gu [76], G. Wahba [149]). Hence, in practice, the exact value of the constant $C^*$ is not needed.

Let us also note that the matrix $G_\varepsilon$ of (2.1) is deduced from the matrix $G$ of the linear system (II–4.1), corresponding to the interpolating spline relative to $A$ and $\beta$, by the relation

$$G_\varepsilon = G + \varepsilon C^* J,$$

where $J$ is obtained from the identity matrix of order $N + \mathfrak{M}$ by replacing the $\mathfrak{M}$ last diagonal elements by 0.

***Theorem* 2.1** – *For any $\varepsilon > 0$, the matrix $G_\varepsilon$ is regular.*

**Proof** – Let us prove that the homogeneous system associated with (2.1) admits only the null solution. To this end, let us suppose that the coefficients $\lambda_a^*$, for $a \in A$, and $c_1, \ldots, c_{\mathfrak{M}}$ are a solution of the system (2.1) for $\beta = 0$. Let

$$u = \sum_{a \in A} \lambda_a^* \delta_a * K_{2m+2s-n} + \sum_{j=1}^{\mathfrak{M}} c_j p_j.$$

Reasoning as in the proof of Theorem II–4.2, we see that $u$ belongs to $X^{m,s}$ and verifies (II–4.2), where $C = (2\pi)^{-2m} C^*$. Likewise, it is clear that

$$\forall v \in X^{m,s}, \ \langle \rho u, \rho v \rangle = \sum_{b \in A} \Big( \sum_{a \in A} \lambda_a^* K_{2m+2s-n}(b-a) + \sum_{j=1}^{\mathfrak{M}} c_j p_j(b) \Big) \overline{v(b)}.$$

Using (2.1) with $\beta = 0$, we obtain

$$\forall v \in X^{m,s}, \ \langle \rho u, \rho v \rangle = -\varepsilon C^* \sum_{b \in A} \lambda_b^* \overline{v(b)}.$$

Taking (II–4.2) into account, we get

$$\forall v \in X^{m,s}, \ \langle \rho u, \rho v \rangle + \varepsilon (u, v)_{m,s} = 0.$$

Thus, $u$ is a solution of problem (1.2) with $\beta = 0$, and so $u = 0$. This implies, by (II–4.2), that $\lambda_a^* = 0$, for any $a \in A$. Finally, since $A$ contains a $P_{m-1}$-unisolvent subset, we derive from (2.1), with $\beta = 0$, that $c_j = 0$, for $j = 1, \ldots, \mathfrak{M}$. The Theorem then follows. □

## 3. CONVERGENCE RESULTS

As in Sections II–5 and II–7, suppose we are given an open set $\Omega \subset \mathbb{R}^n$ with a Lipschitz-continuous boundary (cf. Preliminaries), a function $f \in H^{m+s}(\Omega)$, a set $\mathbb{D} \subset (0, +\infty)$ such that $0 \in \overline{\mathbb{D}}$ and, for any $d \in \mathbb{D}$, a finite subset $A^d$ of $N = N(d)$ points of $\overline{\Omega}$ verifying (II–5.1). For any $d \in \mathbb{D}$, let $\rho^d \in \mathcal{L}(X^{m,s}, \mathbb{C}^N)$ be the operator defined in (II–5.4), and, for any $d \in \mathbb{D}$ and any $\varepsilon > 0$, let us denote by $\sigma_\varepsilon^d$ the smoothing $(m, s)$-spline relative to $A^d$, $\beta^d = \big(f(a)\big)_{a \in A^d}$ and $\varepsilon$, whose existence and uniqueness follow from the fact that $A^d$ contains a $P_{m-1}$-unisolvent subset (cf. Remark II–5.1).

For the sake of simplicity, we shall suppose that $\varepsilon$ *is a function of* $d$ and that $\varepsilon : \mathbb{D} \to (0, +\infty)$ verifies

$$\varepsilon = o(d^{-n}), \ d \to 0. \tag{3.1}$$

From now on, we shall write $\varepsilon$ instead of $\varepsilon(d)$. Let us observe that, from a practical point of view, it is quite reasonable to assume that $\varepsilon$ depends on $d$. This allows us to consider that first we are given the data family $(A^d)_{d \in \mathbb{D}}$ and then we have to choose the corresponding values of $\varepsilon$.

***Theorem* 3.1** – *Suppose that* (II–5.1) *and* (3.1) *hold. Then,*

$$\lim_{d \to 0} \| \sigma_\varepsilon^d - f^\Omega \|_{m,s} = 0,$$

*where $f^{\Omega}$ denotes the unique element of minimal semi-norm $|\cdot|_{m,s}$ in the set $\{v \in X^{m,s} \mid v|_{\Omega} = f\}$.*

**Proof –**

1) Let us prove that, for some $d^* > 0$, the family $(\sigma_\varepsilon^d)_{d\in\mathbb{D}\cap(0,d^*]}$ is bounded in $X^{m,s}$. By definition of $\sigma_\varepsilon^d$ and $f^{\Omega}$ (whose existence and uniqueness have been justified in the proof of Theorem II–5.1), we have

$$\forall d \in \mathbb{D},\ \langle \rho^d \sigma_\varepsilon^d - \beta^d \rangle^2 + \varepsilon |\sigma_\varepsilon^d|_{m,s}^2 \le \langle \rho^d f^{\Omega} - \beta^d \rangle^2 + \varepsilon |f^{\Omega}|_{m,s}^2 = \varepsilon |f^{\Omega}|_{m,s}^2. \tag{3.2}$$

We deduce that

$$\forall d \in \mathbb{D},\ |\sigma_\varepsilon^d|_{m,s} \le |f^{\Omega}|_{m,s} \tag{3.3}$$

and that

$$\forall d \in \mathbb{D},\ \langle \rho^d(\sigma_\varepsilon^d - f^{\Omega}) \rangle^2 \le \varepsilon |f^{\Omega}|_{m,s}^2. \tag{3.4}$$

Let $B_0 = \{b_{01}, \dots, b_{0\mathfrak{M}}\}$ be a $P_{m-1}$-unisolvent subset of $\Omega$ and let $r_0 > 0$ be the constant of Proposition I–2.3. Obviously, there exists $r_0' \in (0, r_0]$ such that

$$\forall j = 1, \dots, \mathfrak{M},\ \overline{B}(b_{0j}, r_0') \subset \overline{\Omega}.$$

It follows from (II–5.1) that

$$\forall d \in \mathbb{D} \cap (0, r_0'),\ \forall j = 1, \dots, \mathfrak{M},\ \overline{B}(b_{0j}, r_0' - d) \subset \bigcup_{a \in A^d \cap \overline{B}(b_{0j}, r_0')} \overline{B}(a, d).$$

Letting $N_j = \operatorname{card}\big(A^d \cap \overline{B}(b_{0j}, r_0')\big)$, we have

$$\forall d \in \mathbb{D} \cap (0, r_0'),\ \forall j = 1, \dots, \mathfrak{M},\ (r_0' - d)^n \le N_j d^n.$$

Thus, for any $d_0 \in (0, r_0')$, we get

$$\forall d \in \mathbb{D} \cap (0, d_0),\ \forall j = 1, \dots, \mathfrak{M},\ N_j \ge (r_0' - d_0)^n d^{-n}.$$

Now, from (3.4) and (3.1), we deduce that

$$\forall j = 1, \dots, \mathfrak{M},\quad \sum_{a \in A^d \cap \overline{B}(b_{0j}, r_0')} |\sigma_\varepsilon^d(a) - f(a)|^2 = o(d^{-n}),\ d \to 0.$$

Likewise, for any $d \in \mathbb{D} \cap (0, r_0')$ and any $j = 1, \dots, \mathfrak{M}$, there exists at least one point $b_j^d \in A^d \cap \overline{B}(b_{0j}, r_0')$ such that

$$|\sigma_\varepsilon^d(b_j^d) - f(b_j^d)| = \min_{a \in A^d \cap \overline{B}(b_{0j}, r_0')} |\sigma_\varepsilon^d(a) - f(a)|.$$

Hence, from the last three relations, we derive

$$\forall j = 1, \dots, \mathfrak{M},\ |\sigma_\varepsilon^d(b_j^d) - f(b_j^d)| = o(1),\ d \to 0. \tag{3.5}$$

For any $d \in \mathbb{D} \cap (0, r_0')$, let $B^d = \{b_1^d, \ldots, b_{\mathfrak{M}}^d\}$. Applying Proposition I–2.3 with $B = B^d$, we derive from (3.3) and (3.5) that

$$\exists C > 0,\ \exists d^* > 0,\ \forall d \in \mathbb{D} \cap (0, d^*],\ \|\sigma_\varepsilon^d\|_{m,s} \leq C,$$

that is, the family $(\sigma_\varepsilon^d)_{d \in \mathbb{D} \cap (0,d^*]}$ is bounded in $X^{m,s}$.

In consequence (cf. Corollary 1 in Preliminaries), there exists a sequence $(\sigma_{\varepsilon_l}^{d_l})_{l \in \mathbb{N}}$, with, for any $l \in \mathbb{N}$, $\varepsilon_l = \varepsilon(d_l)$ and

$$\lim_{l \to +\infty} d_l = \lim_{l \to +\infty} d_l^n \varepsilon_l = 0,$$

extracted from the family $(\sigma_\varepsilon^d)_{d \in \mathbb{D} \cap (0,d^*]}$, and there also exists an element $f^* \in X^{m,s}$ such that, as $l \to +\infty$,

$$\sigma_{\varepsilon_l}^{d_l} \to f^*, \quad \text{weakly in } X^{m,s}. \tag{3.6}$$

2) We shall prove that $f^*|_\Omega = f$, arguing by contradiction. Suppose that $f^*|_\Omega \neq f$. Then, taking into account Corollary I–2.1 and that, by (I–2.2), $H^{m+s}(\Omega)$ is continuously imbedded into $C^0(\overline{\Omega})$ (cf. (3) in Preliminaries), there exists a nonempty open set $\mathcal{O}$ contained in $\Omega$ and a real number $\alpha > 0$ such that

$$\forall x \in \mathcal{O},\ |f^*(x) - f(x)| > \alpha.$$

But, in fact, the imbedding of $H^{m+s}(\Omega)$ into $C^0(\overline{\Omega})$ is compact. Hence, it follows from (3.6) that the sequence $(\sigma_{\varepsilon_l}^{d_l})_{l \in \mathbb{N}}$ converges uniformly over $\overline{\Omega}$ to $f^*$, and so there exists $l_0 \in \mathbb{N}$ such that

$$\forall l \in \mathbb{N},\ l \geq l_0,\ \forall x \in \mathcal{O},\ |\sigma_{\varepsilon_l}^{d_l}(x) - f^*(x)| \leq \alpha/2.$$

Then, for any $l \in \mathbb{N}$, $l \geq l_0$, we have

$$\forall x \in \mathcal{O},\ |\sigma_{\varepsilon_l}^{d_l}(x) - f(x)| \geq |f^*(x) - f(x)| - |\sigma_{\varepsilon_l}^{d_l}(x) - f^*(x)| > \alpha/2. \tag{3.7}$$

Now, a reasoning similar to that in point 1) shows that, for any $l \in \mathbb{N}$ sufficiently large, there exists one point $b^{d_l} \in A^{d_l} \cap \mathcal{O}$ such that

$$|\sigma_{\varepsilon_l}^{d_l}(b^{d_l}) - f(b^{d_l})| = o(1),\ l \to +\infty,$$

which leads to a contradiction with (3.7). Therefore, $f^*|_\Omega = f$.

3) Let us prove that

$$\lim_{l \to +\infty} \|\sigma_{\varepsilon_l}^{d_l} - f^\Omega\|_{m,s} = 0. \tag{3.8}$$

Obviously, we have

$$\forall l \in \mathbb{N},\ |\sigma_{\varepsilon_l}^{d_l} - f^*|_{m,s}^2 = |\sigma_{\varepsilon_l}^{d_l}|_{m,s}^2 + |f^*|_{m,s}^2 - 2\Re(\sigma_{\varepsilon_l}^{d_l}, f^*)_{m,s}.$$

But, from (3.3), the definition of $f^\Omega$ and the relation $f^*|_\Omega = f$, we deduce that

$$\forall l \in \mathbb{N},\ |\sigma_{\varepsilon_l}^{d_l}|_{m,s} \leq |f^\Omega|_{m,s} \leq |f^*|_{m,s}, \tag{3.9}$$

and, hence, by (3.6),

$$\lim_{l \to +\infty} |\sigma_{\varepsilon_l}^{d_l} - f^*|_{m,s} = 0.$$

Likewise, Corollary I–2.1, the compact imbedding of $H^{m+s}(\Omega)$ into $L^2(\Omega)$ (cf. (2) in Preliminaries) and (3.6) imply that

$$\lim_{l \to +\infty} \|\sigma_{\varepsilon_l}^{d_l} - f^*\|_{0,\Omega} = 0.$$

Thus,

$$\lim_{l \to +\infty} \|\sigma_{\varepsilon_l}^{d_l} - f^*\|_{m,s} = 0.$$

Finally, taking limits in (3.9), we deduce that $f^* = f^\Omega$, and, therefore, (3.8) holds.

4) We conclude the proof reasoning again by contradiction. Assume that, under the hypotheses of the Theorem, $\|\sigma_\varepsilon^d - f^\Omega\|_{m,s}$ does not tend to 0 when $d$ does. Then, there exists a real number $\alpha > 0$ and two sequences $(d'_l)_{l \in \mathbb{N}} \subset \mathbb{D}$ and $(\varepsilon'_l)_{l \in \mathbb{N}} \subset (0, +\infty)$, with, for any $l \in \mathbb{N}$, $\varepsilon'_l = \varepsilon(d'_l)$ and

$$\lim_{l \to +\infty} d'_l = \lim_{l \to +\infty} (d'_l)^n \varepsilon'_l = 0,$$

verifying

$$\forall l \in \mathbb{N}, \ \|\sigma_{\varepsilon'_l}^{d'_l} - f^\Omega\|_{m,s} > \alpha. \tag{3.10}$$

But the sequence $(\sigma_{\varepsilon'_l}^{d'_l})_{l \in \mathbb{N}}$ is bounded in $X^{m,s}$. A similar argument to that of points 1), 2) and 3) proves that there exists a subsequence of $(\sigma_{\varepsilon'_l}^{d'_l})_{l \in \mathbb{N}}$ which converges to $f^\Omega$ in $X^{m,s}$, in contradiction with (3.10). The Theorem follows. □

**Remark 3.1** – The following result can also be proved (cf. M. C. López de Silanes and R. Arcangéli [101]):

*Assume that* (II–5.1) *holds and that*

$$N = O(d^{-n}), \ d \to 0. \tag{3.11}$$

*Then, except for the trivial case $f \in P_{m-1}(\Omega)$, the condition* (3.1) *is necessary and sufficient for the convergence, as $d \to 0$, of $\sigma_\varepsilon^d|_\Omega$ to $f$ in $H^{m+s}(\Omega)$.*

Hypothesis (3.11) means that, asymptotically, the points of $A^d$ should be regularly distributed in $\overline{\Omega}$. Let us observe that, by (II–5.1), we have

$$\forall d \in \mathbb{D}, \ \overline{\Omega} \subset \bigcup_{a \in A^d} \overline{B}(a, d),$$

from which we derive

$$\exists C > 0, \ \forall d \in \mathbb{D}, \ N d^n \geq C.$$

Therefore, as $d \to 0$, $N$ is of order $\geq n$ compared with $d^{-1}$. Hypothesis (3.11) implies that $N$ is exactly of order $n$, i.e., as $d \to 0$, $N$ tends to $+\infty$ at the same rate as $d^{-n}$. □

**Remark 3.2** – It is not necessary to suppose that $\varepsilon$ depends on $d$. Theorem 3.1 can be formulated in a more general way: under the only hypothesis (II–5.1), the smoothing $(m, s)$-spline relative to $A^d$, $\big(f(a)\big)_{a\in A^d}$ and $\varepsilon > 0$ converges to $f^\Omega$ in $X^{m,s}$ through the filter basis $\mathcal{B} = \{ B_{\alpha\beta} \mid \alpha > 0,\ \beta > 0\}$, with $B_{\alpha\beta} = \{ (d, \varepsilon) \in \mathbb{D} \times (0, +\infty) \mid d \leq \alpha,\ d^n\varepsilon \leq \beta \}$. □

# 4. ESTIMATES OF THE APPROXIMATION ERROR

Let us suppose again that we are given an open set $\Omega \subset \mathbb{R}^n$ with a Lipschitz-continuous boundary (cf. Preliminaries), a function $f \in H^{m+s}(\Omega)$, a set $\mathbb{D} \in (0, +\infty)$ such that $0 \in \overline{\mathbb{D}}$, and, for any $d \in \mathbb{D}$, a finite subset $A^d$ of $N = N(d)$ points of $\overline{\Omega}$ verifying (II–5.1). For any $d \in \mathbb{D}$ and any $\varepsilon > 0$, we denote by $\sigma_\varepsilon^d$ the smoothing $(m, s)$-spline relative to $A^d$, $\beta^d = \big(f(a)\big)_{a\in A^d}$ and $\varepsilon$. Likewise, we let $u_\varepsilon^d = f - \sigma_\varepsilon^d|_\Omega$ and $\tilde{u}_\varepsilon^d = \tilde{P}u_\varepsilon^d$, where $\tilde{P}$ is the operator defined in (II–6.2).

Let us recall that $\mathfrak{K}$ denotes the dimension of $P_k$, with $k \in \mathbb{N}$ given by (II–6.1), and let $R > 1$ be the constant introduced in Proposition II–6.6. From the proof of this result, we extract the following facts:

- there exists a set $\{\hat{\alpha}_1, \ldots, \hat{\alpha}_{\mathfrak{K}}\} \subset \mathbb{R}^n$ such that $\prod_{j=1}^{\mathfrak{K}} \overline{B}(\hat{\alpha}_j, 1)$ is a compact subset of $(\mathbb{R}^n)^{\mathfrak{K}}$ formed by $P_k$-unisolvent $\mathfrak{K}$-tuples; the constant $R$ is chosen, in fact, so that, for some $\hat{a} \in \mathbb{R}^n$, $\bigcup_{j=1}^{\mathfrak{K}} \overline{B}(\hat{\alpha}_j, 1) \subset \overline{B}(\hat{a}, R)$;

- the closed balls $\mathcal{B}_1, \ldots, \mathcal{B}_{\mathfrak{K}}$ of radius $d$, contained in $\overline{B}(t, Rd)$, that Proposition II–6.6 associates with any $d > 0$ and any $t \in \mathbb{R}^n$ are defined as follows:

$$\forall j = 1, \ldots, \mathfrak{K},\ \mathcal{B}_j = F_t^d\big(\overline{B}(\hat{\alpha}_j, 1)\big),$$

  where $F_t^d$ stands for the invertible affine mapping $F_t^d : x \in \mathbb{R}^n \mapsto t + d(x - \hat{a}) \in \mathbb{R}^n$; any $\mathfrak{K}$-tuple belonging to $\prod_{j=1}^{\mathfrak{K}} \mathcal{B}_j$ is $P_k$-unisolvent.

Let us consider the constants $M > 1$ and $\lambda_0 > 0$ and, for any $\lambda \in (0, \lambda_0]$, the set $T_\lambda$ whose existence is established in Proposition II–6.1. Now, let $d \in \mathbb{D} \cap (0, \lambda_0/R]$ and $t \in T_{Rd}$. It follows from (II–5.1) that there exists a $\mathfrak{K}$-tuple $a_t^d = (a_{1t}^d, \ldots, a_{\mathfrak{K}t}^d) \in \prod_{j=1}^{\mathfrak{K}} (\mathcal{B}_j \cap A^d)$, where $\mathcal{B}_1, \ldots, \mathcal{B}_{\mathfrak{K}}$ are the closed balls associated with $d$ and $t$ by Proposition II–6.6. Since $a_t^d$ is $P_k$-unisolvent, it is possible to define on $H^{m+s}\big(\overline{B}(t, MRd)\big)$ the Lagrange $P_k$-interpolating operator $\Pi_t^d$ given by

$$\Pi_t^d v \in P_k \text{ and, for } j = 1, \ldots, \mathfrak{K},\ \Pi_t^d v(a_{jt}^d) = v(a_{jt}^d). \tag{4.1}$$

***Proposition* 4.1** – *Suppose that* (II–5.1) *holds. Then, there exists* $C > 0$ *such that, for any* $d \in \mathbb{D} \cap (0, \lambda_0/R]$ *and for any* $\varepsilon > 0$, *we have*

$$\forall l = 0, \ldots, k,\ \sum_{t\in T_{Rd}} |\Pi_t^d \tilde{u}_\varepsilon^d|_{l,\overline{B}(t,MRd)}^2 \leq C d^{n-2l} \varepsilon.$$

**Proof** – (Cf. M. C. López de Silanes and R. Arcangéli [100]). Let $l$ be an integer such that $0 \le l \le k$.

1) For any $d \in \mathbb{D} \cap (0, \lambda_0/R]$ and any $t \in T_{Rd}$, the balls $B(\hat{a}, MR)$ and $B(t, MRd)$ are affine-equivalent, since the bijective affine mapping $F_t^d$ transforms the former ball into the latter. By Proposition II–6.4 (with $F = F_t^d$, $\|L^{-1}\| = 1/d$ and $\det L = d^n$), there exists a constant $C = C(l, n)$ such that

$$\begin{gathered} \forall \varepsilon > 0,\ \forall d \in \mathbb{D} \cap (0, \lambda_0/R],\ \forall t \in T_{Rd}, \\ |\Pi_t^d \tilde{u}_\varepsilon^d|_{l,\overline{B}(t,MRd)} \le C d^{n/2-l} |(\Pi_t^d \tilde{u}_\varepsilon^d) \circ F_t^d|_{l,\overline{B}(\hat{a},MR)}. \end{gathered} \tag{4.2}$$

Now, it can be seen that the mapping $v \mapsto \left( \sum_{j=1}^{\mathfrak{K}} |v(\hat{\alpha}_j)|^2 + |v|^2_{m+s,B(\hat{a},MR)} \right)^{1/2}$ is a norm on $H^{m+s}(B(\hat{a}, MR))$ which is equivalent to the norm $\| \cdot \|_{m+s,B(\hat{a},MR)}$ (if $m + s \in \mathbb{N}^*$, see J. Nečas [109, Theorem 2.7.1]; otherwise, one can reason analogously, using the properties of the Sobolev spaces of noninteger order). Therefore, there exists $C > 0$ such that

$$\begin{aligned} &\forall \varepsilon > 0,\ \forall d \in \mathbb{D} \cap (0, \lambda_0/R],\ \forall t \in T_{Rd},\ \|(\Pi_t^d \tilde{u}_\varepsilon^d) \circ F_t^d\|_{m+s,\overline{B}(\hat{a},MR)} \\ &\le C \left( \sum_{j=1}^{\mathfrak{K}} \left|((\Pi_t^d \tilde{u}_\varepsilon^d) \circ F_t^d)(\hat{\alpha}_j)\right|^2 + \left|(\Pi_t^d \tilde{u}_\varepsilon^d) \circ F_t^d\right|^2_{m+s,\overline{B}(\hat{a},MR)} \right)^{1/2}. \end{aligned} \tag{4.3}$$

From the definition (II–6.1) of $k$, we deduce that the second term on the right-hand side of (4.3) is null. Consequently, it follows from (4.2) and (4.3) that

$$\begin{gathered} \exists C > 0,\ \forall \varepsilon > 0,\ \forall d \in \mathbb{D} \cap (0, \lambda_0/R],\ \forall t \in T_{Rd}, \\ |\Pi_t^d \tilde{u}_\varepsilon^d|_{l,\overline{B}(t,MRd)} \le C d^{n/2-l} \left( \sum_{j=1}^{\mathfrak{K}} \left|((\Pi_t^d \tilde{u}_\varepsilon^d) \circ F_t^d)(\hat{\alpha}_j)\right|^2 \right)^{1/2}. \end{gathered} \tag{4.4}$$

2) Let $\widehat{B} = \prod_{j=1}^{\mathfrak{K}} \overline{B}(\hat{\alpha}_j, 1)$. Let us prove that there exists a constant $C > 0$ such that

$$\forall \psi \in P_k,\ \forall \hat{b} = (\hat{b}_1, \ldots, \hat{b}_{\mathfrak{K}}) \in \widehat{B},\ \sum_{j=1}^{\mathfrak{K}} |\psi(\hat{\alpha}_j)|^2 \le C \sum_{j=1}^{\mathfrak{K}} |\psi(\hat{b}_j)|^2. \tag{4.5}$$

For this, we reason as in the proof of Proposition II–6.5. Let $p_1, \ldots, p_{\mathfrak{K}}$ be a basis of $P_k$ and, for any $\hat{b} = (\hat{b}_1, \ldots, \hat{b}_{\mathfrak{K}}) \in \widehat{B}$, let $M(\hat{b}) = \left(p_i(\hat{b}_j)\right)_{1 \le i,j \le \mathfrak{K}}$. The matrix $M(\hat{b})$ is regular, since $\hat{b}$ is $P_k$-unisolvent. Denoting by $m'_{ij}(\hat{b})$ the generic element of the inverse matrix $M(\hat{b})^{-1}$, we have

$$\forall \psi \in P_k,\ \psi = \sum_{i,j=1}^{\mathfrak{K}} \psi(\hat{b}_i) m'_{ij}(\hat{b}) p_j.$$

Since matrix inversion is a continuous operation, each function $m'_{ij}$ is bounded on the compact set $\widehat{B}$. If

$$\gamma_0 = \max_{1 \le i,j \le \mathfrak{K}} \sup_{\hat{b} \in \widehat{B}} |m'_{ij}(\hat{b})| \quad \text{and} \quad \gamma_1 = \max_{1 \le i,j \le \mathfrak{K}} |p_j(\hat{\alpha}_i)|,$$

it is clear that

$$\forall \psi \in P_k,\ \forall \hat{b} = (\hat{b}_1, \ldots, \hat{b}_{\mathfrak{K}}) \in \widehat{B},\ \forall j = 1, \ldots, \mathfrak{K},\ |\psi(\hat{\alpha}_j)| \le \mathfrak{K}\gamma_0\gamma_1 \sum_{i=1}^{\mathfrak{K}} |\psi(\hat{b}_i)|.$$

This implies (4.5).

Now, for any $d \in \mathbb{D} \cap (0, \lambda_0/R]$, $t \in T_{Rd}$ and $j = 1, \ldots, \mathfrak{K}$, let $\hat{a}^d_{jt} = (F^d_t)^{-1}(a^d_{jt})$. By definition of $F^d_t$ and $a^d_t$, the $\mathfrak{K}$-tuple $\hat{a}^d_t = (\hat{a}^d_{1t}, \ldots, \hat{a}^d_{\mathfrak{K}t})$ belongs to $\widehat{B}$. Using (4.5), with $\psi = (\Pi^d_t \tilde{u}^d_\varepsilon) \circ F^d_t$ and $\hat{b} = \hat{a}^d_t$, and taking (4.1) into account, we derive from (4.4) that

$$\exists C > 0,\ \forall \varepsilon > 0,\ \forall d \in \mathbb{D} \cap (0, \lambda_0/R],\ \forall t \in T_{Rd},$$

$$|\Pi^d_t \tilde{u}^d_\varepsilon|_{l, \overline{B}(t, MRd)} \le C d^{n/2 - l} \left( \sum_{j=1}^{\mathfrak{K}} |\tilde{u}^d_\varepsilon(a^d_{jt})|^2 \right)^{1/2}. \tag{4.6}$$

3) Using point (iii) of Proposition (II–6.1), we get from (4.6)

$$\exists C > 0,\ \forall \varepsilon > 0,\ \forall d \in \mathbb{D} \cap (0, \lambda_0/R],\ \sum_{t \in T_{Rd}} |\Pi^d_t \tilde{u}^d_\varepsilon|^2_{l, \overline{B}(t, MRd)}$$

$$\le C d^{n-2l} \sum_{t \in T_{Rd}} \left( \sum_{a \in A^d \cap \overline{B}(t, MRd)} |\tilde{u}^d_\varepsilon(a)|^2 \right) \le C M_1 d^{n-2l} \sum_{a \in A^d} |\tilde{u}^d_\varepsilon(a)|^2.$$

Reasoning as in the proof of Theorem 3.1 to get (3.4), we have

$$\exists C > 0,\ \forall \varepsilon > 0,\ \forall d \in \mathbb{D} \cap (0, \lambda_0/R],\ \sum_{a \in A^d} |\tilde{u}^d_\varepsilon(a)|^2 \le C\varepsilon.$$

The Proposition is then a trivial consequence of the last two relations. □

***Proposition* 4.2** – *Suppose that* (II–5.1) *holds. Then, there exists* $C > 0$ *such that, for any* $d \in \mathbb{D} \cap (0, \lambda_0/R]$ *and for any* $\varepsilon > 0$, *we have*

$$\forall l = 0, \ldots, k,\ \sum_{t \in T_{Rd}} |\tilde{u}^d_\varepsilon - \Pi^d_t \tilde{u}^d_\varepsilon|^2_{l, \overline{B}(t, MRd)} \le C d^{2(m+s-l)} |u^d_\varepsilon|^2_{m+s, \Omega}.$$

**Proof** – (Cf. M. C. López de Silanes and R. Arcangéli [100]). For any $\varepsilon > 0$, $d \in \mathbb{D} \cap (0, \lambda_0/R]$ and $t \in T_{Rd}$, it is clear that $\tilde{u}^d_\varepsilon - \Pi^d_t \tilde{u}^d_\varepsilon$ belongs to $H^{m+s}(\overline{B}(t, MRd))$ and is null in at least one point in each of the balls $\mathcal{B}_1, \ldots, \mathcal{B}_{\mathfrak{K}}$ associated with $d$ and $t$ by Proposition II–6.6. Applying this result, we deduce that there exists a constant $C > 0$, depending only on $M$, $n$ and $m + s$, such that, for any $\varepsilon > 0$, $d \in \mathbb{D} \cap (0, \lambda_0/R]$ and $t \in T_{Rd}$, we have

$$\forall l = 0, \ldots, k,\ |\tilde{u}^d_\varepsilon - \Pi^d_t \tilde{u}^d_\varepsilon|_{l, \overline{B}(t, MRd)} \le C d^{m+s-l} |\tilde{u}^d_\varepsilon - \Pi^d_t \tilde{u}^d_\varepsilon|_{m+s, \overline{B}(t, MRd)}.$$

Then, we proceed as in Theorem II–7.1. □

From now on, for simplicity, we shall write $\sigma_\varepsilon^d$ instead of $\sigma_\varepsilon^d|_\Omega$.

***Theorem* 4.1** – *Suppose that* (II–5.1) *holds. Then, there exists* $C > 0$ *such that, for any* $d \in \mathbb{D} \cap (0, \lambda_0/R]$ *and for any* $\varepsilon > 0$, *we have*

$$\forall l = 0, \ldots, k, \ |f - \sigma_\varepsilon^d|_{l,\Omega} \leq C \left( |f - \sigma_\varepsilon^d|_{m+s,\Omega}\, d^{m+s-l} + d^{n/2-l} \varepsilon^{1/2} \right).$$

**Proof** – (Cf. M. C. López de Silanes and R. Arcangéli [100]). Let $l$ be an integer such that $0 \leq l \leq k$. For any $d \in \mathbb{D} \cap (0, \lambda_0/R]$ and any $\varepsilon > 0$, it follows from Proposition II–6.1 that

$$|u_\varepsilon^d|_{l,\Omega}^2 \leq |\tilde{u}_\varepsilon^d|_{l,\bigcup_{t \in T_{Rd}} \overline{B}(t,MRd)}^2 \leq \sum_{t \in T_{Rd}} |\tilde{u}_\varepsilon^d|_{l,\overline{B}(t,MRd)}^2.$$

In addition, it is obvious that

$$|\tilde{u}_\varepsilon^d|_{l,\overline{B}(t,MRd)}^2 \leq 2 \left( |\tilde{u}_\varepsilon^d - \Pi_t^d \tilde{u}_\varepsilon^d|_{l,\overline{B}(t,MRd)}^2 + |\Pi_t^d \tilde{u}_\varepsilon^d|_{l,\overline{B}(t,MRd)}^2 \right).$$

Applying Propositions 4.1 and 4.2 to bound the right hand side of the last relation, we get

$$|u_\varepsilon^d|_{l,\Omega}^2 \leq C \left( |u_\varepsilon^d|_{m+s,\Omega}^2 d^{2(m+s-l)} + d^{n-2l} \varepsilon \right),$$

from which we deduce the Theorem. □

***Corollary* 4.1** – *Under the conditions of Theorem 3.1, we have*

$$\forall l = 0, \ldots, k, \ |f - \sigma_\varepsilon^d|_{l,\Omega} = o(d^{m+s-l}) + O(d^{n/2-l} \varepsilon^{1/2}), \ d \to 0, \tag{4.7}$$

*and also evidently*

$$|f - \sigma_\varepsilon^d|_{m+s,\Omega} = o(1), \ d \to 0.$$

**Proof** – It is sufficient to apply Theorems 4.1 and 3.1 and Corollary I–2.1. □

The error estimates (4.7) are only significant when the term $d^{n/2-l}\varepsilon^{1/2}$ tends to zero, but they can be improved under supplementary hypotheses. This is what is shown in the following theorem, whose point (i) has been proved by F. Utreras [146] in the case of $D^m$-splines over $\mathbb{R}^n$ (i.e. $(m, 0)$-splines).

***Theorem* 4.2** – *Suppose that* (II–5.1), (3.1) *and* (3.11) *hold, and assume that*

$$\exists C > 0, \ \varepsilon \geq C d^{2m+2s-n}, \ d \to 0. \tag{4.8}$$

*Then, when* $d \to 0$, *we have*

*(i)* $\forall l = 0, \ldots, k, \ |f - \sigma_\varepsilon^d|_{l,\Omega} = O\Big( (\varepsilon/N)^{(m+s-l)/(2m+2s)} \Big)$;

*(ii)* $|f - \sigma_\varepsilon^d|_{m+s,\Omega} = o(1)$.

**Proof** – Taking (3.1) and (4.8) into account, from (4.7) we deduce that

$$|f - \sigma_\varepsilon^d|_{0,\Omega} = O(d^{n/2}\varepsilon^{1/2}),\ d \to 0.$$

Likewise, we have (ii) from Theorem 3.1 and Corollary I–2.1. Then, using (3.11) and a theorem concerning the intermediate semi-norms (cf. R. A. Adams [1, Theorem 4.14, p. 75]), we obtain (i). □

**Remark 4.1** – Condition (4.8) means that, as $d \to 0$, the parameter $\varepsilon$ can either be unbounded, or remain bounded, or even tend to 0 as an infinitesimal function of order $\leq 2m + 2s - n$. □

## 5. CONVERGENCE FOR NOISY DATA

The results obtained in Sections 3 and 4 are results of convergence and error estimates for *exact data*. However, in most real problems, the data are *noisy*. We shall not develop the study of convergence and error estimates in this case. Nevertheless, we shall establish a result, of *deterministic* type, which shows that the problem of smoothing noisy data by $(m,s)$-splines is well posed in the Hadamard sense. Finally, placing us in a *stochastic* situation (where the data are perturbed by a *white noise*), we shall give without proof a Theorem of convergence and error estimates in the sense of *almost sure convergence*.

We keep the notations of the two preceding sections concerning $\Omega$, $f$, $\mathbb{D}$, $A^d$, $N = N(d)$, $\rho^d$ and $\sigma_\varepsilon^d$. For any $d \in \mathbb{D}$, let $\nu^d = (\nu_a^d)_{a\in A^d}$ be any "error vector". For any $\varepsilon > 0$ and for any $d \in \mathbb{D}$, we denote by $\tilde{\sigma}_\varepsilon^d$ the smoothing $(m,s)$-spline relative to $A^d$, $\big(f(a)\big)_{a\in A^d} + \nu^d$ and $\varepsilon$, and by $e_\varepsilon^d$, the smoothing $(m,s)$-spline relative to $A^d$, $\nu^d$ and $\varepsilon$. We evidently have $\tilde{\sigma}_\varepsilon^d = \sigma_\varepsilon^d + e_\varepsilon^d$, since the operator which to any $\beta^d \in \mathbb{R}^N$ assigns the smoothing $(m,s)$-spline relative to $A^d$, $\beta^d$ and $\varepsilon$ is linear.

We assume in the sequel that $\varepsilon$ is a function of $d$.

**Theorem 5.1** – *Suppose that* (II–5.1) *and* (3.1) *hold. Suppose, in addition, that*

$$d = O\big(\inf\{\,\delta(a,b) \mid a \in A^d,\ b \in A^d,\ a \neq b\}\big),\ d \to 0, \tag{5.1}$$

*and that*

$$\sup_{a\in A^d} |\nu_a^d| = o\big((\varepsilon/N)^{1/2}\big),\ d \to 0. \tag{5.2}$$

*Then,*

$$\lim_{d\to 0} \|\tilde{\sigma}_\varepsilon^d - f^\Omega\|_{m,s} = 0,$$

*where $f^\Omega$ denotes the unique element of minimal semi-norm $|\cdot|_{m,s}$ in the set $\{v \in X^{m,s} \mid v|_\Omega = f\}$.*

**Proof** – For any $d \in \mathbb{D}$, the equation (1.2) for the smoothing spline $e_\varepsilon^d$, with $v = e_\varepsilon^d$, can be written as

$$\langle \rho^d e_\varepsilon^d \rangle^2 + \varepsilon |e_\varepsilon^d|_{m,s}^2 = \langle \nu^d, \rho^d e_\varepsilon^d \rangle.$$

Therefore,

$$\langle \rho^d e_\varepsilon^d \rangle \leq \langle \nu^d \rangle,$$

and hence

$$\langle \rho^d e_\varepsilon^d \rangle^2 + \varepsilon |e_\varepsilon^d|_{m,s}^2 \leq \langle \nu^d \rangle^2. \tag{5.3}$$

On the other hand, it follows from (II–5.1) that

$$\forall d \in \mathbb{D},\ \overline{\Omega} \subset \bigcup_{a \in A^d} \overline{B}(a, d),$$

and consequently

$$\exists C > 0,\ \forall d \in \mathbb{D},\ N d^n \geq C.$$

This implies, together with (3.1), that

$$\varepsilon = o(N),\ d \to 0. \tag{5.4}$$

In particular, there exists $C_0 > 0$ and $d_0 > 0$ such that, for any $d \in \mathbb{D} \cap (0, d_0]$, $\varepsilon \leq C_0 N$. Taking (5.3) into account, we have

$$\forall d \in \mathbb{D} \cap (0, d_0],\ \frac{1}{C_0 N} \langle \rho^d e_\varepsilon^d \rangle^2 + |e_\varepsilon^d|_{m,s}^2 \leq \frac{1}{\varepsilon} \langle \nu^d \rangle^2.$$

But, under the hypotheses (II–5.1) and (5.1), there exists $d^*$ such that the mapping

$$v \mapsto \left( \frac{1}{C_0 N} \langle \rho^d v \rangle^2 + |v|_{m,s}^2 \right)^{1/2}$$

is, for any $d \in \mathbb{D} \cap (0, d^*]$, a norm on $X^{m,s}$ uniformly equivalent over $\mathbb{D} \cap (0, d^*]$ to the norm $\|\cdot\|_{m,s}$ (cf. F. Utreras [146, Theorem 3.4]). Thus, we derive from (5.2) that

$$\|e_\varepsilon^d\|_{m,s} = o(1),\ d \to 0.$$

The Theorem then follows from Theorem 3.1. □

In real problems, the hypothesis (5.2) is not acceptable (because it implies, together with (5.4), that $\lim_{d \to 0} \langle \nu^d \rangle = 0$). Due to this fact, the deterministic convergence of $\tilde{\sigma}_\varepsilon^d$ to $f^\Omega$ cannot be ensured in such cases. However it is possible to obtain convergence results of stochastic type. To do this, we restrict the generality of the problem. On the one hand, we limit ourselves to the case of $D^m$-splines over $\mathbb{R}^n$ (i.e. $(m, 0)$-splines). On the other hand, we suppose that we have, not an arbitrary family of data sets, but a *sequence* of such sets, which requires slight changes in the notations.

Suppose that we are given, for any $j \in \mathbb{N}$, an ordered set $A^j$ of $N = N(j)$ distinct points of $\overline{\Omega}$ which contains a $P_{m-1}$-unisolvent subset. Let

$$d = d(j) = \sup_{x \in \Omega} \delta(x, A^j)$$

and assume that

$$\lim_{j \to +\infty} d = 0. \tag{5.5}$$

Likewise, for any $j \in \mathbb{N}$, let $\nu^j \in \mathbb{R}^N$ be a given error vector. We suppose that the sequence $(\nu^j)_{j\in\mathbb{N}}$ satisfies

$$\text{for any } j \in \mathbb{N},\ \nu^j \text{ is a ``white noise''} \tag{5.6}$$

(i.e $\nu^j$ is a Gaussian vector of independent, identically distributed random variables, with null mean and the same positive variance).

Let $f$ now be a function in $H^m(\Omega)$. For any $j \in \mathbb{N}$ and for any $\varepsilon > 0$, we denote by $\tilde{\sigma}_\varepsilon^j$ the smoothing $D^m$-spline relative to $A^j$, $\big(f(a)\big)_{a\in A^j} + \nu^j$ and $\varepsilon$. We suppose that $\varepsilon$ depends on $j$ and we finally formulate the following supplementary hypotheses:

$$\varepsilon = \varepsilon(j) = o(d^{-n}),\ j \to +\infty, \tag{5.7}$$

$$d = O\big(\inf\{\, \delta(a,b) \mid a \in A^j,\ b \in A^j,\ a \neq b\}\big),\ j \to +\infty, \tag{5.8}$$

and

$$\left|\begin{array}{l} \varepsilon = N^{n/(n+2m)}\omega(N), \text{ with } \lim\limits_{j\to+\infty} \omega(N)/j^\theta = +\infty, \text{ for some } \theta > 0, \\ \text{and } \omega(N) = o\Big(N^{2m/(n+2m)}\Big),\ j \to +\infty. \end{array}\right. \tag{5.9}$$

The following theorem generalizes results by D. Ragozin [121] in dimension 1, and by F. Utreras [146] and by M. C. López de Silanes and R. Arcangéli [101] in dimension $n$.

**Theorem 5.2** – *Suppose that* (5.5), (5.6), (5.7), (5.8) *and* (5.9) *hold. Then, we have almost surely*

$$\lim_{j\to+\infty} \|\tilde{\sigma}_\varepsilon^j - f^\Omega\|_{m,0} = 0,$$

*and*

$$\forall l = 0,\dots,m-1,\ |\tilde{\sigma}_\varepsilon^j|_\Omega - f|_{l,\Omega}^2 = O\Big(N^{-2(m-l)/(n+2m)}\big(\omega(N)\big)^{(m-l)/m}\Big),\ j \to +\infty.$$

**Proof** – Cf. R. Arcangéli and B. Ycart [19]. □

**Remark 5.1** – Theorem 5.2 has been obtained, in fact, under the following hypothesis, which is more general than (5.6): for any $j \in \mathbb{N}$, $\nu^j$ is a vector of independent random variables with mean 0 and uniformly bounded moments of any order. In addition, the results established in this theorem are also valid in quadratic mean. □

# CHAPTER IV

# $(m, l, s)$-SPLINES

The $(m, l, s)$-splines have been introduced, under the denomination $L^{m,l,s}$-splines, by A. Bouhamidi [37] and A. Bouhamidi and A. Le Méhauté [39]. They are a natural generalization of the $(m, s)$-splines, which correspond to the case $l = 0$. Moreover, if the semi-norm $|\cdot|_{m,l,s}$, defined later in Section 1, is slightly modified, the $(m, l, s)$-splines also comprise, as a particular case, the thin plate splines under tension (cf. [38], [68], [91] and [132]).

## 1. THE SPACES $X^{m,l,s}$

Let $n \in \mathbb{N}^*$. For any $m \in \mathbb{N}^*$ and any $s \in \mathbb{R}$ satisfying (I–2.2), i.e. the condition

$$-m + \frac{n}{2} < s < \frac{n}{2},$$

and for any $l \in \mathbb{N}$, we write

$$X^{m,l,s} = \{ v \in \mathcal{D}' \mid \forall \alpha \in \mathbb{N}^n,\ m \leq |\alpha| \leq m+l,\ \partial^\alpha v \in \widetilde{H}^s \},$$

where $\widetilde{H}^s$ is the space introduced in Section I–1. We also define a scalar semi-product

$$\forall u, v \in X^{m,l,s},\ (u,v)_{m,l,s} = \sum_{r=0}^{l} \sum_{|\alpha|=m+r} \frac{(m+r)!}{\alpha!} \int_{\mathbb{R}^n} |\xi|^{2s} \mathcal{F}\partial^\alpha u(\xi) \overline{\mathcal{F}\partial^\alpha v(\xi)}\, d\xi$$

(a more general form is available in [37] and [39]) and the corresponding semi-norm

$$\forall v \in X^{m,l,s},\ |v|_{m,l,s} = \big((v,v)_{m,l,s}\big)^{1/2}.$$

Let us note that

$$(u,v)_{m,l,s} = \sum_{r=0}^{l} (u,v)_{m+r,s},$$

where $(\cdot\,,\,\cdot)_{m+r,s}$ is the scalar semi-product introduced in Section I–1.

Reasoning as in Chapter I for the spaces $X^{m,s}$, we can prove, under the hypothesis (I–2.2), the following properties of the spaces $X^{m,l,s}$:

- endowed with the semi-norm $|\cdot|_{m,l,s}$, the space $X^{m,l,s}$ is a semi-Hilbert space (i.e. complete for this semi-norm) that is contained in $\mathcal{S}'$;
- $\mathcal{S} \subset X^{m,l,s}$;

- for any bounded, connected, nonempty, open subset $\Omega^*$ of $\mathbb{R}^n$, the space $X^{m,l,s}$, endowed with the norm

$$\|v\|_{m,l,s}^{\Omega^*} = \left( \int_{\Omega^*} |v(x)|^2 \, dx + |v|_{m,l,s}^2 \right)^{1/2},$$

  is a Hilbert space whose topology is independent of $\Omega^*$; hereafter, we shall assume that $X^{m,l,s}$ is endowed with a norm $\|\cdot\|_{m,l,s}^{\Omega^*}$, which we shall simply write as $\|\cdot\|_{m,l,s}$, without making any reference to a particular open set $\Omega^*$;

- for any open set $\Omega$ in $\mathbb{R}^n$ with a Lipschitz-continuous boundary (cf. Preliminaries), the operator $R_\Omega$ of restriction to $\Omega$ is linear and continuous from $X^{m,l,s}$ onto $H^{m+l+s}(\Omega)$;

- $X^{m,l,s}$ is contained in $C^0(\mathbb{R}^n)$ with continuous injection.

We do not give the proofs of these results and, from now on, we shall suppose that (I–2.2) holds.

## 2. INTERPOLATING $(m, l, s)$-SPLINES

Suppose we are given an ordered subset $A$ of $N$ distinct points of $\mathbb{R}^n$ which contains a $P_{m-1}$-unisolvent subset and an element $\beta$ of $\mathbb{C}^N$. Denoting by $\rho \in \mathcal{L}(X^{m,l,s}, \mathbb{C}^N)$ the operator defined by

$$\rho v = \big(v(a)\big)_{a \in A}$$

and letting

$$\mathcal{K} = \{ v \in X^{m,l,s} \mid \rho v = \beta \},$$

we agree to call *interpolating $(m, l, s)$-spline relative to $A$ and $\beta$* any solution, if any exists, of the problem: find $\sigma$ such that

$$\begin{cases} \sigma \in \mathcal{K}, \\ \forall v \in \mathcal{K}, \ |\sigma|_{m,l,s} \le |v|_{m,l,s} \end{cases} \tag{2.1}$$

(cf. Section II–1). Let us take up the proof of Theorem II–1.1. Since $\mathcal{S} \subset X^{m,l,s}$, it is easily seen that $\mathcal{K}$ is nonempty. Then, defining over $X^{m,l,s}$ a norm (analogous to the norm (I–2.5)) which is equivalent to $\|\cdot\|_{m,l,s}$, one concludes that problem (2.1) admits a unique solution (this fact is established in a different way in [37] and [39]).

Contrary to the case of $(m, s)$-splines, we do not know, in general, how to express the $(m, l, s)$-splines in terms of functions explicitly known. However, A. Bouhamidi and A. Le Méhauté have proved (cf. [37, 39]) that $\sigma$ can be formally written in terms of fundamental solutions $K_{m,l,s}$ of differential or pseudo-differential operators:

$$\forall x \in \mathbb{R}^n, \ \sigma(x) = \sum_{a \in A} \lambda_a K_{m,l,s}(|x-a|) + p(x),$$

where the coefficients $\lambda_a$, with $a \in A$, and the polynomial $p \in P_{m-1}$ are determined by solving a linear system. The explicit expressions of these fundamental solutions are unknown in general, but they are available in many particular cases (cf. A. Le Méhauté and A. Bouhamidi [92]).

We are now interested in the problems of convergence and error estimates. Let $f$ be a given function in $H^{m+l+s}(\Omega)$. Suppose that we are in the situation of Section II–5, changing $X^{m,s}$ into $X^{m,l,s}$. Let us conserve the notations introduced there for $\mathbb{D}$ and $A^d$, and let $\sigma^d$ be now the interpolating $(m, l, s)$-spline relative to $A^d$ and $\big(f(a)\big)_{a \in A^d}$. The following two results (cf. M. C. López de Silanes [97]) generalize Theorem II–5.1 and Corollary II–7.1.

***Theorem* 2.1** – *Suppose that* (II–5.1) *holds. Then,*

$$\lim_{d \to 0} \|\sigma^d - f^\Omega\|_{m,l,s} = 0,$$

*where $f^\Omega$ denotes the unique element of minimal semi-norm $|\cdot|_{m,l,s}$ in the set $\{ v \in X^{m,l,s} \mid v|_\Omega = f \}$.*

***Theorem* 2.2** – *Suppose that* (II–5.1) *holds. Then, we have*

$$\forall i = 0, \dots, k, \ |f - \sigma^d|_{i,\Omega} = o(d^{m+l+s-i}), \ d \to 0,$$

*where, for simplicity, we have written $\sigma^d$ instead of $\sigma^d|_\Omega$, and $k$ denotes the integer part of $m + l + s$ if $m + l + s$ is noninteger, or $m + l + s - 1$, otherwise.*

## 3. SMOOTHING $(m, l, s)$-SPLINES

We keep the definition of $A$, $N$, $\beta$ and $\rho$ given in the preceding section. We denote by $\langle \cdot \rangle$ and $\langle \cdot , \cdot \rangle$, respectively, the hermitian norm and the hermitian scalar product in $\mathbb{C}^N$. For any $\varepsilon$, we write

$$\forall v \in X^{m,l,s}, \ J_\varepsilon(v) = \langle \rho v - \beta \rangle^2 + \varepsilon |v|^2_{m,l,s}.$$

Then, we call *smoothing $(m, l, s)$-spline relative to $A$, $\beta$ and $\varepsilon$* any solution, if any exists, of the problem: find $\sigma_\varepsilon$ such that

$$\begin{cases} \sigma_\varepsilon \in X^{m,l,s}, \\ \forall v \in X^{m,l,s}, \ J_\varepsilon(\sigma_\varepsilon) \leq J_\varepsilon(v) \end{cases} \tag{3.1}$$

(cf. Section III–1). Proceeding as in the proof of Theorem III–1.1, we can easily prove that problem (3.1) admits a unique solution, which is also the unique solution of the problem: find $\sigma_\varepsilon$ such that

$$\begin{cases} \sigma_\varepsilon \in X^{m,l,s}, \\ \forall v \in X^{m,l,s}, \ \langle \rho \sigma_\varepsilon, \rho v \rangle + \varepsilon (\sigma_\varepsilon, v)_{m,l,s} = \langle \beta, \rho v \rangle. \end{cases}$$

Finally, we consider again the problems of convergence and error estimates. Let $f$ be a given function in $H^{m+l+s}(\Omega)$. Suppose that we are in the situation of Section

III–3 changing $X^{m,s}$ into $X^{m,l,s}$ and conserving the notations $\mathbb{D}$, $A^d$ and $N = N(d)$. Let $\sigma_\varepsilon^d$ denote the smoothing $(m,l,s)$-spline relative to $A^d$, $\big(f(a)\big)_{a\in\mathbb{D}}$ and $\varepsilon$. Theorems III–3.1 and III–4.2 are extended by the following results (cf. M. C. López de Silanes [97]).

***Theorem* 3.1** – *Suppose that* (II–5.1) *and* (III–3.1) *hold. Then, we have*

$$\lim_{d\to 0} \|\sigma_\varepsilon^d - f^\Omega\|_{m,l,s} = 0,$$

*where $f^\Omega$ is defined as in Theorem 2.1.*

**Remark 3.1** – Proceeding as in [101], it can be shown that the result of Theorem 3.1 is optimal in the sense that, under (II–5.1) and the additional hypothesis (III–3.11), the condition (III–3.1) is necessary and sufficient for the convergence of $\sigma_\varepsilon^d$ to $f^\Omega$ in $X^{m,l,s}$, except for the trivial case $f \in P_{m-1}(\Omega)$ (cf. Remark III–3.1). □

***Theorem* 3.2** – *Suppose that* (II–5.1), (III–3.1) *and* (III–3.11) *hold, as well as the hypothesis*

$$\exists C > 0, \ \varepsilon \geq C\, d^{2m+2l+2s-n}, \ d \to 0.$$

*Then, when $d \to 0$, we have*

*(i)* $\forall i = 0, \ldots, k, \ |f - \sigma_\varepsilon^d|_{i,\Omega} = O\Big((\varepsilon/N)^{(m+l+s-i)/(2m+2l+2s)}\Big),$

*(ii)* $|f - \sigma_\varepsilon^d|_{m+l+s,\Omega} = o(1),$

*where we have written $\sigma_\varepsilon^d$ instead of $\sigma_\varepsilon^d|_\Omega$, and $k$ denotes the constant defined in Theorem 2.2.*

# PART B

## *$D^m$-SPLINES OVER A BOUNDED DOMAIN OF $\mathbb{R}^n$*

# *INTRODUCTION*

The introduction of $D^m$-splines over a bounded domain of $\mathbb{R}^n$ (i.e. those which minimize the semi-norm $|\cdot|_{m,\Omega}$) is credited to M. Attéia [22, 23]. As in Part A, we consider in the following chapters the model problem of *Lagrange interpolation*. In Chapter V we study the existence, the uniqueness, the characterization, the regularity and the convergence of interpolating and smoothing $D^m$-splines.

Contrary to $(m, s)$-splines, $D^m$-splines over a bounded domain of $\mathbb{R}^n$ are not explicitly known in terms of linear combinations of known functions (except in the case of dimension $n = 1$). It is then natural to approximate them, which we shall do by discretizing them by means of the Finite Element Method. In this way, we obtain the *discrete $D^m$-splines*. Chapter VI is devoted to the study of interpolating and smoothing discrete $D^m$-splines. Questions about convergence and error estimates are specially examined. Part C will show the interest which these (or analogous) splines present for various problems of approximation of functions.

Finally, Chapter VII treats the $D^m$-splines in dimension $n = 1$, whose content is classical.

All spaces considered in this part are real.

# CHAPTER V

# $D^m$-SPLINES OVER $\Omega$

Throughout this chapter we shall implicitly assume that $m$ and $n$ are two positive integers such that $m > n/2$. Likewise, we shall always denote by $\Omega$ an open subset of $\mathbb{R}^n$ with a Lipschitz-continuous boundary (cf. Preliminaries).

## 1. INTERPOLATING $D^m$-SPLINES

Let $A$ be an ordered set of $N$ distinct points of $\overline{\Omega}$ which contains a $P_{m-1}$-unisolvent subset. We denote by $\rho \in \mathcal{L}(H^m(\Omega), \mathbb{R}^N)$ the operator defined by

$$\rho v = \big(v(a)\big)_{a \in A},$$

whose continuity follows from the imbedding $H^m(\Omega) \hookrightarrow C^0(\overline{\Omega})$ (cf. (4) in Preliminaries).

Let $\beta \in \mathbb{R}^N$. We consider the affine linear variety

$$\mathcal{K} = \{ v \in H^m(\Omega) \mid \rho v = \beta \}$$

as well as the associated vector subspace

$$\mathcal{K}_0 = \{ v \in H^m(\Omega) \mid \rho v = 0 \}.$$

Then, we pose the following model problem: find $\sigma$ solution of

$$\begin{cases} \sigma \in \mathcal{K}, \\ \forall v \in \mathcal{K}, \ |\sigma|_{m,\Omega} \le |v|_{m,\Omega}. \end{cases} \tag{1.1}$$

Every solution of (1.1), if any exists, will be called *interpolating $D^m$-spline over $\Omega$ relative to $A$ and $\beta$*. In the sequel, if nothing else is specified, the term *$D^m$-spline* will always refer to the $D^m$-splines over $\Omega$.

**Remark 1.1** – It is possible to generalize without difficulty problem (1.1) by modifying the definition of the interpolating operator $\rho$. For example, one can include the values of derivatives of order $k$, for some $k \in \mathbb{N}^*$ such that $m > n/2 + k$, or replace the point values by local mean values. In the latter case, it is possible to develop for the $D^m$-splines over a bounded domain a study analogous to that done in Section II–3 for the $(1,0)$-splines defined by local mean values (cf. D. Apprato, R. Arcangéli and J. Gaches [9]). □

Let us begin with two results of norm equivalence in $H^m(\Omega)$ which we shall need in the sequel (they are analogous to Propositions I–2.2 and I–2.3 relative to the space

$X^{m,s}$). For any finite subset $E$ of $\overline{\Omega}$, we agree to denote by $[\![\cdot]\!]_{E,m,\Omega}$ the mapping from $H^m(\Omega)$ onto $[0,+\infty)$ defined by

$$\forall v \in H^m(\Omega),\ [\![v]\!]_{E,m,\Omega} = \left(\sum_{a\in E} |v(a)|^2 + |v|_{m,\Omega}^2\right)^{1/2}. \tag{1.2}$$

***Proposition* 1.1** – *Let $E$ be a finite subset of $\overline{\Omega}$. Then, if $E$ contains a $P_{m-1}$-unisolvent subset, the mapping $[\![\cdot]\!]_{E,m,\Omega}$ is a Hilbertian norm on $H^m(\Omega)$ equivalent to the norm $\|\cdot\|_{m,\Omega}$. Conversely, if $[\![\cdot]\!]_{E,m,\Omega}$ is a norm on $H^m(\Omega)$, the set $E$ contains a $P_{m-1}$-unisolvent subset and $[\![\cdot]\!]_{E,m,\Omega}$ is equivalent to $\|\cdot\|_{m,\Omega}$.*

**Proof** –

1) Suppose that $E$ contains a $P_{m-1}$-unisolvent subset. It is clear that $[\![\cdot]\!]_{E,m,\Omega}$ is a semi-norm on $H^m(\Omega)$. Now, if $[\![v]\!]_{E,m,\Omega} = 0$ for some $v \in H^m(\Omega)$, we have the relations $v = 0$ on $E$ and $|v|_{m,\Omega} = 0$. Taking account of the connectedness of $\Omega$, it follows from the latter relation that $v$ belongs to $P_{m-1}(\Omega)$. Then, by hypothesis on $E$, the former relation implies that $v = 0$ on $\Omega$. We conclude that, in fact, $[\![\cdot]\!]_{E,m,\Omega}$ is a norm, obviously associated with a scalar product. The equivalence of $[\![\cdot]\!]_{E,m,\Omega}$ and $\|\cdot\|_{m,\Omega}$ (and hence the completeness of $H^m(\Omega)$ with the norm $[\![\cdot]\!]_{E,m,\Omega}$) follows from Theorem 2.7.1 of J. Nečas [109].

2) The converse is obtained as in point 2) of Proposition I–2.2. □

***Proposition* 1.2** – *Let $B_0 = \{b_{01},\dots,b_{0\mathfrak{M}}\}$ be a $P_{m-1}$-unisolvent subset of $\overline{\Omega}$. For any $r > 0$, we denote by $\mathcal{B}_r$ the family of all subsets $B = \{b_1,\dots,b_{\mathfrak{M}}\}$ of $\overline{\Omega}$ which satisfy the following condition*

$$\forall j = 1,\dots,\mathfrak{M},\ |b_j - b_{0j}| \le r.$$

*Then, there exists $r_0 > 0$ such that the family $\mathcal{B}_{r_0}$ is formed by $P_{m-1}$-unisolvent subsets and the mapping $[\![\cdot]\!]_{B,m,\Omega}$ is, for every $B \in \mathcal{B}_{r_0}$, a norm on $H^m(\Omega)$, uniformly equivalent over $\mathcal{B}_{r_0}$ to the norm $\|\cdot\|_{m,\Omega}$.*

**Proof** – This proof is similar to that of Proposition I–2.3, replacing the spaces $X^{m,s}$ and $H^{m+s}(\Omega)$ by $H^m(\Omega)$. □

***Theorem* 1.1** – *Problem* (1.1) *admits a unique solution $\sigma$.*

**Proof** – Using a corollary of Urysohn's Theorem (cf. L. Hörmander [80, Theorem 1.2.2, p. 4]), one can find real functions $\varphi_a \in \mathcal{D}(\Omega)$ such that

$$\forall a, b \in A,\ \varphi_a(b) = \begin{cases} 1, & b = a, \\ 0, & b \neq a. \end{cases} \tag{1.3}$$

Hence, if $\beta = (\beta_a)_{a\in A}$, the function $\sum_{a\in A} \beta_a \varphi_a$ belongs to $\mathcal{K}$. It is then clear that $\mathcal{K}$ is a nonempty, closed, convex subset of $H^m(\Omega)$.

Now, since problem (1.1) is obviously equivalent to the problem

$$\begin{cases} \sigma \in \mathcal{K}, \\ \forall v \in \mathcal{K},\ [\![\sigma]\!]_{A,m,\Omega} \leq [\![v]\!]_{A,m,\Omega}, \end{cases}$$

we conclude, taking Proposition 1.1 into account, that (1.1) admits a unique solution, namely the element of minimal norm $[\![\cdot]\!]_{A,m,\Omega}$ in $\mathcal{K}$. □

***Proposition* 1.3** – *The solution $\sigma$ of problem* (1.1) *is characterized by*

$$\begin{cases} \sigma \in \mathcal{K}, \\ \forall w \in \mathcal{K}_0,\ (\sigma, w)_{m,\Omega} = 0. \end{cases} \tag{1.4}$$

**Proof** – The element $\sigma$ is the projection of the null element of $H^m(\Omega)$ onto $\mathcal{K}$. Thus, $\sigma$ is characterized by the relations

$$\begin{cases} \sigma \in \mathcal{K}, \\ \forall v \in \mathcal{K},\ [\![0-\sigma, v-\sigma]\!]_{A,m,\Omega} \leq 0, \end{cases}$$

where $[\![\cdot\,,\,\cdot]\!]_{A,m,\Omega}$ is the scalar product associated with the norm $[\![\cdot]\!]_{A,m,\Omega}$. The Proposition is then a simple consequence. □

***Proposition* 1.4** – *There exists one and only one pair $(\sigma, \lambda) \in H^m(\Omega) \times \mathbb{R}^N$, with $\lambda = (\lambda_a)_{a\in A}$, which is the solution of*

$$\begin{cases} \sigma \in \mathcal{K}, \\ \forall v \in H^m(\Omega),\ (\sigma, v)_{m,\Omega} = \sum_{a\in A} \lambda_a v(a), \end{cases} \tag{1.5}$$

*where $\sigma$ is just the solution of problem* (1.1).

**Proof** – If $(\sigma, \lambda)$ is a solution of (1.5), then $\sigma \in \mathcal{K}$ and, for any $w \in \mathcal{K}_0$, $(\sigma, w)_{m,\Omega} = 0$. Hence, $\sigma$ is the solution of (1.1) and $\sigma$ is unique. Now, if $(\sigma, \lambda')$ and $(\sigma, \lambda'')$ are two solutions of (1.5), with $\lambda' = (\lambda'_a)_{a\in A}$ and $\lambda'' = (\lambda''_a)_{a\in A}$, we deduce that, for any $v \in H^m(\Omega)$, $\sum_{a\in A}(\lambda'_a - \lambda''_a)v(a) = 0$, which implies that $\lambda' = \lambda''$. Therefore, there exists, at most, one solution of (1.5).

On the other hand, it is clear that, for any $v \in H^m(\Omega)$, the function $w = v - \sum_{a\in A} v(a)\varphi_a$ belongs to $\mathcal{K}_0$, where, for any $a \in A$, $\varphi_a$ is any function in $\mathcal{D}(\Omega)$ verifying (1.3). Then, (1.5) follows from (1.4) if we take, for any $a \in A$, $\lambda_a = (\sigma, \varphi_a)_{m,\Omega}$. This completes the proof. □

The preceding proof shows that the constants $\lambda_a$ in (1.5) depend on $\sigma$. But, against what its definition seems to indicate, they are in fact independent of the choice of the functions $\varphi_a$.

The vector $-2\lambda$, where $\lambda = (\lambda_a)_{a\in A}$ is the vector introduced in (1.5), is just the *Lagrange multiplier* of problem (1.1).

## 2. THE SPACE OF $D^m$-SPLINES

We write

$$S = \{ s \in H^m(\Omega) \mid \forall w \in \mathcal{K}_0,\ (s,w)_{m,\Omega} = 0 \}.$$

***Proposition* 2.1** – *The set $S$ is a subspace of dimension $N$ of $H^m(\Omega)$. Moreover, the restriction $\rho_\perp$ of $\rho$ to $S$ is an isomorphism from $S$ onto $\mathbb{R}^N$ and, for any $\beta \in \mathbb{R}^N$, $\rho_\perp^{-1}(\beta)$ is just the interpolating $D^m$-spline relative to $A$ and $\beta$.*

**Proof** – The set $S$ is obviously a subspace (the orthogonal complement of $\mathcal{K}_0$ in $(H^m(\Omega), [\![\cdot]\!]_{A,m,\Omega})$). From Proposition 1.3, we have

$$\forall\beta \in \mathbb{R}^N,\ \exists!\sigma \in S,\ \rho\sigma = \beta.$$

Thus, we deduce that $\rho_\perp$ is a linear bijection from $S$ onto $\mathbb{R}^N$ and hence $\dim S = N$. The Proposition then follows. □

For any $\beta \in \mathbb{R}^N$, the subspace $S$ contains the interpolating $D^m$-spline relative to $A$ and $\beta$. We shall show in Section 3 that, for any $\beta \in \mathbb{R}^N$ and for any $\varepsilon > 0$, $S$ also contains the *smoothing $D^m$-spline over $\Omega$ relative to $A$, $\beta$ and $\varepsilon$*. We say that $S$ is the *space of the $D^m$-splines over $\Omega$ relative to $A$*.

**Remark 2.1** – Contrary to the case of the $D^m$-splines over $\mathbb{R}^n$, *the expression of the $D^m$-splines over $\Omega$ is not explicitly known* (except for $n = 1$, cf. Chapter VII). However, the Finite Element Method can be used to construct real numerical approximants, called *discrete $D^m$-splines*. We shall show this in Chapter VI. □

For any $j = 1, \ldots, N$, we denote by $\sigma_j$ the interpolating $D^m$-spline relative to $A$ and $e_j$, where $e_j$ denotes the $j$th element of the canonical basis of $\mathbb{R}^N$, i.e. $\sigma_j = \rho_\perp^{-1}(e_j)$. It is an immediate consequence of Proposition 2.1 that $\{\sigma_1, \ldots, \sigma_N\}$ is a basis of the space $S$. These functions are called *basis $D^m$-splines*.

The following results show the regularity properties of the elements of $S$.

***Proposition* 2.2** – *Let $P$ be the operator in partial derivatives defined by $Pv = (-1)^m \sum_{|\alpha|=m} \partial^{2\alpha} v$ and, for any $a \in A$, let $\delta_a$ be the Dirac measure at the point $a \in A$. Then, for any $s \in S$, there exists a unique real vector $\lambda = (\lambda_a)_{a\in A}$ such that*

$$Ps = \sum_{a\in A} \lambda_a \delta_a \ \textit{in}\ \mathcal{D}'(\Omega) \tag{2.1}$$

*and*

$$\forall p \in P_{m-1}(\Omega),\ \sum_{a\in A} \lambda_a p(a) = 0. \tag{2.2}$$

**Proof** – Let $s \in S$. A reasoning similar to that in the proof of Proposition 1.4 shows that there exists a unique vector $\lambda = (\lambda_a)_{a\in A}$ such that

$$\forall v \in H^m(\Omega),\ (s,v)_{m,\Omega} = \sum_{a\in A} \lambda_a v(a). \tag{2.3}$$

In fact, for any $a \in A$, the constant $\lambda_a$ is equal to $(s, \varphi_a)_{m,\Omega}$, where $\varphi_a$ is any function in $\mathcal{D}(\Omega)$ satisfying (1.3). Now, for any $\varphi \in \mathcal{D}(\Omega)$, we have

$$Ps.\varphi = (-1)^m \sum_{|\alpha|=m} \partial^{2\alpha} s.\varphi = \sum_{|\alpha|=m} \partial^\alpha s.\partial^\alpha \varphi = (s, \varphi)_{m,\Omega}.$$

This relation, together with (2.3), implies (2.1). Finally, (2.2) is a trivial consequence of (2.3), since, for any $p \in P_{m-1}(\Omega)$, $(s,p)_{m,\Omega} = 0$. □

***Theorem* 2.1** – *Every element $s \in S$ is an analytic function in the open set $\Omega' = \Omega \setminus A$. Moreover, for all $\theta > 0$, $s \in H_{\text{loc}}^{2m-n/2-\theta}(\Omega)$ and therefore $s \in C^{2m-n-1}(\Omega)$.*

**Proof** – Let $s$ be any element of $S$. It is clear from (2.1) that $Ps = 0$ on $\Omega'$. As $P$ is an elliptic operator with analytic coefficients (because these coefficients are constant), the Analytic Regularity Theorem (cf. L. Hörmander [80, p. 178]) proves that $s$ is analytic in $\Omega'$.

Likewise, using the definition, by means of the Fourier transform, of the Sobolev spaces of noninteger order over $\mathbb{R}^n$, and taking into account that, for all $a, \xi \in \mathbb{R}^n$, $|\hat{\delta}_a(\xi)| = 1$, one can verify that

$$\forall \theta > 0, \ \sum_{a \in A} \lambda_a \delta_a \in H^{-n/2-\theta}(\mathbb{R}^n).$$

Then, since $P$ is elliptic of order $2m$, it follows from (2.1) and Friedrichs' Theorem (cf. J. L. Lions and E. Magenes [94, Theorem 3.2, p. 138]) that, for any $\theta > 0$, $s \in H_{\text{loc}}^{2m-n/2-\theta}(\Omega)$.

Finally, for any $x \in \Omega$, there exists $r > 0$ and $\varphi \in \mathcal{D}(\Omega)$ such that $\overline{B}(x,r) \subset \Omega$ and $\varphi = 1$ on $\overline{B}(x,r)$. Since, for any $\theta \in (0,1)$, $2m - n/2 - \theta > 2m - n - 1$ and $s\varphi \in H^{2m-n/2-\theta}(\Omega)$, Sobolev's Continuous Imbedding Theorem (cf. (4) in Preliminaries) proves that $s \in C^{2m-n-1}(\overline{B}(x,r))$. Therefore, $s$ belongs to $C^{2m-n-1}(\Omega)$. □

**Remark 2.2** – Since $S \subset H^m(\Omega)$, Sobolev's Continuous Imbedding Theorem implies that $S \subset C^k(\overline{\Omega})$, where $k = m - n/2 - 1$, if $n$ is even, or $m - n/2 - 1/2$, if $n$ is odd. In fact, by Theorem 2.1, $S$ is a subspace of $C^k(\overline{\Omega}) \cap C^{2m-n-1}(\Omega)$. Therefore, Theorem 2.1 constitutes a result of *super-regularity* for the space $S$ of $D^m$-splines over $\Omega$. Let us observe that the $D^m$-splines over $\mathbb{R}^n$ are also functions of class $C^{2m-n-1}$ (cf. Corollary II–2.1).

The result of Theorem 2.1 can be slightly improved in dimension $n = 1$. We shall see in Chapter VII that the elements of $S$ are piecewise polynomial functions of degree $\leq 2m - 1$. Hence, in this case, $S \subset C^{2m-2}(\overline{\Omega})$. □

## 3. *SMOOTHING $D^m$-SPLINES*

We keep the definition of $A$, $N$, $\rho$, $\beta$ and $\mathcal{K}_0$ given in Section 1. Likewise, we denote by $\langle \cdot \rangle$ (respectively by $\langle \cdot , \cdot \rangle$) the Euclidean norm (respectively, the Euclidean scalar product) in $\mathbb{R}^N$. For any $\varepsilon > 0$, we put

$$\forall v \in H^m(\Omega),\ J_\varepsilon(v) = \langle \rho v - \beta \rangle^2 + \varepsilon |v|^2_{m,\Omega}.$$

Then, we call *smoothing $D^m$-spline over $\Omega$ relative to $A$, $\beta$ and $\varepsilon$* any solution, if any exists, of the problem: find $\sigma_\varepsilon$ such that

$$\begin{cases} \sigma_\varepsilon \in H^m(\Omega), \\ \forall v \in H^m(\Omega),\ J_\varepsilon(\sigma_\varepsilon) \le J_\varepsilon(v). \end{cases} \tag{3.1}$$

***Theorem* 3.1** – *Problem* (3.1) *admits a unique solution $\sigma_\varepsilon$, which is also the unique solution of the variational problem: find $\sigma_\varepsilon$ such that*

$$\begin{cases} \sigma_\varepsilon \in H^m(\Omega), \\ \forall v \in H^m(\Omega),\ \langle \rho\sigma_\varepsilon, \rho v \rangle + \varepsilon(\sigma_\varepsilon, v)_{m,\Omega} = \langle \beta, \rho v \rangle. \end{cases} \tag{3.2}$$

*In addition, the smoothing $D^m$-spline $\sigma_\varepsilon$ belongs to the space $S$ of $D^m$-splines.*

**Proof** – Let us endow $H^m(\Omega)$ with the norm $[\![\cdot]\!]_{A,m,\Omega}$, defined in (1.2) with $E = A$, for which, by Proposition 1.1, $H^m(\Omega)$ is a Hilbert space. Given that $\rho \in \mathcal{L}(H^m(\Omega), \mathbb{R}^N)$, the mapping

$$(u, v) \mapsto \langle \rho u, \rho v \rangle + \varepsilon(u, v)_{m,\Omega}$$

is a continuous symmetric bilinear form on $H^m(\Omega) \times H^m(\Omega)$, since

$$\forall u, v \in H^m(\Omega),\ |\langle \rho u, \rho v \rangle + \varepsilon(u, v)_{m,\Omega}| \le \max(1, \varepsilon)[\![u]\!]_{A,m,\Omega}[\![v]\!]_{A,m,\Omega},$$

and it is also $H^m(\Omega)$-elliptic, since

$$\forall v \in H^m(\Omega),\ \langle \rho v \rangle^2 + \varepsilon |v|^2_{m,\Omega} \ge \min(1, \varepsilon)[\![v]\!]^2_{A,m,\Omega}.$$

Likewise, the mapping $v \mapsto \langle \beta, \rho v \rangle$ is a continuous linear form on $H^m(\Omega)$. Then, the Lax-Milgram Lemma (cf. P. G. Ciarlet [45, Theorem 1.1.3]) proves that (3.1) and (3.2) admit the same unique solution $\sigma_\varepsilon$. Finally, it follows from (3.2) that, for any $v \in \mathcal{K}_0$, $(\sigma_\varepsilon, v)_{m,\Omega} = 0$, and hence $\sigma_\varepsilon \in S$. □

The following result shows that the smoothing $D^m$-spline relative to $A$, $\beta$ and $\varepsilon$ is an *approximant*, as $\varepsilon \to 0$, of the smoothing $D^m$-spline $\sigma$ relative to $A$ and $\beta$. The proof is similar to that of Theorem III–1.2.

***Theorem* 3.2** – *As $\varepsilon \to 0$, we have $\|\sigma_\varepsilon - \sigma\|_{m,\Omega} = O(\varepsilon)$.*

## 4. CONVERGENCE AND ERROR ESTIMATES

The $D^m$-splines over a bounded domain possess the same properties as the $(m, s)$-splines concerning convergence and approximation errors. Thus, for example, one can prove that, given a function $f \in H^m(\Omega)$, the interpolating $D^m$-spline relative to $A$ and $(f(a))_{a \in A}$ converges to $f$ in $H^m(\Omega)$ as $N = \operatorname{card} A \to +\infty$, assuming, of course, some additional hypotheses.

For the sake of brevity, in this section we shall limit ourselves to state, but not prove, some results about convergence and error estimates for $D^m$-splines. The reader can be sure that, in order to get explicit proofs, it suffices to adapt the theory of Sections II–5, II–6, II–7, III–3 and III–4, replacing the spaces $X^{m,s}$ and $H^{m+s}(\Omega)$ by $H^m(\Omega)$ and formally putting $s = 0$. As in these sections, we suppose that we are given a subset $\mathbb{D}$ of $(0, +\infty)$ such that $0 \in \overline{\mathbb{D}}$ and, for any $d \in \mathbb{D}$, an ordered set $A^d$ of $N = N(d)$ distinct points in $\overline{\Omega}$ verifying condition (II–5.1), i.e.

$$\sup_{x \in \Omega} \delta(x, A^d) = d.$$

We recall that the left-hand member of this condition is the Hausdorff distance between $A^d$ and $\overline{\Omega}$, which tends to 0 as $d$ does.

The following result is the counterpart of Proposition II–5.1, which would be needed in the proof of Theorem 4.1 (see below).

**Proposition 4.1** – *Suppose that* (II–5.1) *holds. Then, there exists* $\eta > 0$ *and, for any* $d \in \mathbb{D}$, *a* $P_{m-1}$*-unisolvent subset* $A_0^d$ *of* $A^d$ *such that the mapping* $[\![ \cdot ]\!]_{A_0^d, m, \Omega}$, *defined by* (1.2) *with* $E = A_0^d$, *is, for any* $d \in \mathbb{D} \cap (0, \eta]$, *a norm on* $H^m(\Omega)$, *uniformly equivalent over* $\mathbb{D} \cap (0, \eta]$ *to the norm* $\| \cdot \|_{m,\Omega}$.

Since, by (II–5.1), $\mathbb{D}$ is bounded, we may assume, without loss of generality, that $\eta = \sup \mathbb{D}$. Thus, for any $d \in \mathbb{D}$, $A^d$ contains a $P_{m-1}$-unisolvent subset. This fact guarantees, for any $d \in \mathbb{D}$, the existence and the uniqueness of the interpolating $D^m$-spline over $\Omega$ relative to $A^d$ and $(f(a))_{a \in A^d}$, that we shall denote by $\sigma^d$, as well as, for any $d \in D$ and for any $\varepsilon > 0$, the existence and the uniqueness of the smoothing $D^m$-spline $\sigma_\varepsilon^d$ relative to $A^d$, $(f(a))_{a \in A^d}$ and $\varepsilon$. As in Section III–3, we assume that $\varepsilon$ is a function of $d$ and that $\varepsilon : \mathbb{D} \to (0, +\infty)$ satisfies (III–3.1).

The next two theorems establish the convergence of interpolating and smoothing $D^m$-splines.

**Theorem 4.1** – *Suppose that* (II–5.1) *holds. Then,*

$$\lim_{d \to 0} \|\sigma^d - f\|_{m,\Omega} = 0.$$

**Theorem 4.2** – *Suppose that* (II–5.1) *and* (III–3.1) *hold. Then,*

$$\lim_{d \to 0} \|\sigma_\varepsilon^d - f\|_{m,\Omega} = 0.$$

**Remark 4.1** – From a numerical point of view, Theorems 4.1 and 4.2 are merely abstract results. They lack real utility, insofar as the $D^m$-splines over $\Omega$ cannot be explicitly known (except in the case $n = 1$, cf. Chapter VII). □

We conclude with the results relative to error estimates.

**Theorem 4.3** – *Suppose that* (II–5.1) *holds. Then, we have*

$$\forall l = 0, \ldots, m-1, \; |f - \sigma^d|_{l,\Omega} = o(d^{m-l}), \; d \to 0.$$

**Theorem 4.4** – *Suppose that* (II–5.1), (III–3.1) *and* (III–3.11) *hold and that*

$$\exists C > 0, \; \varepsilon \geq C d^{2m-n}, \; d \to 0. \tag{4.1}$$

*Then, we have*

$$\forall l = 0, \ldots, m-1, \; |f - \sigma_\varepsilon^d|_{l,\Omega} = O\Big((\varepsilon/N)^{(m-l)/(2m)}\Big), \; d \to 0.$$

The reader is invited to review the appropriate sections of Chapters II and III. Most of the comments made there about the counterparts of the above results and their corresponding hypotheses are still valid in the present context.

# *CHAPTER VI*

# *DISCRETE $D^m$-SPLINES*

We have already mentioned several times that, for $n > 1$, the $D^m$-splines over a bounded domain cannot be explicitly expressed in terms of a finite number of known functions. Hence, for practical purposes, the $D^m$-splines always need to be approximated. To this end, the Finite Element Method, in short F. E. M., constitutes a basic tool, since it provides an easy way to discretize the elliptic variational problems of which the $D^m$-splines are the solutions. Therefore, we are led to consider variational problems (or their equivalent minimization problems) analogous to those introduced in Chapter V, but posed in suitable finite-dimensional spaces. The corresponding solutions are called *discrete $D^m$-splines.*

As in Chapter V, hereafter we assume that $m$ and $n$ are two positive integers such that $m > n/2$, and also that $\Omega$ is an open subset of $\mathbb{R}^n$ with a Lipschitz-continuous boundary (cf. Preliminaries).

## *1. THE FINITE ELEMENT FRAMEWORK* †

Throughout this chapter we shall implicitly assume that we are given

- a bounded subset $\mathbb{H}$ of $(0, +\infty)$ such that $0 \in \overline{\mathbb{H}}$,
- a bounded polyhedral open subset $\widetilde{\Omega}$ of $\mathbb{R}^n$ such that $\Omega \subset \widetilde{\Omega}$,
- for any $h \in \mathbb{H}$, a triangulation $\widetilde{\mathcal{T}}_h$ of $\overline{\widetilde{\Omega}}$ made up of $n$-simplices or $n$-rectangles of diameter $h_K \leq h$ and a finite element space $\widetilde{V}_h$, constructed on $\widetilde{\mathcal{T}}_h$, such that

$$\widetilde{V}_h \text{ is a finite-dimensional subspace of } H^m(\widetilde{\Omega}). \tag{1.1}$$

For any $h \in \mathbb{H}$, we shall denote by $\Omega_h$ the open set defined by

$$\Omega_h \text{ is the interior of the union of elements } K \in \widetilde{\mathcal{T}}_h \text{ such that } K \cap \Omega \neq \emptyset \tag{1.2}$$

(cf. Figure 1). It can be seen that the family $(\Omega_h)_{h \in \mathbb{H}}$ verifies the relations

$$\forall h \in \mathbb{H},\ \Omega \subset \Omega_h \subset \widetilde{\Omega} \tag{1.3}$$

and

$$\lim_{h \to 0} \operatorname{meas}(\Omega_h \setminus \Omega) = 0. \tag{1.4}$$

In addition, for any $h \in \mathbb{H}$ small enough, $\Omega_h$ has a Lipchitz-continuous boundary. We assume, without loss of generality, that, in fact, this property holds for any $h \in \mathbb{H}$.

†For the notations, definitions and classical results concerning the theory of the Finite Element Method, see, for example, P. G. Ciarlet [45].

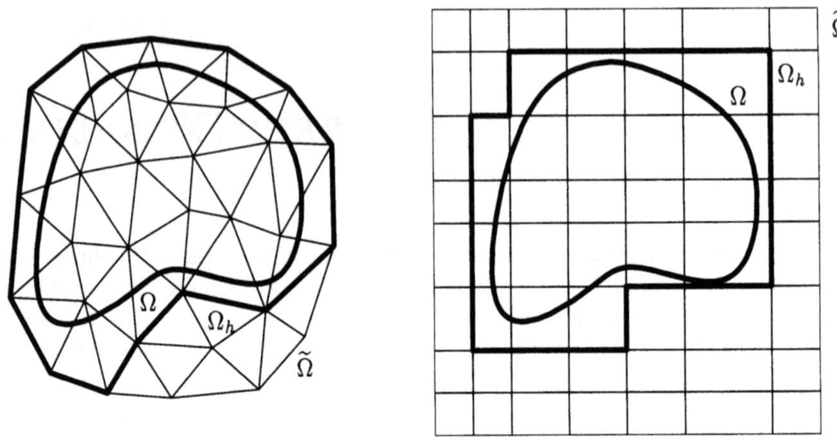


Figure 1: Two examples of open sets $\Omega$, $\widetilde{\Omega}$ and $\Omega_h$.

For any $h \in \mathbb{H}$, we also define the space $V_h$ by

$$\left| \begin{array}{l} V_h \text{ is the space of restrictions to } \Omega_h \text{ of functions in } \widetilde{V}_h, \text{ endowed with} \\ \text{the topology induced by that of } H^m(\Omega_h). \end{array} \right. \tag{1.5}$$

It is clear that $V_h$ is a finite-dimensional subspace of $H^m(\Omega_h)$. We denote by $M = M(h)$ the dimension of $V_h$.

To prove convergence results about discrete $D^m$-splines, for any $h \in \mathbb{H}$, we shall need to approximate functions in $H^q(\widetilde{\Omega})$ by functions in $\widetilde{V}_h$, $q$ being a suitable integer $\geq m$. At a first glance, it seems that, for such purposes, it may suffice to consider, for any $h \in \mathbb{H}$, the exact $\widetilde{V}_h$-interpolating operator $\Pi_h$ (cf. P. G. Ciarlet [45]). But this operator can be applied to functions in $H^q(\widetilde{\Omega})$ only if the imbedding $H^q(\widetilde{\Omega}) \hookrightarrow C^s(\overline{\widetilde{\Omega}})$ is ensured, $s$ being the greatest order of derivation occurring in the definition of the degrees of freedom of the generic finite element of the family $(\widetilde{V}_h)_{h\in\mathbb{H}}$. Unfortunately, the preceding regularity condition may be too restrictive in many particular cases.

The works of G. Strang [136], P. Clément [47] or L. R. Scott and S. Zhang [133], among others, show that it is possible to define $\widetilde{V}_h$-approximation operators $\widetilde{\Pi}_h$ under weaker constraints, but still preserving the same optimal degree of approximation provided by the operator $\Pi_h$. Therefore, we may assume that the family $(\widetilde{V}_h)_{h\in\mathbb{H}}$ of finite element spaces is constructed in such a way that the following result, obtained by P. Clément [47, Theorem 1], is verified:

Let $q$ be an integer such that $P_{q-1}(K) \subset P_K \subset H^q(K)$. Then, there exists a constant $C > 0$ and, for any $h \in \mathbb{H}$, a linear operator $\widetilde{\Pi}_h : L^2(\widetilde{\Omega}) \to \widetilde{V}_h$ such that

$$\forall l = 0, \ldots, q, \ \forall v \in H^q(\widetilde{\Omega}), \ \left( \sum_{K \in \widetilde{\mathcal{T}}_h} |v - \widetilde{\Pi}_h v|_{l,K}^2 \right)^{1/2} \leq C h^{q-l} |v|_{q,\widetilde{\Omega}}. \tag{1.6}$$

Here, to state precisely the hypothesis on $q$, $K$ and $P_K$ are the first and second components of the triple $(K, P_K, \Sigma_K)$ which denotes the generic finite element of the family $(\widetilde{V}_h)_{h\in\mathbb{H}}$. We shall alternatively assume the following variant of the above result, which requires a slightly different condition on $q$ (see again [47, Theorem 1]):

> Let $q$ be an integer such that $P_q(K) \subset P_K \subset H^q(K)$. Then, (1.6) holds and, in addition,
>
> $$\forall v \in H^q(\widetilde{\Omega}),\ \lim_{h\to 0}\left(\sum_{K\in\widetilde{\mathcal{T}}_h} |v - \widetilde{\Pi}_h v|^2_{q,K}\right)^{1/2} = 0. \tag{1.7}$$

The value of $q$ for which (1.6) or (1.7) should hold will be indicated in every convergence result.

**Remark 1.1** – In [47], P. Clément limited his study to the usual Hermite triangular elements, for which the degrees of freedom $\phi_i$ are of the type $v \to (D_i v)(b_i)$, where $D_i$ is an homogeneous differential operator associated with the node $b_i$. Nevertheless, his result is also applicable to other finite elements, which may be defined, for example, on $n$-simplices or $n$-rectangles. To obtain (1.6) or (1.7), P. Clément assumes that

$$(\widetilde{\mathcal{T}}_h)_{h\in\mathbb{H}} \text{ is regular (in the P. G. Ciarlet and P.-A. Raviart [46] sense)} \tag{1.8}$$

and that, for any $h \in \mathbb{H}$, the basis functions $\tilde{w}_1, \ldots, \tilde{w}_{\widetilde{M}}$ of $\widetilde{V}_h$ satisfy the following "uniformity" property, introduced in a different form by G. Strang in [136]:

$$\left|\begin{array}{l}\exists C > 0,\ \forall i = 1, \ldots, \widetilde{M},\ \text{with } s_i \le q-1,\ \forall h \in \mathbb{H},\ \forall K \in \widetilde{\mathcal{T}}_h,\ \forall l = 0, \ldots, q,\\ \forall v \in C^\infty(\overline{\widetilde{\Omega}}),\ |\phi_i(v)\tilde{w}_i|_{l,K} \le C\, h_K^{s_i-l}(\operatorname{meas} K)^{1/2} \max_{|\alpha|=s_i} |\partial^\alpha v(b_i)|,\end{array}\right. \tag{1.9}$$

where, for any $i = 1, \ldots, \widetilde{M}$, $s_i$ indicates the order of the differential operator $D_i$ intervening in the definition of the degree of freedom $\phi_i$, and $\tilde{w}_i$ is the basis function associated with $\phi_i$. For $n = 2$, condition (1.9) is fulfilled, for example, by the Argyris and Bell triangles and the Bogner-Fox-Schmit rectangle (see Section VIII–3 for their corresponding definitions).

We conclude by recalling, for any $h \in \mathbb{H}$, the explicit definition of the operator $\widetilde{\Pi}_h$. For any $i = 1, \ldots, \widetilde{M}$, let $S_i$ be the support of the basis function $\tilde{w}_i$. Then, $\widetilde{\Pi}_h$ is given by

$$\forall v \in H^q(\widetilde{\Omega}),\ \widetilde{\Pi}_h v = \sum_{i=1}^{\widetilde{M}} \phi_i(\Pi_i v)\tilde{w}_i, \tag{1.10}$$

where $\Pi_i v$ stands for the $L^2(\widetilde{\Omega})$-orthogonal projection of $v$ on $P_{q-1}(S_i)$, i.e. the unique polynomial defined by the relations

$$\begin{cases}\Pi_i v \in P_{q-1}(S_i),\\ \forall p \in P_{q-1}(S_i),\ (v - \Pi_i v, p)_{0,S_i} = 0.\quad \square\end{cases} \tag{1.11}$$

Hereafter, every time that hypotheses (1.6) or (1.7) are explicitly stated, we shall implicitly suppose that (1.8) and (1.9) hold, as well as the precise condition on $q$ that (1.6) and (1.7) contain.

# 2. DISCRETE INTERPOLATING $D^m$-SPLINES

## 2.1. DEFINITION AND PROPERTIES

Let $A$ be an ordered set of $N$ distinct points of $\overline{\Omega}$ which contains a $P_{m-1}$-unisolvent subset. For any $h \in \mathbb{H}$, we denote by $\rho_h \in \mathcal{L}(V_h, \mathbb{R}^N)$ the operator defined by

$$\rho_h v = \big(v(a)\big)_{a \in A},$$

whose continuity follows from (1.5) and the imbedding $H^m(\Omega_h) \hookrightarrow C^0(\overline{\Omega}_h)$ (cf. (4) in Preliminaries). Let us observe that, in fact, the definition of $\rho_h$ does not depend on $h$. Therefore, to simplify notations, we shall write $\rho$ instead of $\rho_h$.

Let $\beta \in \mathbb{R}^N$. For any $h \in \mathbb{H}$, we introduce the affine linear variety

$$\mathcal{K}_h = \{\, v_h \in V_h \mid \rho v_h = \beta \,\}$$

as well as the associated vector subspace

$$\mathcal{K}_{0h} = \{\, v_h \in V_h \mid \rho v_h = 0 \,\}.$$

Then, we consider the following problem: find $\sigma_h$ such that

$$\begin{cases} \sigma_h \in \mathcal{K}_h, \\ \forall v_h \in \mathcal{K}_h, \ |\sigma_h|_{m,\Omega_h} \le |v_h|_{m,\Omega_h}. \end{cases} \tag{2.1}$$

Every solution of (2.1), if any exists, will be called *$V_h$-discrete interpolating $D^m$-spline relative to $A$ and $\beta$.*

***Theorem* 2.1** – *Let $h \in \mathbb{H}$. Suppose that* (1.1), (1.2) *and* (1.5) *hold. Then, if $\mathcal{K}_h$ is nonempty, problem* (2.1) *admits a unique solution, characterized by*

$$\begin{cases} \sigma_h \in \mathcal{K}_h, \\ \forall w_h \in \mathcal{K}_{0h}, \ (\sigma_h, w_h)_{m,\Omega_h} = 0. \end{cases} \tag{2.2}$$

**Proof** – The closed, convex and nonempty subset $\mathcal{K}_h$ of $V_h$ has a unique element $\sigma_h$ of minimal norm $[\![ \cdot ]\!]_{A,m,\Omega_h}$, characterized by (2.2), where $[\![ \cdot ]\!]_{A,m,\Omega_h}$ is defined by (V–1.2) with $E = A$ and $\Omega_h$ instead of $\Omega$. □

It is clear that (2.1) (resp. (2.2)) constitutes a discretization of (V–1.1) (resp. (V–1.4)).

Let us note that, for any $h \in \mathbb{H}$, $\mathcal{K}_h$ is nonempty if the following condition is satisfied:

$$\text{the points of } A \text{ are Lagrange nodes of the space } V_h. \tag{2.3}$$

This hypothesis obviously implies that $N \le M$.

**Remark 2.1** – Let $h \in \mathbb{H}$ and suppose that (2.3) holds. Let $a_1, \ldots, a_N$ be the points of $A$ and let $w_1, \ldots, w_M$ be the basis functions of the space $V_h$, numbered so that, for all $j = 1, \ldots, N$, $w_j(a_j) = 1$. If $\beta = (\beta_j)_{1 \le j \le N}$, it is clear that $\sum_{j=1}^N \beta_j w_j$ belongs to $K_h$ and that $\{w_{N+1}, \ldots, w_M\}$ is a basis of $K_{0h}$. Therefore, the solution $\sigma_h$ of (2.1) and (2.2) can be expressed as

$$\sigma_h = \sum_{j=1}^{N} \beta_j w_j + \sum_{j=N+1}^{M} \alpha_j w_j,$$

with $\alpha_{N+1}, \ldots, \alpha_M \in \mathbb{R}$. In addition, it follows from (2.2) that the unknown coefficients $\alpha_j$ are the solution of the linear system

$$\sum_{j=N+1}^{M} (w_j, w_i)_{m,\Omega_h} \alpha_j = -\sum_{j=1}^{N} \beta_j (w_j, w_i)_{m,\Omega_h}, \quad N+1 \le i \le M,$$

whose matrix is regular, since it is the Gram matrix of a basis of $\mathcal{K}_{0h}$ in the space $(V_h, [\![ \cdot ]\!]_{A,m,\Omega_h})$, with $[\![ \cdot ]\!]_{A,m,\Omega_h}$ defined as in the proof of Theorem 2.1. □

**Remark 2.2** – In many applications, one has to construct an interpolant of class $C^k$, with $k = 1$ or 2, of a bivariate function $f : \overline{\Omega} \to \mathbb{R}$ from the Lagrange data set $\{ f(a) \mid a \in A \}$. In this case, one must use finite element spaces $\widetilde{V}_h$ constructed from a generic Hermite finite element $(K, P_K, \Sigma_K)$ of class $C^{k'}$, with $k' \ge k$, and such that $P_K \subset H^m(K)$ and $m \le k' + 1$. Once the value of $k$ is fixed, for reasons of computational cost, one typically sets $k' = k$ and $m = k + 1$.

Section VIII–3 discusses several questions about finite elements. We only mention here that the implementation of rectangular elements is usually considerably less expensive than that of triangular elements, so, from this point of view, the former should be preferred to the latter whenever possible. However, due to hypothesis (2.3), rectangular elements are only valid for gridded data and hence, for general interpolation problems, triangular elements are the unique choice. To overcome this restriction, one can attempt to introduce Lagrange multipliers (for a similar approach, see Section VIII–5, in particular, (VIII–5.5) and Remark VIII–5.1). But then new difficulties arise, tied, for example, to the condition $\mathcal{K}_h \ne \emptyset$. The problem remains open. □

**Remark 2.3** – Let $h \in \mathbb{H}$ and assume that $K_h \ne \emptyset$. One can easily prove a result analogous to Proposition V–2.1:

*The set*

$$S_h = \{ s_h \in V_h \mid \forall w_h \in \mathcal{K}_{0h},\ (s_h, w_h)_{m,\Omega_h} = 0 \}$$

*is a subspace of dimension $N$ of $V_h$. Moreover, the restriction $\rho_\perp$ of $\rho$ to $S_h$ is an isomorphism from $S_h$ onto $\mathbb{R}^N$ and, for any $\beta \in \mathbb{R}^N$, $\rho_\perp^{-1}(\beta)$ is just the $V_h$-discrete interpolating $D^m$-spline relative to $A$ and $\beta$.*

Thus, for any $\beta \in \mathbb{R}^N$, the subspace $S_h$ contains the $V_h$-discrete interpolating $D^m$-spline relative to $A$ and $\beta$, and also the so-called *$V_h$-discrete smoothing $D^m$-spline relative to $A$, $\beta$ and $\varepsilon$* (cf. Section 3). Hence, $S_h$ is the *space of $V_h$-discrete $D^m$-splines relative to $A$.* □

## 2.2. CONVERGENCE AND ERROR ESTIMATES

Suppose we are given a set $\mathbb{D} \subset (0, +\infty)$ such that $0 \in \overline{\mathbb{D}}$, and, for any $d \in \mathbb{D}$, a finite subset $A^d$ of $N = N(d)$ points of $\overline{\Omega}$ verifying (II–5.1), i.e.

$$\sup_{x \in \Omega} \delta(x, A^d) = d$$

(see the comments about this hypothesis in Section II–5). By Proposition V–4.1, we may assume that, for any $d \in \mathbb{D}$, $A^d$ contains a $P_{m-1}$-unisolvent subset $A_0^d$. Likewise, for any $d \in \mathbb{D}$ and for any function $v$ defined on $A^d$, we write

$$\rho^d v = \big(v(a)\big)_{a \in A^d}. \tag{2.4}$$

In this way, we have introduced an operator $\rho^d$ which can be considered, depending on the context, either as an element of $\mathcal{L}(H^m(\Omega), \mathbb{R}^N)$, or as an element of $\mathcal{L}(H^m(\widetilde{\Omega}), \mathbb{R}^N)$, or, for any $h \in \mathbb{H}$, as an element of $\mathcal{L}(V_h, \mathbb{R}^N)$.

To study the convergence of the discrete interpolating $D^m$-splines, we need a hypothesis analogous to (2.3), but valid for the families $(A^d)_{d \in \mathbb{D}}$ and $(V_h)_{h \in \mathbb{H}}$. We shall suppose that the set

$$\mathcal{E} = \{\, (d,h) \in \mathbb{D} \times \mathbb{H} \mid \text{the points of } A^d \text{ are Lagrange nodes of } V_h \,\}$$

verifies the condition

$$\mathcal{E} \text{ is nonempty and admits } (0,0) \text{ as an accumulation point.} \tag{2.5}$$

We denote by $\widetilde{M} = \widetilde{M}(h)$ the dimension of $\widetilde{V}_h$. It is clear that, for any $(d,h) \in \mathcal{E}$, $N \leq \widetilde{M}$.

**Lemma 2.1** – *Suppose that* (1.1) *and* (2.5) *hold and that* (1.6) *is satisfied for* $q = m+1$. *Then, for any* $v \in H^{m+1}(\widetilde{\Omega})$ *and for any* $(d,h) \in \mathcal{E}$, *there exists* $v_h^d \in \widetilde{V}_h$ *such that* $\rho^d v_h^d = \rho^d v$ *and that*

$$\lim_{\substack{(d,h) \to (0,0) \\ (d,h) \in \mathcal{E}}} \|v_h^d - v\|_{m,\widetilde{\Omega}} = 0.$$

**Proof** –

1) Let $(d,h) \in \mathcal{E}$. Let us denote by $a_1, \ldots, a_N$ the points of $A^d$, and assume that the basis functions $\tilde{w}_1, \ldots, \tilde{w}_{\widetilde{M}}$ of the space $\widetilde{V}_h$ are numbered so that, for $i = 1, \ldots, N$, $\tilde{w}_i(a_i) = 1$. Finally, for $i = 1, \ldots, \widetilde{M}$, let $S_i$ and $\phi_i$ be, respectively, the support of the basis function $\tilde{w}_i$ and its associated degree of freedom. Obviously, for $i = 1, \ldots, N$ and for any function $v$ defined on $A^d$, one has $\phi_i(v) = v(a_i)$.

For any $(d,h) \in \mathcal{E}$ and any $v \in H^{m+1}(\widetilde{\Omega})$, we put

$$v_h^d = \widetilde{\Pi}_h v + \sum_{i=1}^{N} \phi_i(v - \Pi_i v)\tilde{w}_i,$$

where $\widetilde{\Pi}_h$ is the Clément operator introduced in (1.6) and defined in (1.10), and $\Pi_i v$ is the polynomial given by (1.11) with $q = m+1$. One has, by definition of $v_h^d$,

$$\forall (d,h) \in \mathcal{E},\ \forall a \in A^d,\ v_h^d(a) = v(a),$$

and, by (1.6),

$$\lim_{h \to 0} \|v - \widetilde{\Pi}_h v\|_{m,\widetilde{\Omega}} = 0.$$

Therefore, proving the Lemma is equivalent to showing that

$$\lim_{\substack{(d,h)\to(0,0)\\(d,h)\in\mathcal{E}}} \left\| \sum_{i=1}^{N} \phi_i (v - \Pi_i v) \tilde{w}_i \right\|_{m,\widetilde{\Omega}} = 0.$$

2) Let $(d,h) \in \mathcal{E}$ and $i \in \{1,\ldots,N\}$. Taking into account that $S_i$ is star-shaped with respect to the point $a_i$, it follows from Taylor's formula of order $m$ with integral remainder that

$$\forall u \in C^{m+1}(\overline{\widetilde{\Omega}}),\ \forall x \in S_i,\ u(a_i) = \sum_{l=0}^{m} \frac{1}{l!} D^l u(x)(a_i - x)^l$$
$$+ \frac{1}{m!} \int_0^1 (1-t)^m D^{m+1} u\big(x + t(a_i - x)\big)(a_i - x)^{m+1}\, dt,$$

where $D^l u$ denotes the $l$th total derivative of $u$. Thus,

$$\forall u \in C^{m+1}(\overline{\widetilde{\Omega}}),\ \forall x \in S_i,\ |u(a_i)| \leq \sum_{l=0}^{m} \frac{(\operatorname{diam} S_i)^l}{l!} \|D^l u(x)\|$$
$$+ \frac{(\operatorname{diam} S_i)^{m+1}}{m!} \int_0^1 (1-t)^m \|D^{m+1} u\big(x + t(a_i - x)\big)\|\, dt.$$

Taking the $L_x^2(S_i)$ norms on both members of the above inequality, we have (cf. C. B. Morrey [108] or R. Arcangéli and J. L. Gout [16, Proposition 1-1])

$$\forall u \in C^{m+1}(\overline{\widetilde{\Omega}}),\ (\operatorname{meas} S_i)^{1/2} |u(a_i)| \leq \sum_{l=0}^{m} \frac{(\operatorname{diam} S_i)^l}{l!} \left( \int_{S_i} \|D^l u(x)\|\, dx \right)^{1/2}$$
$$+ \frac{(\operatorname{diam} S_i)^{m+1}}{m!(m+1-n/2)} \left( \int_{S_i} \|D^{m+1} u(x)\|\, dx \right)^{1/2}.$$

It is well known that, for any $l = 0,\ldots,m+1$, there exists a constant $C_l$, which only depends on $l$ and $n$, such that

$$\forall u \in C^{m+1}(\overline{\widetilde{\Omega}}),\ \forall x \in S_i,\ \|D^l u(x)\|^2 \leq C_l \sum_{|\alpha|=l} |\partial^\alpha u(x)|^2.$$

Since, for $i = 1,\ldots,N$, $C^{m+1}(S_i)$ is a dense subspace of $H^{m+1}(S_i)$, we finally conclude that there exists a constant $C > 0$, independent of $d$ and $h$, such that

$$\forall (d,h) \in \mathcal{E},\ \forall i = 1,\ldots,N,\ \forall u \in H^{m+1}(S_i),$$
$$|u(a_i)| \leq C (\operatorname{meas} S_i)^{-1/2} \sum_{l=0}^{m+1} (\operatorname{diam} S_i)^l |u|_{l,S_i}. \tag{2.6}$$

3) Thanks to a result by P. Clément (cf. [47, Lemma 1]), we obtain

$$\begin{aligned}&\exists C>0,\ \forall(d,h)\in\mathcal{E},\ \forall i=1,\dots,N,\ \forall l=0,\dots,m+1,\\&\qquad\forall v\in H^{m+1}(\widetilde{\Omega}),\ |v-\Pi_i v|_{l,S_i}\le C(\operatorname{diam}S_i)^{m+1-l}|v|_{m+1,S_i}.\end{aligned}\tag{2.7}$$

Likewise, from (1.8) it follows that

$$\exists C>0,\ \forall(d,h)\in\mathcal{E},\ \forall i=1,\dots,N,\ \forall K\in\widetilde{\mathcal{T}}_h \text{ with } K\subset S_i,\ \operatorname{diam}S_i\le Ch_K \tag{2.8}$$

(to verify (2.8) in the case of triangular finite elements, one can use *Zlamal's condition* (cf. [45])). Taking $u=v-\Pi_i v$ in (2.6), we derive from this relation, (2.7) and (2.8) that

$$\begin{aligned}&\exists C>0,\ \forall(d,h)\in\mathcal{E},\ \forall i=1,\dots,N,\ \forall K\in\widetilde{\mathcal{T}}_h \text{ with } K\subset S_i,\\&\qquad\forall v\in H^{m+1}(\widetilde{\Omega}),\ |v(a_i)-(\Pi_i v)(a_i)|\le C(\operatorname{meas}S_i)^{-1/2}h_K^{m+1}|v|_{m+1,S_i}.\end{aligned}\tag{2.9}$$

Now, we deduce from (1.9) that

$$\begin{aligned}&\exists C>0,\ \forall(d,h)\in\mathcal{E},\ \forall i=1,\dots,N,\ \forall K\in\widetilde{\mathcal{T}}_h,\ \forall l=0,\dots,m,\\&\forall v\in H^{m+1}(\widetilde{\Omega}),\ |\phi_i(v-\Pi_i v)\tilde{w}_i|_{l,K}\le C|v(a_i)-(\Pi_i v)(a_i)|(\operatorname{meas}K)^{1/2}h_K^{-l}.\end{aligned}\tag{2.10}$$

Thus, combining (2.9) and (2.10), we get

$$\begin{aligned}&\exists C>0,\ \forall(d,h)\in\mathcal{E},\ \forall i=1,\dots,N,\ \forall K\in\widetilde{\mathcal{T}}_h \text{ with } K\subset S_i,\\&\forall l=0,\dots,m,\ \forall v\in H^{m+1}(\widetilde{\Omega}),\ |\phi_i(v-\Pi_i v)\tilde{w}_i|_{l,K}\le Ch^{m+1-l}|v|_{m+1,S_i}.\end{aligned}\tag{2.11}$$

It is clear that

$$\exists\nu_1\in\mathbb{N},\ \forall(d,h)\in\mathcal{E},\ \forall K\in\widetilde{\mathcal{T}}_h,\ \operatorname{card}I_K\le\nu_1 \tag{2.12}$$

where $I_K=\{i\in\{1,\dots,N\}\mid S_i\supset K\}$ (in fact, $\nu_1$ can be taken as $\operatorname{card}\Sigma_K$, if $(K,P_K,\Sigma_K)$ denotes the generic finite element of the family $(\widetilde{V}_h)_{h\in\mathbb{H}}$). Hence,

$$\forall(d,h)\in\mathcal{E},\ \forall K\in\widetilde{\mathcal{T}}_h,\ \forall v\in H^{m+1}(\widetilde{\Omega}),\ \sum_{i\in I_K}|v|_{m+1,S_i}\le\nu_1|v|_{m+1,S(K)},$$

where $S(K)=\bigcup_{i\in I_K}S_i$. Then, it follows from (2.11) that

$$\begin{aligned}&\exists C>0,\ \forall(d,h)\in\mathcal{E},\ \forall K\in\widetilde{\mathcal{T}}_h,\ \forall l=0,\dots,m,\\&\qquad\forall v\in H^{m+1}(\widetilde{\Omega}),\ \left|\sum_{i=1}^{N}\phi_i(v-\Pi_i v)\tilde{w}_i\right|_{l,K}\le Ch^{m+1-l}|v|_{m+1,S(K)}.\end{aligned}$$

But relations (1.8) and (2.12) imply that

$$\exists\nu_2\in\mathbb{N},\ \forall(d,h)\in\mathcal{E},\ \forall K\in\widetilde{\mathcal{T}}_h,\ \operatorname{card}\{K^*\in\widetilde{\mathcal{T}}_h\mid K^*\subset S(K)\}\le\nu_2.$$

We deduce that, for any $v \in H^{m+1}(\widetilde{\Omega})$ and for any $(d,h) \in \mathcal{E}$,

$$\sum_{K \in \widetilde{\mathcal{T}}_h} |v|^2_{m+1,S(K)} = \sum_{K \in \widetilde{\mathcal{T}}_h} \int_{\widetilde{\Omega}} \chi_{S(K)} \left( \sum_{|\alpha|=m+1} |\partial^\alpha v(x)|^2 \right) dx$$
$$= \int_{\widetilde{\Omega}} \left( \sum_{K \in \widetilde{\mathcal{T}}_h} \chi_{S(K)} \right) \left( \sum_{|\alpha|=m+1} |\partial^\alpha v(x)|^2 \right) dx \leq \nu_2 |v|^2_{m+1,\widetilde{\Omega}},$$

where $\chi_{S(K)}$ stands for the characteristic function of $S(K)$. Therefore,

$$\lim_{\substack{(d,h)\to(0,0)\\(d,h)\in\mathcal{E}}} \left\| \sum_{i=1}^N \phi_i (v - \Pi_i v)\tilde{w}_i \right\|_{m,\widetilde{\Omega}} = 0,$$

and the Lemma follows. □

To simplify notations, for any $h \in \mathbb{H}$ and for any function $v_h \in V_h$, from now on we shall write $v_h$ instead of $v_h|_\Omega$.

***Theorem 2.2*** – *Suppose that* (II–5.1), (1.1), (1.2), (1.5) *and* (2.5) *hold and that* (1.6) *is satisfied for $q = m+1$. Let $f$ be a given function in $H^{m+1}(\Omega)$ and, for any $(d,h) \in \mathcal{E}$, let $\sigma_h^d$ be the $V_h$-discrete interpolating $D^m$-spline relative to $A^d$ and $\rho^d f$. Then,*

$$\lim_{\substack{(d,h)\to(0,0)\\(d,h)\in\mathcal{E}}} \|\sigma_h^d - f\|_{m,\Omega} = 0. \tag{2.13}$$

**Proof** –

1) We first remark that, for any $(d,h) \in \mathcal{E}$, $\sigma_h^d$ exists and is unique, since $A^d$ contains a $P_{m-1}$-unisolvent subset and, by definition of $\mathcal{E}$, $\mathcal{K}_h^d = \{ v_h \in V_h \mid \rho^d v_h = \rho^d f \}$ is nonempty. Likewise, by (4) and (5) in Preliminaries, there exists $\tilde{f} \in H^{m+1}(\widetilde{\Omega})$ such that $\tilde{f}|_{\overline{\Omega}} = f$. It follows from (1.3) and (1.4) that

$$|\tilde{f}|_{m,\Omega_h} = |f|_{m,\Omega} + o(1), \ h \to 0. \tag{2.14}$$

2) For any $(d,h) \in \mathcal{E}$, let $\tilde{f}_h^d$ be the function in $\widetilde{V}_h$ that Lemma 2.1 associates with $\tilde{f}$ and let $f_h^d = \tilde{f}_h^d|_{\Omega_h}$. Hence,

$$\forall (d,h) \in \mathcal{E}, \ f_h^d \in \mathcal{K}_h^d \tag{2.15}$$

and

$$\lim_{\substack{(d,h)\to(0,0)\\(d,h)\in\mathcal{E}}} \|\tilde{f}_h^d - \tilde{f}\|_{m,\widetilde{\Omega}} = 0. \tag{2.16}$$

From (2.14) and (2.16), we deduce that

$$\lim_{\substack{(d,h)\to(0,0)\\(d,h)\in\mathcal{E}}} |f_h^d|_{m,\Omega_h} = |f|_{m,\Omega}. \tag{2.17}$$

Likewise, by definition of $\sigma_h^d$ and (2.15), we have

$$\forall (d,h) \in \mathcal{E},\ |\sigma_h^d|_{m,\Omega_h} \le |f_h^d|_{m,\Omega_h}. \tag{2.18}$$

This inequality, combined with (2.15), implies that, for any $(d,h) \in \mathcal{E}$,

$$\begin{aligned}
[\![\sigma_h^d]\!]^2_{A_0^d,m,\Omega} &\le \sum_{a\in A_0^d} |\sigma_h^d(a)|^2 + |\sigma_h^d|^2_{m,\Omega_h} \\
&\le \sum_{a\in A_0^d} |f_h^d(a)|^2 + |f_h^d|^2_{m,\Omega_h} = [\![f]\!]^2_{A_0^d,m,\Omega} + \left(|f_h^d|^2_{m,\Omega_h} - |f|^2_{m,\Omega}\right),
\end{aligned}$$

where $A_0^d$ is the $P_{m-1}$-unisolvent subset of $A^d$ given by Proposition V–4.1 and the mapping $[\![\,\cdot\,]\!]_{A_0^d,m,\Omega}$ is the norm defined by (V–1.2) with $E = A_0^d$. Therefore, by (2.17) and Proposition V–4.1, there exists $d_0 > 0$ and $h_0 > 0$ such that the family $(\sigma_h^d)_{(d,h)\in\mathcal{E},\, d\le d_0,\, h\le h_0}$ is bounded in $H^m(\Omega)$. Since $\mathbb{D}$ and $\mathbb{H}$ are bounded, we may assume that, in fact, $d_0 = \sup \mathbb{D}$ and $h_0 = \sup \mathbb{H}$.

By Corollary 1 in Preliminaries and (2.5), there exists a sequence $(\sigma_{h_l}^{d_l})_{l\in\mathbb{N}}$, with $\lim_{l\to+\infty}(d_l,h_l) = (0,0)$, extracted from the family $(\sigma_h^d)_{(d,h)\in\mathcal{E}}$, and there also exists an element $f^* \in H^m(\Omega)$ such that, as $l \to +\infty$,

$$\sigma_{h_l}^{d_l} \to f^*, \quad \text{weakly in } H^m(\Omega).$$

3) We now proceed as in the proof of Theorem II–5.1. Let us first prove that $f^* = f$. Let $x$ be any point in $\Omega$. Hypothesis (II–5.1) implies that

$$\exists (x_l)_{l\in\mathbb{N}},\ (\forall l \in \mathbb{N},\ x_l \in A^{d_l}) \text{ and } (x = \lim_{l\to+\infty} x_l).$$

Then, taking into account that, for any $l \in \mathbb{N}$, $\rho^{d_l}\sigma_{h_l}^{d_l} = \rho^{d_l} f$, we have

$$\forall l \in \mathbb{N},\ f(x) - \sigma_{h_l}^{d_l}(x) = \left(f(x) - f(x_l)\right) + \left(\sigma_{h_l}^{d_l}(x_l) - \sigma_{h_l}^{d_l}(x)\right).$$

Since $H^m(\Omega)$ is compactly imbedded into $C^0(\overline{\Omega})$ (cf. (3) in Preliminaries), we find that

$$\lim_{l\to+\infty} \sigma_{h_l}^{d_l}(x) = f^*(x)$$

and that

$$\lim_{l\to+\infty} f(x_l) = f(x).$$

Likewise, from Sobolev's Hölder Imbedding Theorem for the space $H^m(\Omega)$ (cf. (1) in Preliminaries), we derive

$$\lim_{l\to+\infty} \left(\sigma_{h_l}^{d_l}(x_l) - \sigma_{h_l}^{d_l}(x)\right) = 0.$$

The conjunction of the last four relations implies that $f^* = f$.

4) From (3) in Preliminaries and the weak convergence of $(\sigma_{h_l}^{d_l})_{l\in\mathbb{N}}$ to $f$ in $H^m(\Omega)$, it follows that

$$\lim_{l\to+\infty} \|\sigma_{h_l}^{d_l} - f\|_{m-1,\Omega} = 0.$$

On the other hand, by (2.18), we have

$$\begin{aligned}\forall l \in \mathbb{N},\ |\sigma_{h_l}^{d_l} - f|_{m,\Omega}^2 &= |\sigma_{h_l}^{d_l}|_{m,\Omega}^2 + |f|_{m,\Omega}^2 - 2(\sigma_{h_l}^{d_l}, f)_{m,\Omega} \\ &\le |f_{h_l}^{d_l}|_{m,\Omega_h}^2 + |f|_{m,\Omega}^2 - 2(\sigma_{h_l}^{d_l}, f)_{m,\Omega},\end{aligned}$$

from where, taking (2.17) into account, we derive

$$\lim_{l\to+\infty} |\sigma_{h_l}^{d_l} - f|_{m,\Omega} = 0.$$

In consequence, we deduce that

$$\lim_{l\to+\infty} \|\sigma_{h_l}^{d_l} - f\|_{m,\Omega} = 0.$$

5) To conclude the proof, we argue by contradiction. Assume that $\|\sigma_h^d - f\|_{m,\Omega}$ does not tend to 0 when $(d,h) \mapsto (0,0)$. Then, there exists a real number $\alpha > 0$ and a sequence $(d_l', h_l')_{l\in\mathbb{N}} \subset \mathcal{E}$, convergent to $(0,0)$, such that

$$\forall l \in \mathbb{N},\ \|\sigma_{h_l'}^{d_l'} - f\|_{m,\Omega} > \alpha. \tag{2.19}$$

But the sequence $(\sigma_{h_l'}^{d_l'})_{l\in\mathbb{N}}$ is bounded. The reasoning of points 2), 3) and 4) shows that there exists a subsequence of $(\sigma_{h_l'}^{d_l'})_{l\in\mathbb{N}}$ which converges to $f$ in $H^m(\Omega)$, in contradiction with (2.19). This completes the proof. □

**Remark 2.4** – The proof of Theorem 2.2, like that of Theorem V–4.1, does not really use hypothesis (II–5.1), but the condition (II–5.2), i.e.

$$\lim_{d\to 0} \sup_{x\in\Omega} \delta(x, A^d) = 0$$

(cf. Section II–5). □

**Remark 2.5** – When $\Omega$ is polyhedral, one can take $\widetilde{\Omega} = \Omega$ and, for any $h \in \mathbb{H}$, $\Omega_h = \Omega$. In this case, the proof of Theorem 2.2 can be simplified as follows.

For any $(d,h) \in \mathcal{E}$, considering (V–1.4) with $\sigma^d$ and $\sigma_h^d - f_h^d$ instead of $\sigma$ and $w$ and replacing $\sigma_h$ and $w_h$ in (2.2) by $\sigma_h^d$ and $\sigma_h^d - f_h^d$, respectively, we obtain the relation

$$(\sigma^d - \sigma_h^d, \sigma_h^d - f_h^d)_{m,\Omega} = 0,$$

which can be written as

$$(\sigma^d - \sigma_h^d, \sigma_h^d - \sigma^d + \sigma^d - f_h^d)_{m,\Omega} = 0,$$

from which we deduce that

$$|\sigma_h^d - \sigma^d|_{m,\Omega} \le |\sigma^d - f_h^d|_{m,\Omega} \le |\sigma^d - f|_{m,\Omega} + |f - f_h^d|_{m,\Omega}.$$

This inequality implies that

$$\begin{aligned}|\sigma_h^d - f|_{m,\Omega} &\le |\sigma_h^d - \sigma^d|_{m,\Omega} + |\sigma^d - f|_{m,\Omega} \\ &\le 2|\sigma^d - f|_{m,\Omega} + |f - f_h^d|_{m,\Omega},\end{aligned}$$

and then, using Proposition V–4.1, Theorem V–4.1 and Lemma 2.1, we finally conclude that (2.13) holds. □

With respect to the approximation error, it is possible to establish the following result.

***Theorem* 2.3** – *Under the conditions of Theorem 2.2, we have*

$$\forall l = 0, \ldots, m, \ |f - \sigma_h^d|_{l,\Omega} = o(d^{m-l}), \ (d, h) \to (0, 0).$$

**Proof** – Cf. M. C. López de Silanes and D. Apprato [98]. □

Let us note the remarkable fact that these error estimates are *independent of h*.

# *3. DISCRETE SMOOTHING $D^m$-SPLINES*

## *3.1. DEFINITION AND PROPERTIES*

We keep the definition of $A$, $N$, $\rho$ and $\beta$ given in Subsection 2.1. Likewise, we denote by $\langle \cdot \rangle$ (respectively by $\langle \cdot , \cdot \rangle$) the Euclidean norm (respectively, the Euclidean scalar product) in $\mathbb{R}^N$. For any $\varepsilon > 0$, we put

$$\forall v_h \in V_h, \ J_{\varepsilon h}(v_h) = \langle \rho v_h - \beta \rangle^2 + \varepsilon |v_h|^2_{m,\Omega_h}.$$

Then, we call *$V_h$-discrete smoothing $D^m$-spline relative to $A$, $\beta$ and $\varepsilon$* any solution, if any exists, of the problem: find $\sigma_{\varepsilon h}$ such that

$$\begin{cases}\sigma_{\varepsilon h} \in V_h, \\ \forall v_h \in V_h, \ J_{\varepsilon h}(\sigma_{\varepsilon h}) \le J_{\varepsilon h}(v_h).\end{cases} \tag{3.1}$$

***Theorem* 3.1** – *Let $h \in \mathbb{H}$. Suppose that* (1.1), (1.2) *and* (1.5) *hold. Then, problem* (3.1) *admits a unique solution $\sigma_{\varepsilon h}$ which is also the unique solution of the variational problem: find $\sigma_{\varepsilon h}$ such that*

$$\begin{cases}\sigma_{\varepsilon h} \in V_h, \\ \forall v_h \in V_h, \ \langle \rho\sigma_{\varepsilon h}, \rho v_h \rangle + \varepsilon(\sigma_{\varepsilon h}, v_h)_{m,\Omega_h} = \langle \beta, \rho v_h \rangle.\end{cases} \tag{3.2}$$

*In addition, the discrete smoothing $D^m$-spline $\sigma_{\varepsilon h}$ belongs to the space $S_h$ of discrete $D^m$-splines (cf. Remark 2.3).*

**Proof** – Analogous to the proof of Theorem V–3.1, replacing $H^m(\Omega)$ by $V_h$. □

It is clear that (3.1) (resp. (3.2)) constitutes a discretization of (V–3.1) (resp. (V–3.2)).

**Remark 3.1** – Let $h \in \mathbb{H}$, let $a_1, \dots, a_N$ be the points of $A$, and let $w_1, \dots, w_M$ be the basis functions of the space $V_h$. The solution $\sigma_{\varepsilon h}$ of (3.1) and (3.2) is expressed as

$$\sigma_{\varepsilon h} = \sum_{j=1}^{M} \alpha_j w_j,$$

with $\alpha_1, \dots, \alpha_M \in \mathbb{R}$. Introducing the matrices

$$\mathcal{A} = \big(w_j(a_i)\big)_{1 \le i \le N,\, 1 \le j \le M} \quad \text{and} \quad \mathcal{R} = \big((w_j, w_i)_{m,\Omega_h}\big)_{1 \le i,\, j \le M},$$

one sees that (3.2) is equivalent to the problem: find $\alpha = (\alpha_j)_{1 \le j \le M}$ solution of

$$\begin{cases} \alpha \in \mathbb{R}^M, \\ (\mathcal{A}^T \mathcal{A} + \varepsilon \mathcal{R})\alpha = \mathcal{A}^T \beta, \end{cases} \tag{3.3}$$

where $\mathcal{A}^T$ denotes the transpose matrix of $\mathcal{A}$. Let us note that $\mathcal{A}^T\mathcal{A} + \varepsilon\mathcal{R}$ is a positive definite symmetric matrix whose dimension is exactly that of the space $V_h$.

We remark that problem (3.3) is just the *regularized equation in the Tikhonov sense* of the equation $\mathcal{A}\alpha = \beta$ (cf. A. Tikhonov and V. Arsenine [138]). □

**Remark 3.2** – Contrary to the case of discrete interpolating $D^m$-splines, the computation of discrete smoothing $D^m$-splines is not constrained by a condition like (2.3). For $n = 2$, this means that rectangular elements can be systematically employed to construct the spaces $V_h$, since, as already pointed out in Remark 2.2, those elements are preferable to triangular elements due to reasons of computational cost and ease of implementation. But one pays for this freedom by introducing the smoothing parameter $\varepsilon$, whose value is not obvious at all.

For a fixed $h \in \mathbb{H}$, the choice of $\varepsilon$ can be made by *trial and error*, starting with a large value and following some strategy (progressive decrease, bisection search,...) until a satisfactory fitting is achieved. During this process, the matrices $\mathcal{A}$ and $\mathcal{R}$ introduced in the preceding remark remain constant, so the search for $\varepsilon$ can be made efficiently by incremental modifications of the linear system in (3.3). There also exist methods for an automatic choice of $\varepsilon$, mainly based on statistical considerations, as the *generalized cross validation* and the *generalized maximum likelihood* methods (cf. P. Craven and G. Wahba [49], C. Gu [76], G. Wahba [149]). □

**Remark 3.3** – Let us give a brief account of the generalized cross validation (GCV) method, that we shall use in the numerical examples of Part C.

For a fixed $h \in \mathbb{H}$, the GCV method consists in choosing the parameter $\varepsilon$ so as to minimize the GCV function $\mathcal{V}$, defined, with the notations of Remark 3.1, by

$$\mathcal{V}(\varepsilon) = \frac{\frac{1}{N}\langle (I_N - Q_\varepsilon)\beta \rangle^2}{\left(\frac{1}{N}\operatorname{tr}(I_N - Q_\varepsilon)\right)^2}, \tag{3.4}$$

where $I_N$ is the identity matrix of dimension $N$, $Q_\varepsilon$ is the so-called *influence matrix*, i.e. the $N \times N$ matrix for which $\big(\sigma_{\varepsilon h}(a_i)\big)_{1 \le i \le N} = Q_\varepsilon \beta$, and $\operatorname{tr}(I_N - Q_\varepsilon)$ is the trace of the matrix $I_N - Q_\varepsilon$. It is readily seen that, in the present framework, $Q_\varepsilon = \mathcal{A}(\mathcal{A}^T\mathcal{A} + \varepsilon\mathcal{R})^{-1}\mathcal{A}^T$.

In order to be able to evaluate $\mathcal{V}$ at a low cost, it is a common practice to replace the term $\mathrm{tr}(I_N - Q_\varepsilon)$ by a Monte-Carlo approximant, as, for example, $u^T(I_N - Q_\varepsilon)u$, where $u$ is a single realization of a vector of $N$ independent random variables with zero mean and common variance 1 (cf. D. Girard and P.-J. Laurent [70]). In our implementation of the GCV method, the elements of $u$ take the values 1 and $-1$ with probability 1/2, as suggested by M. F. Hutchinson [83]. This approach leads to minimize the function $\overline{\mathcal{V}}$ defined by

$$\overline{\mathcal{V}}(\varepsilon) = \frac{\frac{1}{N}\langle (I_N - Q_\varepsilon)\beta\rangle^2}{\left(\frac{1}{N}u^T(I_N - Q_\varepsilon)u\right)^2} \tag{3.5}$$

instead of the GCV function. Let us observe that, if $\alpha$ and $\tilde{\alpha}$ are, respectively, the solutions of the systems $(\mathcal{A}^T\mathcal{A} + \varepsilon\mathcal{R})\alpha = \mathcal{A}^T\beta$ and $(\mathcal{A}^T\mathcal{A} + \varepsilon\mathcal{R})\tilde{\alpha} = \mathcal{A}^T u$ (which can be solved simultaneously), we have

$$\overline{\mathcal{V}}(\varepsilon) = N\,\frac{\langle\beta\rangle^2 - 2\alpha^T\mathcal{A}^T\beta + \alpha^T\mathcal{A}^T\mathcal{A}\alpha}{(N - \tilde{\alpha}^T\mathcal{A}^T u)^2}. \tag{3.6}$$

Using the above expression, the search for the optimal value of $\varepsilon$ can then be done by a global minimization method. □

## 3.2. CONVERGENCE AND ERROR ESTIMATES

As in Subsection 2.2, suppose we are given a set $\mathbb{D} \subset (0, +\infty)$ such that $0 \in \overline{\mathbb{D}}$, and, for any $d \in \mathbb{D}$, a finite subset $A^d$ of $N = N(d)$ points of $\overline{\Omega}$ verifying (II–5.1). Likewise, for any $d \in \mathbb{D}$, let $\rho^d$ be the operator defined by (2.4).

We suppose that the families $(A^d)_{d\in\mathbb{D}}$ and $(\widetilde{\mathcal{T}}_h)_{h\in\mathbb{H}}$ are linked by the relation

$$\exists C > 0,\ \forall (d,h) \in \mathbb{D}\times\mathbb{H},\ \forall K \in \widetilde{\mathcal{T}}_h,\ \frac{\mathrm{card}(A^d \cap K)}{\mathrm{meas}\,K} \le C d^{-n}. \tag{3.7}$$

This hypothesis translates a property of "asymptotic regularity" of the density of the points of $A^d$ over the elements $K$ of $\widetilde{\mathcal{T}}_h$. It is close to hypothesis (III–3.11), which concerns the global distribution of the points of $A^d$ over $\overline{\Omega}$. In fact, (3.7) implies (III–3.11), since

$$N = \mathrm{card}\,A^d \le \sum_{K\in\widetilde{\mathcal{T}}_h} \mathrm{card}(A^d \cap K) \le C d^{-n} \sum_{K\in\widetilde{\mathcal{T}}_h} \mathrm{meas}\,K = C d^{-n}\,\mathrm{meas}\,\widetilde{\Omega}.$$

**Lemma 3.1** – *Suppose that* (1.6), *with* $q > n/2$, *and* (3.7) *hold. Then, there exists* $C > 0$ *such that*

$$\forall (d,h) \in \mathbb{D}\times\mathbb{H},\ \forall v \in H^q(\widetilde{\Omega}),\ \langle \rho^d(\widetilde{\Pi}_h v - v)\rangle^2 \le C\frac{h^{2q}}{d^n}|v|^2_{q,\widetilde{\Omega}},$$

*where, for any* $h \in \mathbb{H}$, $\widetilde{\Pi}_h$ *denotes the operator introduced in* (1.6).

**Proof –**

1) Let $h \in \mathbb{H}$ and $K \in \widetilde{\mathcal{T}}_h$. Since $K$ is convex, $K$ is star-shaped with respect to any point $a \in K$. Then, reasoning as in point 2) of the proof of Lemma 2.1 with $q$, $K$ and $a$ instead of $m+1$, $S_i$ and $a_i$, we deduce that there exists a constant $C > 0$, independent of $h$, such that

$$\forall h \in \mathbb{H},\ \forall K \in \widetilde{\mathcal{T}}_h,\ \forall a \in K,\ \forall u \in H^q(K),$$
$$|u(a)| \leq C(\operatorname{meas} K)^{-1/2} \sum_{l=0}^{q} h^l |u|_{l,K}. \tag{3.8}$$

2) It is clear that, for any $(d,h) \in \mathbb{D} \times \mathbb{H}$ and for any $v \in H^q(\widetilde{\Omega})$,

$$\begin{aligned}\langle \rho^d(\widetilde{\Pi}_h v - v)\rangle^2 &\leq \sum_{K \in \widetilde{\mathcal{T}}_h} \sum_{a \in A^d \cap K} |(\widetilde{\Pi}_h v - v)(a)|^2 \\ &\leq \sum_{K \in \widetilde{\mathcal{T}}_h} (\operatorname{card}(A^d \cap K)) \max_{a \in K} |(\widetilde{\Pi}_h v - v)(a)|^2.\end{aligned}$$

Hence, taking $u = (\widetilde{\Pi}_h v - v)|_K$ in (3.8), we derive that

$$\exists C > 0,\ \forall (d,h) \in \mathbb{D} \times \mathbb{H},\ \forall v \in H^q(\widetilde{\Omega}),$$
$$\langle \rho^d(\widetilde{\Pi}_h v - v)\rangle^2 \leq C \sum_{K \in \widetilde{\mathcal{T}}_h} \frac{\operatorname{card}(A^d \cap K)}{\operatorname{meas} K} \sum_{l=0}^{q} h^{2l} |\widetilde{\Pi}_h v - v|_{l,K}^2.$$

This inequality, combined with (3.7), implies that

$$\exists C > 0,\ \forall (d,h) \in \mathbb{D} \times \mathbb{H},\ \forall v \in H^q(\widetilde{\Omega}),$$
$$\langle \rho^d(\widetilde{\Pi}_h v - v)\rangle^2 \leq C d^{-n} \sum_{l=0}^{q} h^{2l} \sum_{K \in \widetilde{\mathcal{T}}_h} |\widetilde{\Pi}_h v - v|_{l,K}^2.$$

We then deduce from (1.6) that

$$\exists C > 0,\ \forall (d,h) \in \mathbb{D} \times \mathbb{H},\ \forall v \in H^q(\widetilde{\Omega}),\ \langle \rho^d(\widetilde{\Pi}_h v - v)\rangle^2 \leq C d^{-n} \sum_{l=0}^{q} h^{2q} |v|_{q,\widetilde{\Omega}}^2,$$

and the Lemma follows. □

Let $f$ be a given function of $H^{m'}(\Omega)$, where $m'$ is an integer $\geq m$. We recall that, by Proposition V–4.1, we may assume that any set in the family $(A^d)_{d \in \mathbb{D}}$ contains a $P_{m-1}$-unisolvent subset. Thus, for any $d \in \mathbb{D}$ and any $\varepsilon > 0$, the $V_h$-discrete smoothing $D^m$-spline relative to $A^d$, $\rho^d f$ and $\varepsilon$, denoted by $\sigma^d_{\varepsilon h}$, exists and is unique.

From now on we shall suppose that $\varepsilon$ *and* $h$ *are functions of* $d$. Therefore, one can consider that the family $(A^d)_{d \in \mathbb{D}}$ is first given and then one can choose $\varepsilon$ and the families $(\widetilde{\mathcal{T}}_h)_{h\in\mathbb{H}}$ and $(V_h)_{h\in\mathbb{H}}$ accordingly. More precisely, we shall assume that the functions $\varepsilon : \mathbb{D} \to (0, +\infty)$ and $h : \mathbb{D} \to \mathbb{H}$ verify the relation

$$\frac{h^{2m'}}{d^n \varepsilon} = o(1), \; d \to 0, \tag{3.9}$$

and that (III–3.1) also holds. From (III–3.1) and (3.9), we derive that $d \to 0$ implies that $h \to 0$. Let us observe that (3.9) is related to the *size* of the elements of every triangulation $\widetilde{\mathcal{T}}_h$, whereas (3.7) influences their *distribution* in $\overline{\widetilde{\Omega}}$. To simplify notations, hereafter we shall write $\varepsilon$ and $h$ instead of $\varepsilon(d)$ and $h(d)$.

The following theorem constitutes a general result of convergence of discrete smoothing $D^m$-splines. Let us remember that, for any $h \in \mathbb{H}$ and for any function $v_h \in V_h$, we agree to write $v_h$ instead of $v_h|_\Omega$.

***Theorem 3.2*** – *Suppose that* (II–5.1), (III–3.1), (1.1), (1.2), (1.5), (3.7) *and* (3.9) *hold. If $m' > m$, we also assume that* (1.6) *is verified for $q = m'$, whereas, if $m' = m$, we suppose that* (1.7) *holds with $q = m$. Then,*

$$\lim_{d\to 0} \|\sigma^d_{\varepsilon h} - f\|_{m,\Omega} = 0.$$

**Proof** –

1) Let $\tilde{f} \in H^{m'}(\widetilde{\Omega})$ be an extension of $f$, whose existence is justified by (5) in Preliminaries. It follows from (1.3) and (1.4) that

$$|\tilde{f}|_{m,\Omega_h} = |f|_{m,\Omega} + o(1), \; h \to 0. \tag{3.10}$$

Let $(\widetilde{\Pi}_h)_{h\in\mathbb{H}}$ be the family of operators introduced in (1.6). For any $h \in \mathbb{H}$, we write $f_h = (\widetilde{\Pi}_h \tilde{f})|_{\Omega_h}$. Taking into account that

$$\forall h \in \mathbb{H}, \; ||f_h|_{m,\Omega_h} - |\tilde{f}|_{m,\Omega_h}| \le |\widetilde{\Pi}_h \tilde{f} - \tilde{f}|_{m,\widetilde{\Omega}},$$

we derive from (3.10), combined with (1.6), if $m' > m$, or (1.7), if $m' = m$, that

$$|f_h|_{m,\Omega_h} = |f|_{m,\Omega} + o(1), \; h \to 0. \tag{3.11}$$

On the other hand, by Lemma 3.1, we have

$$\forall d \in \mathbb{D}, \; \langle \rho^d(\widetilde{\Pi}_h \tilde{f} - \tilde{f})\rangle^2 \le C \frac{h^{2m'}}{d^n} |\tilde{f}|^2_{m',\widetilde{\Omega}}, \tag{3.12}$$

with $C$ independent of $d$ (and $h$). Likewise, for any $d \in \mathbb{D}$, we deduce from the definition of $\sigma^d_{\varepsilon h}$ that

$$\langle \rho^d \sigma^d_{\varepsilon h} - \rho^d f\rangle^2 + \varepsilon |\sigma^d_{\varepsilon h}|^2_{m,\Omega_h} \le \langle \rho^d f_h - \rho^d f\rangle^2 + \varepsilon |f_h|^2_{m,\Omega_h}.$$

The conjunction of this inequality, (3.9), (3.11) and (3.12) implies the relations

$$|\sigma^d_{\varepsilon h}|_{m,\Omega_h} \le |f|_{m,\Omega} + o(1), \; d \to 0, \tag{3.13}$$

and

$$\langle \rho^d(\sigma^d_{\varepsilon h} - f)\rangle^2 = O(\varepsilon),\ d \to 0. \tag{3.14}$$

2) Let $B_0 = \{b_{01}, \dots, b_{0\mathfrak{M}}\}$ be a $P_{m-1}$-unisolvent subset of $\Omega$ and let $r_0 > 0$ be the constant of Proposition V–1.2. Obviously, there exists $r_0' \in (0, r_0]$ such that

$$\forall j = 1, \dots, \mathfrak{M},\ \overline{B}(b_{0j}, r_0') \subset \overline{\Omega}.$$

It follows from (II–5.1) that

$$\forall d \in \mathbb{D} \cap (0, r_0'),\ \forall j = 1, \dots, \mathfrak{M},\ \overline{B}(b_{0j}, r_0' - d) \subset \bigcup_{a \in A^d \cap \overline{B}(b_{0j}, r_0')} \overline{B}(a, d).$$

Letting $N_j = \operatorname{card}\big(A^d \cap \overline{B}(b_{0j}, r_0')\big)$, we have

$$\forall d \in \mathbb{D} \cap (0, r_0'),\ \forall j = 1, \dots, \mathfrak{M},\ (r_0' - d)^n \le N_j d^n.$$

Thus, for any $d_0 \in (0, r_0')$, we get

$$\forall d \in \mathbb{D} \cap (0, d_0),\ \forall j = 1, \dots, \mathfrak{M},\ N_j \ge (r_0' - d_0)^n d^{-n}.$$

Now, from (III–3.1) and (3.14), we deduce that

$$\forall j = 1, \dots, \mathfrak{M},\quad \sum_{a \in A^d \cap \overline{B}(b_{0j}, r_0')} |\sigma^d_{\varepsilon h}(a) - f(a)|^2 = o(d^{-n}),\ d \to 0.$$

Likewise, for any $d \in \mathbb{D} \cap (0, r_0')$ and any $j = 1, \dots, \mathfrak{M}$, there exists at least one point $b_j^d \in A^d \cap \overline{B}(b_{0j}, r_0')$ such that

$$|\sigma^d_{\varepsilon h}(b_j^d) - f(b_j^d)| = \min_{a \in A^d \cap \overline{B}(b_{0j}, r_0')} |\sigma^d_{\varepsilon h}(a) - f(a)|.$$

Hence, from the last three relations, we derive

$$\forall j = 1, \dots, \mathfrak{M},\ |\sigma^d_{\varepsilon h}(b_j^d) - f(b_j^d)| = o(1),\ d \to 0. \tag{3.15}$$

For any $d \in \mathbb{D} \cap (0, r_0')$, let $B^d = \{b_1^d, \dots, b_{\mathfrak{M}}^d\}$. Applying Proposition V–1.2 with $B = B^d$, it follows from (3.13) and (3.15) that

$$\exists C > 0,\ \exists d^* > 0,\ \forall d \in \mathbb{D} \cap (0, d^*],\ \|\sigma^d_{\varepsilon h}\|_{m,\Omega} \le C,$$

that is, the family $(\sigma^d_{\varepsilon h})_{d \in \mathbb{D} \cap (0, d^*]}$ is bounded in $H^m(\Omega)$. Therefore, by Corollary 1 in Preliminaries, there exists a sequence $(\sigma^{d_l}_{\varepsilon_l h_l})_{l \in \mathbb{N}}$, with, for any $l \in \mathbb{N}$, $\varepsilon_l = \varepsilon(d_l)$, $h_l = h(d_l)$ and

$$\lim_{l \to +\infty} d_l = \lim_{l \to +\infty} d_l^n \varepsilon_l = \lim_{l \to +\infty} \frac{h_l^{2m'}}{d_l^n \varepsilon_l} = 0,$$

extracted from the family $(\sigma^d_{\varepsilon h})_{d \in \mathbb{D} \cap (0, d^*]}$, and there also exists an element $f^* \in H^m(\Omega)$ such that, as $l \to +\infty$,

$$\sigma^{d_l}_{\varepsilon_l h_l} \to f^*,\ \text{weakly in } H^m(\Omega). \tag{3.16}$$

3) We shall prove that $f^* = f$, arguing by contradiction. Suppose that $f^* \neq f$. Then, taking into account that $H^m(\Omega)$ is continuously imbedded into $C^0(\overline{\Omega})$ (cf. (3) in Preliminaries), there exists a nonempty open set $\mathcal{O}$ contained in $\Omega$ and a real number $\alpha > 0$ such that

$$\forall x \in \mathcal{O}, \ |f^*(x) - f(x)| > \alpha.$$

In fact, $H^m(\Omega)$ is compactly imbedded into $C^0(\overline{\Omega})$. Hence, we deduce from (3.16) that the sequence $(\sigma^{d_l}_{\varepsilon_l h_l})_{l \in \mathbb{N}}$ converges uniformly over $\overline{\Omega}$ to $f^*$, and so there exists $l_0 \in \mathbb{N}$ such that

$$\forall l \in \mathbb{N}, \ l \geq l_0, \ \forall x \in \mathcal{O}, \ |\sigma^{d_l}_{\varepsilon_l h_l}(x) - f^*(x)| \leq \alpha/2.$$

Thus, for any $l \geq l_0$, we have

$$\forall x \in \mathcal{O}, \ |\sigma^{d_l}_{\varepsilon_l h_l}(x) - f(x)| \geq |f^*(x) - f(x)| - |\sigma^{d_l}_{\varepsilon_l h_l}(x) - f^*(x)| > \alpha/2. \qquad (3.17)$$

Now, a reasoning analogous to that in point 2) shows that, for any $l \in \mathbb{N}$ sufficiently large, there exists one point $b^{d_l} \in A^{d_l} \cap \mathcal{O}$ such that

$$|\sigma^{d_l}_{\varepsilon_l h_l}(b^{d_l}) - f(b^{d_l})| = o(1), \ l \to +\infty$$

which contradicts (3.17). Therefore, $f^* = f$.

4) Let us prove that

$$\lim_{l \to +\infty} \|\sigma^{d_l}_{\varepsilon_l h_l} - f\|_{m,\Omega} = 0. \qquad (3.18)$$

For any $l \in \mathbb{N}$, it is clear that

$$|\sigma^{d_l}_{\varepsilon_l h_l} - f|^2_{m,\Omega} = |\sigma^{d_l}_{\varepsilon_l h_l}|^2_{m,\Omega} + |f|^2_{m,\Omega} - 2(\sigma^{d_l}_{\varepsilon_l h_l}, f)_{m,\Omega}.$$

Then, from (3.13) and the weak convergence of $(\sigma^{d_l}_{\varepsilon_l h_l})_{l \in \mathbb{N}}$ to $f$ in $H^m(\Omega)$, it follows that

$$\lim_{l \to +\infty} |\sigma^{d_l}_{\varepsilon_l h_l} - f|_{m,\Omega} = 0.$$

On the other hand, the compact imbedding of $H^m(\Omega)$ into $H^{m-1}(\Omega)$ (cf. (2) in Preliminaries) and (3.16) imply that $f$ is the strong limit of $(\sigma^{d_l}_{\varepsilon_l h_l})_{l \in \mathbb{N}}$ in $H^{m-1}(\Omega)$. We deduce that (3.18) holds.

5) We conclude the proof reasoning again by contradiction. Assume that, under the hypotheses of the Theorem, $\|\sigma^d_{\varepsilon h} - f\|_{m,\Omega}$ does not tend to 0 when $d \to 0$. Then, there exists $\alpha > 0$ and three sequences $(d'_l)_{l \in \mathbb{N}} \subset \mathbb{D}$, $(\varepsilon'_l)_{l \in \mathbb{N}}$ and $(h'_l)_{l \in \mathbb{N}}$, with, for any $l \in \mathbb{N}$, $\varepsilon'_l = \varepsilon(d'_l)$, $h'_l = h(d'_l)$ and

$$\lim_{l \to +\infty} d'_l = \lim_{l \to +\infty} (d'_l)^n \varepsilon'_l = \lim_{l \to +\infty} \frac{(h'_l)^{2m'}}{(d'_l)^n \varepsilon'_l} = 0,$$

verifying

$$\forall l \in \mathbb{N}, \ \|\sigma^{d'_l}_{\varepsilon'_l h'_l} - f\|_{m,\Omega} > \alpha. \qquad (3.19)$$

But the sequence $(\sigma^{d'_l}_{\varepsilon'_l h'_l})_{l \in \mathbb{N}}$ is bounded in $H^m(\Omega)$. A similar argument to that of points 1), 2), 3) and 4) proves that there exists a subsequence of $(\sigma^{d'_l}_{\varepsilon'_l h'_l})_{l \in \mathbb{N}}$ which converges to $f$ in $H^m(\Omega)$, in contradiction with (3.19). The Theorem follows. □

**Remark 3.4** – Theorem 3.2 can be formulated in a more general way. Suppose that (II–5.1), (1.1), (1.2), (1.5) and (3.7) hold. If $m' > m$, assume also that (1.6) is verified with $q = m'$, whereas, if $m' = m$, suppose that (1.7) holds with $q = m$. Then, the $V_h$-discrete smoothing $D^m$-spline $\sigma^d_{\varepsilon h}$ relative to $A^d$, $\rho^d f$ and $\varepsilon$ converges to $f$ in $H^m(\Omega)$ through the filter basis $\mathcal{B} = \{B_{\alpha\beta\gamma} \mid \alpha > 0,\ \beta > 0,\ \gamma > 0\}$, with $B_{\alpha\beta\gamma} = \{(d, \varepsilon, h) \in \mathbb{D} \times (0, +\infty) \times \mathbb{H} \mid d \le \alpha,\ d^n\varepsilon \le \beta,\ h^{2m'}/(d^n\varepsilon) \le \gamma\}$. □

We can also obtain estimates of the approximation error by discrete smoothing $D^m$-splines. The reader is referred to M. C. López de Silanes and D. Apprato [98] for the corresponding proofs, analogous to those of Corollary III–4.1 and Theorem III–4.2.

***Theorem* 3.3** – *Under the conditions of Theorem 3.2, we have*

$$\forall l = 0, \dots, m,\ |f - \sigma^d_{\varepsilon h}|_{l,\Omega} = o(d^{m-l}) + O(d^{n/2-l}\varepsilon^{1/2}),\ d \to 0.$$

***Theorem* 3.4** – *Under the conditions of Theorem 3.2, if* (V–4.1) *holds, then, when* $d \to 0$, *we have*

*(i)* $\forall l = 0, \dots, m-1,\ |f - \sigma^d_{\varepsilon h}|_{l,\Omega} = O\Big((\varepsilon/N)^{(m-l)/(2m)}\Big)$,

*(ii)* $|f - \sigma^d_{\varepsilon h}|_{m,\Omega} = o(1)$.

Let us note that, as in the case of discrete interpolating $D^m$-splines, the above error estimates are independent of $h$.

**Remark 3.5** – There also exist results about convergence and error estimates of smoothing $D^m$-splines in the case of noisy data. Although not mentioned in Chapter V, Theorems III–5.1 and III–5.2 can be extended to smoothing $D^m$-splines over a bounded domain (cf. M. C. López de Silanes, D. Apprato and R. Arcangéli [00]). Unfortunately, for discrete smoothing $D^m$-splines, we are able to proof only the counterpart of Theorem III–5.1. In the case of data perturbed by a white noise, the problem of convergence remains open. □

**Remark 3.6** – We have cited in Remark 3.2 some practical methods to set the value of the smoothing parameter $\varepsilon$. From a theoretical point of view, the choice of $\varepsilon$ should conform with (III–3.1) and also, if error estimates are desired to hold, with (V–4.1) (cf. Theorems 3.2 and 3.4). These hypotheses imply the existence of two positive constants $C$ and $C^*$ such that, for any $d \in \mathbb{D}$ sufficiently small, $Cd^{2m-n} \le \varepsilon \le C^* d^{-n}$. Hence, as $d \to 0$, $\varepsilon$ can either remain between 0 and $+\infty$, or tend to 0, or even tend to $+\infty$, but, in the latter two cases, not too fast. □

**Remark 3.7** – It is clear that, for any $d \in \mathbb{D}$, the choice of the parameter $h$ must be done prior to the choice of $\varepsilon$ (given a data set, one first fixes the finite element space and then finds a suitable value of the smoothing parameter). In practice, since $\varepsilon$ and $h$ are tied by (3.9), it is necessary to replace this condition by another that only involves $d$ and $h$. To this end, let us assume that all the hypotheses of Theorem 3.2 (except (3.9)) hold and that, in addition, (V–4.1) and the "inverse assumption" (cf. P. G. Ciarlet [45]), i.e.

$$\exists \nu > 0,\ \forall h \in \mathbb{H},\ \forall K \in \widetilde{\mathcal{T}}_h,\ \frac{h}{h_K} \le \nu, \tag{3.20}$$

are satisfied. It follows from (II–5.1), (1.8) (implicitly assumed by (1.6) and (1.7)) and (3.20) that there exists a constant $C > 0$ such that

$$\forall d \in \mathbb{D}, \ \frac{h^{m'}}{d^m} \leq C \left( \frac{N^m}{M^{m'}} \right)^{1/n}.$$

Therefore, using (V–4.1), we conclude that a sufficient condition for (3.9) to hold is that

$$\frac{N^m}{M^{m'}} = o(1), \ d \to 0. \tag{3.21}$$

This fact has a remarkable consequence, which becomes essential in those applications where $N$ is large: *if $m' > m$, the approximation of a function $f \in H^{m'}(\Omega)$ by discrete smoothing $D^m$-splines can be performed in a space $V_h$ whose dimension $M$ is much lower than the number $N$ of data points.* To see this, it suffices to take $M$ in (3.21) such that $N^{m/m'} < M < N$, with $M$ close to $N^{m/m'}$, and observe that, since $\lim_{d\to 0} N = +\infty$ (cf. Remark III–3.1), the quantity $N^{m/m'}$ is much smaller than $N$ as $d \to 0$.

Let us consider an example. Suppose that $n = 2$ and that the finite element spaces $\widetilde{V}_h$ have been constructed from the Bogner-Fox-Schmit rectangle of class $C^1$ (cf. Section VIII–3). The conditions $m > n/2$ and (1.1) imply that necessarily $m = 2$. Likewise, from the conditions $m' > m$ and (1.6) (which should hold with $q = m'$), we deduce that $m' = 3$ or 4. If $N = 10^6$, for example, then $N^{m/m'} = 10000$ or 1000, respectively. Thus, to fit a function in $H^{m'}(\Omega)$, we can use a space $V_h$ with a dimension $M$ clearly smaller than $N$. □

# *CHAPTER VII*

# *UNIVARIATE $D^m$-SPLINES* †

## 1. *CHARACTERIZATION AND EXPLICIT FORM*

Let us consider again the framework introduced in Chapter V, but restricted to dimension $n = 1$. Hence, the notations $m$, $\Omega$ and $A$ stand here, respectively, for a positive integer, a bounded open interval $(a, b)$ of $\mathbb{R}$, and an ordered set of $N$ distinct points of $\overline{\Omega} = [a, b]$ which contains a $P_{m-1}$-unisolvent subset. Since $\dim P_{m-1} = m$, this condition on $A$ is equivalent to the inequality $N \geq m$.

For convenience, we denote by $x_1, \ldots, x_N$ the points of $A$ and we assume that they are numbered in increasing order, i.e. $x_1 < x_2 < \cdots < x_{N-1} < x_N$. Likewise, we consider again the linear continuous operator $\rho : H^m(\Omega) \to \mathbb{R}^N$, given by

$$\rho v = \big(v(x_i)\big)_{1 \leq i \leq N}, \tag{1.1}$$

as well as the vector subspace

$$\mathcal{K}_0 = \{\, v \in H^m(\Omega) \mid \rho v = 0 \,\}.$$

In this situation, it is clear that all the results of Chapter V are valid. For example, given $\beta \in \mathbb{R}^N$, there exists one and only one interpolating $D^m$-spline over $\Omega$ relative to $A$ and $\beta$ (cf. Theorem V–1.1). Analogously, given $\beta \in \mathbb{R}^N$ and $\varepsilon > 0$, by Theorem V–3.1, there exists a unique smoothing $D^m$-spline over $\Omega$ relative to $A$, $\beta$ and $\varepsilon$.

Let us turn our attention to the space $S$ of $D^m$-splines over $\Omega$ relative to $A$, defined in Section V–2 by

$$S = \{\, s \in H^m(\Omega) \mid \forall w \in \mathcal{K}_0,\ (s, w)_{m,\Omega} = 0 \,\}.$$

In the univariate case, the elements of $S$ can be explicitly expressed in terms of a finite number of known functions. In fact, any $D^m$-spline over $\Omega$ is a piecewise polynomial function of degree $2m-1$. For simplicity, we prove this statement under the condition $A \subset \Omega$, which can be easily removed, as shown later, for $m = 2$, in Subsection 3.1.

***Theorem* 1.1** – *Suppose that $A \subset \Omega$ (i.e. $a < x_1$ and $x_N < b$). Then, a function $s \in H^m(\Omega)$ belongs to $S$ if and only if the following four conditions are fulfilled:*

$$s \in C^{2m-2}(\overline{\Omega}), \tag{1.2}$$

$$\forall i = 1, \ldots, N-1,\ s \in P_{2m-1}\big((x_i, x_{i+1})\big), \tag{1.3}$$

$$s \in P_{m-1}\big((a, x_1)\big), \tag{1.4}$$

$$s \in P_{m-1}\big((x_N, b)\big). \tag{1.5}$$

†Cf. [2], [33], [74], [88], [120], [128], [129].

*Moreover, $s$ belongs to $S$ if and only if there exist unique vectors $(\alpha_0, \dots, \alpha_{m-1}) \in \mathbb{R}^m$ and $(d_1, \dots, d_N) \in \mathbb{R}^N$ such that*

$$\forall x \in \Omega,\ s(x) = \sum_{k=0}^{m-1} \alpha_k x^k + \sum_{i=1}^{N} d_i \frac{(x - x_i)_+^{2m-1}}{(2m-1)!} \tag{1.6}$$

*and*

$$\forall k = 0, \dots, m-1,\ \sum_{i=1}^{N} d_i x_i^k = 0, \tag{1.7}$$

*where, for any $i = 1, \dots, N$, $(x - x_i)_+ = x - x_i$, if $x \geq x_i$, and $0$, otherwise.*

**Proof** – First, let us remark that $C^{2m-2}(\overline{\Omega}) \subset C^{m-1}(\overline{\Omega})$, since $2m-2 \geq m-1$. Then it follows that conditions (1.2), (1.3), (1.4) and (1.5) define a subspace of $H^m(\Omega)$. This justifies the formulation of the Theorem .

From now on, we shall write $x_0 = a$ and $x_{N+1} = b$.

1) Suppose that $s \in S$ and let us prove that (1.2), (1.3), (1.4) and (1.5) hold.

Let $i \in \{0, \dots, N\}$. It is clear that any function $\varphi$ in $\mathcal{D}(\Omega)$ having its support in $(x_i, x_{i+1})$ belongs to $\mathcal{K}_0$ and hence, by definition of $S$, $(s, \varphi)_{m,\Omega} = 0$. But

$$(s, \varphi)_{m,\Omega} = (s, \varphi)_{m,(x_i, x_{i+1})} = \langle s^{(m)}, \varphi^{(m)} \rangle = (-1)^m \langle s^{(2m)}, \varphi \rangle,$$

where $\langle \cdot\,, \cdot \rangle$ denotes here the duality between $\mathcal{D}'\big((x_i, x_{i+1})\big)$ and $\mathcal{D}\big((x_i, x_{i+1})\big)$. It follows that

$$s^{(2m)} = 0 \text{ in } \mathcal{D}'\big((x_i, x_{i+1})\big).$$

Since $(x_i, x_{i+1})$ is connected, we deduce that

$$s \in P_{2m-1}\big((x_i, x_{i+1})\big).$$

This proves (1.3) and also that

$$s \in P_{2m-1}\big((a, x_1)\big) \text{ and } s \in P_{2m-1}\big((x_N, b)\big). \tag{1.8}$$

Now, by Theorem V–2.1, we have $S \subset C^{2m-2}(\Omega)$. This relation, together with (1.8), yields (1.2).

To prove (1.4), let us consider any function $\varphi \in \mathcal{D}(\overline{\Omega})$ with support in $[a, x_1)$. Obviously, $\varphi$ belongs to $\mathcal{K}_0$. Hence, taking account of (1.8) and integrating by parts, we get

$$\begin{aligned} 0 = (s, \varphi)_{m,\Omega} &= (s, \varphi)_{m,(a,x_1)} = \int_a^{x_1} s^{(m)}(x) \varphi^{(m)}(x)\, dx \\ &= \sum_{l=0}^{m-1} (-1)^l \big[s^{(m+l)}(x) \varphi^{(m-l-1)}(x)\big]_a^{x_1} = \sum_{j=0}^{m-1} (-1)^{m-j} s^{(2m-1-j)}(a) \varphi^{(j)}(a). \end{aligned}$$

Since, for any $j = 0, \ldots, m-1$, we can choose $\varphi$ so that $\varphi^{(j)}(a) \neq 0$ and

$$\forall k = 0, \ldots, m-1,\ k \neq j,\ \varphi^{(k)}(a) = 0,$$

we deduce that

$$\forall l = m, \ldots, 2m-1,\ s^{(l)}(a) = 0,$$

which implies (1.4). A similar reasoning on $(x_N, b]$ shows that (1.5) also holds.

2) Conversely, let us prove that, if $s$ verifies (1.2), (1.3), (1.4) and (1.5), then $s \in S$.

First, since $2m-2 \geq m-1$, by (1.2), $s$ belongs to $C^{m-1}(\overline{\Omega})$. Then, it follows from (1.3), (1.4) and (1.5) that $s \in H^m(\Omega)$. It remains to prove that, for any $v \in \mathcal{K}_0$, $(s, v)_{m,\Omega} = 0$.

Now, for any $v \in \mathcal{K}_0$, we have

$$(s, v)_{m,\Omega} = \sum_{i=0}^{N} \int_{x_i}^{x_{i+1}} s_i^{(m)}(x) v^{(m)}(x)\, dx,$$

where, for $i = 0, \ldots, N$, $s_i = s|_{(x_i, x_{i+1})}$. Integrating by parts, which is justified, since, for $i = 0, \ldots, N$ and $l = 0, \ldots, m$, $v^{(m-l)}$ is integrable on $(x_i, x_{i+1})$ and $s_i^{(m+l)}$ is absolutely continuous, and taking into account that, by (1.3), (1.4) and (1.5), $s_i^{(2m)} = 0$, we deduce that, for any $m \geq 2$ (for $m = 1$ the result is simplified)

$$\begin{aligned}(s, v)_{m,\Omega} = &\sum_{i=0}^{N} \sum_{l=0}^{m-2} (-1)^l \left[s_i^{(m+l)}(x) v^{(m-l-1)}(x)\right]_{x_i}^{x_{i+1}} \\ &+ \sum_{i=0}^{N} (-1)^{m-1} \left[s_i^{(2m-1)}(x) v(x)\right]_{x_i}^{x_{i+1}}.\end{aligned} \tag{1.9}$$

It follows from (1.2) and the continuous imbedding of $H^m(\Omega)$ into $C^0(\overline{\Omega})$ that, for $l = 0, \ldots, m-2$, $s_i^{(m+l)} v^{(m-l-1)}$ is continuous on $\overline{\Omega}$. Taking (1.4) and (1.5) into account, we conclude that the first term on the right-hand side of (1.9) is null. The second term also vanishes, since $v \in \mathcal{K}_0$, $s^{(2m-1)}(a) = 0$ and $s^{(2m-1)}(b) = 0$. Therefore, $(s, v)_{m,\Omega} = 0$.

3) Let us now assume that there exist unique vectors $(\alpha_0, \ldots, \alpha_{m-1}) \in \mathbb{R}^m$ and $(d_1, \ldots, d_N) \in \mathbb{R}^N$ such that (1.6) and (1.7) hold. We have to see that $s \in S$.

It is clear that (1.2) holds, since $s$ is, by (1.6), a linear combination of functions in $C^{2m-2}(\overline{\Omega})$. Likewise, it is a straightforward consequence of (1.6) that (1.3) and (1.4) are satisfied. Finally, if we denote by $\tilde{s}_N$ the polynomial function over $\mathbb{R}$ which coincides with $s$ over $(x_N, b)$, it follows from (1.6) and (1.7) that

$$\forall l = m, \ldots, 2m-1,\ \tilde{s}_N^{(l)}(0) = \frac{(-1)^{2m-1-l}}{(2m-1-l)!} \sum_{i=1}^{N} d_i x_i^{2m-l-1} = 0,$$

which implies (1.5). Therefore, by the first part of the Theorem, $s$ belongs to $S$.

4) Finally, let us prove the converse of the relation established in 3). We suppose that $s \in S$, or equivalently, that (1.2), (1.3), (1.4) and (1.5) are satisfied.

For $i = 0, \ldots, N$, let $\tilde{s}_i$ be the polynomial function over $\mathbb{R}$ which coincides with $s$ over $(x_i, x_{i+1})$. By (1.4), there exists a unique vector $(\alpha_0, \ldots, \alpha_{m-1}) \in \mathbb{R}^m$, such that

$$\forall x \in \mathbb{R},\ \tilde{s}_0(x) = \sum_{k=0}^{m-1} \alpha_k x^k.$$

Likewise, for $i = 1, \ldots, N$, let

$$d_i = s^{(2m-1)}(x_i^+) - s^{(2m-1)}(x_i^-)$$

(i.e. $d_i$ is the jump of $s^{(2m-1)}$ at the point $x_i$). On the one hand, it is clear that

$$\forall i = 1, \ldots, N,\ \tilde{s}_i^{(2m-1)}(x_i) = \tilde{s}_{i-1}^{(2m-1)}(x_i) + d_i.$$

On the other hand, it follows from (1.2) that

$$\forall i = 1, \ldots, N,\ \forall l = 0, \ldots, 2m-2,\ \tilde{s}_i^{(l)}(x_i) = \tilde{s}_{i-1}^{(l)}(x_i).$$

Since Taylor's Formula shows that, for $i = 1, \ldots, N$ and for any $x \in \mathbb{R}$,

$$\tilde{s}_{i-1}(x) = \sum_{l=0}^{2m-1} \frac{\tilde{s}_{i-1}^{(l)}(x_i)}{l!}(x - x_i)^l \quad \text{and} \quad \tilde{s}_i(x) = \sum_{l=0}^{2m-1} \frac{\tilde{s}_i^{(l)}(x_i)}{l!}(x - x_i)^l,$$

we deduce that

$$\forall i = 1, \ldots, N,\ \forall x \in \mathbb{R},\ \tilde{s}_i(x) = \tilde{s}_{i-1}(x) + d_i \frac{(x - x_i)^{2m-1}}{(2m-1)!}.$$

By recurrence, we obtain

$$\forall i = 1, \ldots, N,\ \forall x \in [x_i, x_{i+1}], s(x) = \tilde{s}_i(x) = \sum_{k=0}^{m-1} \alpha_k x^k + \sum_{j=1}^{i} d_j \frac{(x - x_j)^{2m-1}}{(2m-1)!},$$

from which we derive (1.6). The relation (1.5) implies that, for $l = m, \ldots, 2m-1$, $\tilde{s}_N^{(l)}(0) = 0$, and so we get (1.7). Finally, the coefficients $d_1, \ldots, d_N$ are unique, since they are necessarily the jumps of $s^{(2m-1)}$ at the points $x_1, \ldots, x_N$. □

**Remark 1.1** – The space of piecewise polynomial functions over $\overline{\Omega}$ that satisfy (1.2), (1.3), (1.4) and (1.5) is usually called the *space of natural polynomial splines of order* $2m$ *with simple knots* $x_1, \ldots, x_N$ (cf. L. L. Schumaker [129, Section 8.2]). Theorem 1.1 shows that this space is, in fact, the space $S$ of $D^m$-splines over $\Omega$ relative to $A$. □

**Remark 1.2** – Let $\widetilde{S}$ be the space of $D^m$-splines over $\mathbb{R}$ (i.e. $(m, 0)$-splines) relative to $A$. We can show that $\tilde{s} \in X^{m,0}$ belongs to $\widetilde{S}$ if and only if

$$\tilde{s} \in C^{2m-2}(\mathbb{R}), \tag{1.10}$$

$$\forall i = 1, \ldots, N-1,\ \tilde{s} \in P_{2m-1}\big((x_i, x_{i+1})\big), \tag{1.11}$$

$$\tilde{s} \in P_{m-1}\big((-\infty, x_1)\big) \text{ and } \tilde{s} \in P_{m-1}\big((x_N, +\infty)\big). \tag{1.12}$$

The key to the proof of this result is to identify, using Plancherel's Theorem (cf. K. Yosida [151, p. 153]), the space $X^{m,0}$ and its semi-norm $|\cdot|_{m,0}$ with the space $\{v \in \mathcal{D}' \mid v^{(m)} \in L^2\}$ and the usual semi-norm $|\cdot|_{m,\mathbb{R}}$. Then one reasons in a way similar to that of points 1) and 2) in the proof of Theorem 1.1. Let us observe that (1.10) is a consequence of Corollary II–2.1. Likewise, given $\tilde{s} \in \widetilde{S}$, (1.12) follows from the condition $\tilde{s} \in X^{m,0}$ and the fact that $\tilde{s}$ is a polynomial of degree $2m-1$ in $(-\infty, x_1)$ and $(x_N, +\infty)$.

The above result, combined with Theorem 1.1, immediately implies that the space $S$ of $D^m$-splines over $\Omega$ relative to $A$ is just the space of restrictions to $\Omega$ of the elements of $\widetilde{S}$. □

## 2. COMPUTATION OF $D^m$-SPLINES

We keep the notations of Section 1 and we assume that $A \subset \Omega$.

Let $\beta = (\beta_1, \ldots, \beta_N) \in \mathbb{R}^N$. By Theorem 1.1, the interpolating $D^m$-spline $\sigma$ over $\Omega$ relative to $A$ and $\beta$ can be written as

$$\sigma(x) = \sum_{k=0}^{m-1} \alpha_k x^k + \sum_{i=1}^{N} d_i \frac{(x-x_i)_+^{2m-1}}{(2m-1)!},$$

with $d_1, \ldots, d_N$ satisfying (1.7). The unknown $N+m$ coefficients $\alpha_0, \ldots, \alpha_{m-1}$ and $d_1, \ldots, d_N$ can be determined imposing the interpolation conditions

$$\forall i = 1, \ldots, N, \ \sigma(x_i) = \beta_i.$$

This leads to the linear system

$$\begin{cases} \displaystyle\sum_{i=1}^{j} d_i \frac{(x_j - x_i)^{2m-1}}{(2m-1)!} + \sum_{k=0}^{m-1} \alpha_k x_j^k = \beta_j, \quad j = 1, \ldots, N, \\ \displaystyle\sum_{i=1}^{N} d_i x_i^k = 0, \quad k = 0, \ldots, m-1. \end{cases} \tag{2.1}$$

Let $G$ be the coefficient matrix of this linear system. We remark that, contrary to the matrix of the system (II–4.1), $G$ is not symmetric.

***Theorem* 2.1** – *The matrix $G$ is regular.*

**Proof** – Let us prove that the homogeneous system associated with (2.1) admits only the null solution. We suppose that $(d_1^*, \ldots, d_N^*, \alpha_0^*, \alpha_1^*, \ldots, \alpha_{m-1}^*)$ is a solution of (2.1) for $\beta = 0$, and we denote by $u$ the function defined over $\Omega$ by

$$u(x) = \sum_{k=0}^{m-1} \alpha_k^* x^k + \sum_{i=1}^{N} d_i^* \frac{(x-x_i)_+^{2m-1}}{(2m-1)!}.$$

It is clear that $\rho u = 0$ and that, by Theorem 1.1, $u$ belongs to $S$. Since the null function over $\Omega$ also belongs to $S$ and satisfies the same interpolation conditions as $u$, Proposition V–2.1 implies that $u$ is the null function. The uniqueness of the vectors $(\alpha_0, \ldots, \alpha_{m-1})$ and $(d_1, \ldots, d_N)$ corresponding to $u$, established in Theorem 1.1, implies that $\alpha_0^* = \alpha_1^* = \cdots = \alpha_{m-1}^* = d_1^* = \cdots = d_N^* = 0$. □

Given $\varepsilon > 0$, let $\sigma_\varepsilon$ be the smoothing $D^m$-spline over $\Omega$ relative to $A$, $\beta$ and $\varepsilon$. We recall that $\sigma_\varepsilon$ verifies the variational equation (V–3.2) and belongs to $S$, so it can be expressed, like $\sigma$, as

$$\sigma_\varepsilon(x) = \sum_{k=0}^{m-1} \alpha_k x^k + \sum_{i=1}^{N} d_i \frac{(x - x_i)_+^{2m-1}}{(2m-1)!},$$

with $d_1, \ldots, d_N$ satisfying (1.7).

For any $j = 1, \ldots, N$, let $\varphi_j$ be a function in $\mathcal{D}(\Omega)$ such that $\varphi_j(x_j) = 1$ and, for any $i = 1, \ldots, N$, with $i \neq j$, $\varphi_j(x_i) = 0$ (cf. the proof of Theorem V–1.1). Taking $v = \varphi_j$ in (V–3.2), we derive

$$\forall j = 1, \ldots, N, \;\; \sigma_\varepsilon(x_j) + (-1)^m \varepsilon \langle \sigma_\varepsilon^{(2m)}, \varphi_j \rangle = \beta_j,$$

where $\langle \cdot\,, \cdot \rangle$ denotes the duality between $\mathcal{D}'(\Omega)$ and $\mathcal{D}(\Omega)$. But, from the expression of $\sigma_\varepsilon$, it follows that

$$\sigma_\varepsilon^{(2m)} = \sum_{i=1}^{N} d_i \delta_{x_i}, \text{ in } \mathcal{D}'(\Omega).$$

Hence, we have

$$\forall j = 1, \ldots, N, \;\; \sigma_\varepsilon(x_j) + (-1)^m \varepsilon d_j = \beta_j.$$

This relation, together with (1.7), yields the following linear system, which provides the $N + m$ unknown coefficients of $\sigma_\varepsilon$:

$$\begin{cases} (-1)^m \varepsilon d_j + \displaystyle\sum_{i=1}^{j} d_i \frac{(x_j - x_i)^{2m-1}}{(2m-1)!} + \sum_{k=0}^{m-1} \alpha_k x_j^k = \beta_j, \quad j = 1, \ldots, N, \\ \displaystyle\sum_{i=1}^{N} d_i x_i^k = 0, \quad k = 0, \ldots, m-1. \end{cases} \tag{2.2}$$

Let us note that the matrix $G_\varepsilon$ of (2.2) is deduced from the matrix $G$ of the linear system (2.1), corresponding to the interpolating spline, by the relation

$$G_\varepsilon = G + (-1)^m \varepsilon J,$$

where $J$ is obtained from the identity matrix of order $N + m$ replacing the last $m$ diagonal terms by 0.

***Theorem* 2.2** – *For any $\varepsilon > 0$, the matrix $G_\varepsilon$ is regular.*

**Proof** – Let us prove that the homogeneous system associated with (2.2) admits only the null solution. To this end, let us suppose that $(d_1^*, \ldots, d_N^*, \alpha_0^*, \alpha_1^*, \ldots, \alpha_{m-1}^*)$ is a solution of (2.2) for $\beta = 0$, and we denote by $u$ the function defined over $\Omega$ by

$$u(x) = \sum_{k=0}^{m-1} \alpha_k^* x^k + \sum_{i=1}^{N} d_i^* \frac{(x - x_i)_+^{2m-1}}{(2m-1)!}.$$

Let us observe that, for $j = 1, \ldots, N$,

$$(-1)^m \varepsilon d_j^* + u(x_j) = (-1)^m \varepsilon d_j^* + \sum_{i=1}^{j} d_i^* \frac{(x_j - x_i)^{2m-1}}{(2m-1)!} + \sum_{k=0}^{m-1} \alpha_k^* x_j^k = 0. \tag{2.3}$$

Now, by Theorem 1.1, $u$ belongs to $S$. From (V–2.1), (V–2.3) and the expression of $u$, it follows that

$$\forall v \in H^m(\Omega),\ (u, v)_{m,\Omega} = \sum_{j=1}^{N} (-1)^m d_j^* v(x_j).$$

This relation, combined with (2.3), implies that

$$\forall v \in H^m(\Omega),\ \sum_{j=1}^{N} u(x_j) v(x_j) + \varepsilon (u, v)_{m,\Omega} = \sum_{j=1}^{N} (u(x_j) + (-1)^m \varepsilon d_j^*) v(x_j) = 0,$$

i.e. $u$ is the solution of (V–3.2) with $\beta = 0$. We conclude, as in the previous theorem, that $u = 0$ and, therefore, that $\alpha_0^* = \alpha_1^* = \ldots = \alpha_{m-1}^* = d_1^* = \ldots = d_N^* = 0$. □

For the numerical computation of interpolating or smoothing univariate $D^m$-splines, there exists specialized software that implements more powerful algorithms than those based on the direct resolution of systems (2.1) or (2.2), which generally present ill-conditioned matrices. One has to resort to the piecewise polynomial form of the $D^m$-splines or use basis of *B-splines* (cf. Section 3 for a brief account on this subject).

## *3. CUBIC SPLINES*

It is possible to define $D^m$-splines satisfying conditions more general than those which have been considered in the two preceding sections: Lagrange conditions on $\overline{\Omega}$, Hermite or Fourier conditions, conditions defined by local mean values, etc. On the other hand, in order to derive fast, robust algorithms for the computation and evaluation of $D^m$-splines, one can use basis of the space $S$ composed by $D^m$-splines having a local support, called $B$-splines. We shall content ourselves with giving a notion of these questions in a simple example, that of $D^2$-splines, or *cubic splines*, over $\Omega$. For a more general study, one can consult P.-J. Laurent [88] and L. L. Schumaker [129], among others.

## *3.1. NATURAL CUBIC SPLINES*

In the preceding sections, we have supposed that the set $A = \{x_i\}_{1\leq i\leq N}$ of interpolation points, with $x_1 < \cdots < x_N$, is contained in the interval $\Omega = (a, b)$, i.e. $a < x_1$ and $x_N < b$. This restriction has been imposed only for the sake of simplicity. In fact, the theory of $D^m$-splines has been developed in Chapter V under the more general condition $A \subset \overline{\Omega}$. We next show, for $m = 2$, how to extend the first part of Theorem 1.1, covering also the cases $a = x_1$ and $x_N < b$, $a < x_1$ and $x_N = b$, and $a = x_1$ and $x_N = b$.

**Theorem 3.1** – *Let $\nu(a) = 1$, if $a < x_1$, or 2, if $a = x_1$. Let $\nu(b) = N$, if $x_N < b$, or $N - 1$, if $x_N = b$. Then, a function $s \in H^2(\Omega)$ belongs to $S$ if and only if the following conditions are fulfilled:*

$$s \in C^2(\overline{\Omega}), \tag{3.1}$$

$$\forall i = \nu(a), \ldots, \nu(b) - 1, \;\; s \in P_3\big((x_i, x_{i+1})\big), \tag{3.2}$$

$$s \in P_3\big((a, x_{\nu(a)})\big) \text{ and } s \in P_3\big((x_{\nu(b)}, b)\big), \tag{3.3}$$

$$s''(x_1) = s''(x_N) = 0, \tag{3.4}$$

$$s'''(x_1^-) = 0, \text{ if } a < x_1, \text{ and } s'''(x_N^+) = 0, \text{ if } x_N < b. \tag{3.5}$$

**Proof** –

1) Suppose that $s \in S$. Reasoning as in point 1) of the proof of Theorem 1.1, we easily see that (3.1), (3.2) and (3.3) hold.

Now, let $\varphi$ be a function in $\mathcal{D}(\overline{\Omega})$ with support in $[a, x_{\nu(a)})$. Taking account of (3.3) and integrating by parts, we get

$$\begin{aligned}(s, \varphi)_{2,\Omega} &= \int_a^{x_{\nu(a)}} s''(x)\varphi''(x)\,dx \\ &= \big[s''(x)\varphi'(x)\big]_a^{x_{\nu(a)}} - \big[s'''(x)\varphi(x)\big]_a^{x_{\nu(a)}} = -s''(a)\varphi'(a) + s'''(a)\varphi(a).\end{aligned}$$

We can choose $\varphi$ so as $\varphi(a) = 0$ and $\varphi'(a) \neq 0$. Then, $(s, \varphi)_{2,\Omega} = 0$, since $\varphi \in \mathcal{K}_0$, and hence $s''(a) = 0$. If $a = x_1$, we obviously have $s''(x_1) = 0$. If $a < x_1$, we can also take $\varphi$ such that $\varphi(a) \neq 0$ and $\varphi'(a) = 0$. Once again, $\varphi \in \mathcal{K}_0$, and so $s'''(a) = 0$. Since $s''(a) = s'''(a) = 0$, by (3.3), $s|_{(a,x_1)}$ is a polynomial of degree 1. Therefore, $s''(x_1) = s'''(x_1^-) = 0$. Using similar arguments in the interval $(x_{\nu(b)}, b)$, we finally conclude that (3.4) and (3.5) are satisfied.

2) Assume that (3.1)–(3.5) hold. Now, for any $v \in \mathcal{K}_0$, integrating by parts, we have

$$\begin{aligned}(s, v)_{2,\Omega} =& \big[s''(x)v'(x) - s'''(x)v(x)\big]_a^{x_{\nu(a)}} \\ &+ \sum_{i=\nu(a)}^{\nu(b)-1} \big[s''(x)v'(x) - s'''(x)v(x)\big]_{x_i}^{x_{i+1}} + \big[s''(x)v'(x) - s'''(x)v(x)\big]_{x_{\nu(b)}}^{b}.\end{aligned}$$

Since $s''v'$ is continuous on $\overline{\Omega}$ and $v \in \mathcal{K}_0$, we get

$$(s, v)_{2,\Omega} = -s''(a)v'(a) + s'''(a)v(a) + s''(b)v'(b) - s'''(b)v(b).$$

If $a < x_1$, the relations (3.3), (3.4) and (3.5) imply that $s''(a) = s'''(a) = 0$. If $a = x_1$, then $s''(a) = 0$ and $v(a) = 0$. Hence, the first two terms on the right-hand side of the above relation are null. The last two terms also vanish for similar reasons. We conclude that $s \in S$. □

## 3.2. CLAMPED CUBIC SPLINES

The operator $\rho$ defined in (1.1) only contains Lagrange interpolation conditions. In many circumstances, one may be interested in introducing additional conditions, for example, of Hermite type. To fix ideas, suppose we are given an interval $\Omega = (a, b)$ and a finite set $A = \{x_i\}_{1 \le i \le N}$ such that $a = x_1 < \cdots < x_N = b$. We now define the operator $\rho : H^2(\Omega) \to \mathbb{R}^{N+2}$ by

$$\rho v = \big(v'(x_1), v(x_1), \ldots, v(x_N), v'(x_N)\big). \tag{3.6}$$

Let us observe that $\rho$ is well defined, since $H^2(\Omega) \hookrightarrow C^1(\overline{\Omega})$. In comparison with (1.1), the expression of $\rho$ includes here two first order Hermite conditions at the end points of $\Omega$.

With the new operator $\rho$ in hand, one can consider again the interpolation and smoothing problems posed in Chapter V. Given $\beta = (\beta_1', \beta_1, \ldots, \beta_N, \beta_N') \in \mathbb{R}^{N+2}$, we introduce, as usual, the affine linear variety

$$\mathcal{K} = \{\, v \in H^2(\Omega) \mid \rho v = \beta \,\}$$

as well as the associated vector subspace

$$\mathcal{K}_0 = \{\, v \in H^2(\Omega) \mid \rho v = 0 \,\}.$$

Then, one easily sees that problem (V–1.1), with $m = 2$, has one and only one solution $\sigma$, called interpolating $D^2$-spline over $\Omega$ relative to $\rho$ and $\beta$ and equivalently characterized by (V–1.4). Likewise, given $\beta \in \mathbb{R}^{N+2}$ and $\varepsilon > 0$, problem (V–3.1) admits a unique solution $\sigma_\varepsilon$, which also satisfies the variational equation (V–3.2). The function $\sigma_\varepsilon$ is called smoothing $D^2$-spline over $\Omega$ relative to $\rho$, $\beta$ and $\varepsilon$.

Proposition V–2.1 can be easily modified to cope with this new framework. Hence, one proves that the set

$$S = \{\, s \in H^2(\Omega) \mid \forall w \in \mathcal{K}_0,\ (s, w)_{2,\Omega} = 0 \,\}$$

is a subspace of dimension $N + 2$ of $H^2(\Omega)$. In addition, the restriction $\rho_\perp$ of $\rho$ to $S$ is an isomorphism from $S$ onto $\mathbb{R}^{N+2}$ and, for any $\beta \in \mathbb{R}^{N+2}$, $\rho_\perp^{-1}(\beta)$ is the interpolating $D^2$-spline relative to $\rho$ and $\beta$. The space $S$ also contains any smoothing $D^2$-spline and so $S$ is known as the *space of $D^2$-splines relative to $\rho$*. In the literature, $S$ is also referred as the space of *clamped cubic splines*. Reasoning as in the proof of Theorem 3.1, one proves that

| $S$ is the space of functions $s$ of class $C^2$ on $\overline{\Omega}$ such that, for $i = 1, \ldots, N-1$, $s \in P_3\big((x_i, x_{i+1})\big)$. (3.7)

## 3.3. THE BASIS OF CUBIC B-SPLINES

We keep the notations of the preceding subsection. It follows from (3.7) that any clamped cubic spline can be expressed as a piecewise polynomial function of degree 3. This may be convenient in some situations. But it turns out that, in many computational applications, it is preferable to write $s$ as a linear combination of a basis of $S$. In this way, for example, one only needs to save the $N+2$ corresponding coefficients in order to store $s$ in a computer. This compares favourably with the $4(N-1)$ memory positions required by the piecewise polynomial representation.

The most widely used basis of $S$ is, of course, that of *normalized cubic B-splines*, which, for simplicity, we present here only in the case of equally spaced nodes.

In addition to the points $x_1, \ldots, x_N$ in $A$, we introduce six supplementary nodes $x_i$, with $i \in \{-2, -1, 0, N+1, N+2, N+3\}$, and we assume that, for $i = -2, \ldots, N+3$, $x_i = a + (i-1)h$, where $h = (b-a)/(N-1)$. For $i = 0, \ldots, N+1$, let $\widetilde{\mathcal{N}}_i$ be the real function defined over $\mathbb{R}$ by

$$\widetilde{\mathcal{N}}_i(x) = \frac{1}{6}\widetilde{\mathcal{N}}\left(\frac{x - x_i}{h}\right),$$

with $\widetilde{\mathcal{N}} : \mathbb{R} \to \mathbb{R}$ given by

$$\widetilde{\mathcal{N}}(x) = \begin{cases} (x+2)^3, & -2 \le x < -1, \\ 4 - 6x^2 - 3x^3, & -1 \le x < 0, \\ 4 - 6x^2 + 3x^3, & 0 \le x < 1, \\ (2-x)^3, & 1 \le x \le 2, \\ 0, & \text{otherwise.} \end{cases}$$

For any $i = 0, \ldots, N+1$, it can be verified without difficulty that

$$\widetilde{\mathcal{N}}_i \in C^2(\mathbb{R}), \tag{3.8}$$

$$\widetilde{\mathcal{N}}_i(x_j) = \begin{cases} 2/3, & j = i, \\ 1/6, & j = i \pm 1, \\ 0, & \text{otherwise,} \end{cases} \tag{3.9}$$

and

$$\widetilde{\mathcal{N}}_i'(x_{i-1}) = \frac{1}{2h}, \ \widetilde{\mathcal{N}}_i'(x_i) = 0, \ \widetilde{\mathcal{N}}_i'(x_{i+1}) = -\frac{1}{2h}. \tag{3.10}$$

For $i = 0, \ldots, N+1$, we denote by $\mathcal{N}_i$ the restriction of $\widetilde{\mathcal{N}}_i$ to $\overline{\Omega}$.

***Theorem* 3.2** – *The set $\{\mathcal{N}_0, \ldots, \mathcal{N}_{N+1}\}$ is a basis of the space $S$ of clamped cubic splines.*

**Proof** –

1) For $i = 0, \ldots, N+1$, it is clear that, by (3.8), $\mathcal{N}_i$ belongs to $C^2(\overline{\Omega})$ and that, for $j = 1, \ldots, N-1$, $\mathcal{N}_i$ is, by definition, a polynomial of degree 3 on every interval

$(x_j, x_{j+1})$ (in fact, the null polynomial if $j < i-2$ or $j > i+2$). If follows from (3.7) that $\mathcal{N}_i$ belongs to $S$.

2) Taking the preceding point into account, the set $\{\mathcal{N}_0, \dots, \mathcal{N}_{N+1}\}$ is a basis of $S$ if and only if

$$\forall s \in S,\ \exists! \,(\alpha_0, \dots, \alpha_{N+1}) \in \mathbb{R}^{N+2},\ s = \sum_{i=0}^{N+1} \alpha_i \mathcal{N}_i. \tag{3.11}$$

We have mentioned before that Proposition V–2.1 can be modified to show that the restriction $\rho_\perp$ of the operator $\rho$ to $S$ is an isomorphism from $S$ onto $\mathbb{R}^{N+2}$. Thus, instead of (3.11), we can prove that

$$\forall \beta \in \mathbb{R}^{N+2},\ \exists! \,(\alpha_0, \dots, \alpha_{N+1}) \in \mathbb{R}^{N+2},\ \rho\left(\sum_{i=0}^{N+1} \alpha_i \mathcal{N}_i\right) = \beta.$$

But, writing $\alpha = (\alpha_0, \dots, \alpha_{N+1})$ and using (3.9) and (3.10), this amounts to seeing that, for any $\beta \in \mathbb{R}^{N+2}$, the matrix equation $\mathcal{A}\alpha = \beta$, with

$$\mathcal{A} = \frac{1}{6}\begin{pmatrix} -3/h & 0 & 3/h & 0 & \cdots\cdots & 0 \\ 1 & 4 & 1 & 0 & \cdots\cdots & 0 \\ 0 & 1 & 4 & 1 & \cdots\cdots & 0 \\ \vdots & & \ddots & \ddots & \ddots & \vdots \\ 0 & \cdots\cdots & 1 & 4 & 1 & 0 \\ 0 & \cdots\cdots & 0 & 1 & 4 & 1 \\ 0 & \cdots\cdots & 0 & -3/h & 0 & 3/h \end{pmatrix},$$

has a unique solution, which, in turn, is equivalent to verifying that $\det \mathcal{A} \neq 0$. To do this, we add the first column to the third and the $(N+2)$th column to the $N$th one. After developing the resulting determinant by the first row and then by the last one, we get $\det \mathcal{A} = -\Delta/(4h^2)$, where $\Delta$ is the determinant of a strictly dominant diagonal (and so regular) matrix. Therefore, $\det \mathcal{A} \neq 0$. This completes the proof. □

Normalized $B$-splines enjoy many properties that make them so interesting from a computational point of view (positivity, local support of minimal length, partition of unity, recursion formulae,...). The theory of $B$-splines is generalized (in dimension 1) to the case of $B$-splines of arbitrary degree, but only the $B$-splines of odd degree are also $D^m$-splines (cf. P. M. Prenter [120] or L. L. Schumaker [129], for instance).

## 4. TENSOR PRODUCT OF UNIVARIATE $D^m$-SPLINES

We conclude this chapter by giving a brief idea of the interpolation by tensor product of univariate $D^m$-splines.

Let us consider an open rectangle $\Omega = (a, b) \times (c, d)$ and two sets $X = \{x_1, \dots, x_K\}$ and $Y = \{y_1, \dots, y_L\}$ such that

$$a = x_1 < x_2 < \cdots < x_K = b \quad \text{and} \quad c = y_1 < y_2 < \cdots < y_L = d.$$

Given a real matrix $\beta = (\beta_{ij})_{1\le i\le K,\,1\le j\le L}$, we can pose the problem of finding a function $\tilde{\sigma}$, defined on $\overline{\Omega}$, such that

$$\forall i = 1,\ldots,K,\ \forall j = 1,\ldots,L,\ \tilde{\sigma}(x_i, y_j) = \beta_{ij}. \tag{4.1}$$

Of course, there exist many different solutions to this problem, but one of them can be easily obtained as follows.

Let $\{\varphi_i \mid 1 \le i \le K\}$ be the set of basis $D^m$-splines over $(a,b)$ relative to $X$ (i.e. for $i = 1,\ldots,K$, $\varphi_i$ is the interpolating $D^m$-spline relative to $X$ and the $i$th element of the canonical basis of $\mathbb{R}^K$). Likewise, let $\{\psi_j \mid 1 \le j \le L\}$ be the set of basis $D^m$-splines over $(c,d)$ relative to $Y$. Then, it suffices to define $\tilde{\sigma}$ as

$$\tilde{\sigma} = \sum_{i=1}^{K}\sum_{j=1}^{L} \beta_{ij}\,\varphi_i \otimes \psi_j,$$

where $\varphi_i \otimes \psi_j$ denotes the tensor product of $\varphi_i$ and $\psi_j$. It is clear that $\tilde{\sigma}$ satisfies the interpolation conditions (4.1).

The function $\tilde{\sigma}$ has good properties. It belongs to $C^{2m-2}(\overline{\Omega})$. Moreover, although nothing seemed to indicate that $\tilde{\sigma}$ is a spline function associated with a minimization problem in a Hilbert space, this is certainly so (cf. D. Apprato [5]).

When $m = 2$, we remark that, according to Theorem 3.1, the functions $v$ in the interpolation space (i.e. the space spanned by $\{\varphi_i \otimes \psi_j\}_{1\le i\le K,\,1\le j\le L}$) verify the relations

$$\forall y \in (c,d),\ \frac{\partial^2 v}{\partial x^2}(a,y) = \frac{\partial^2 v}{\partial x^2}(b,y) = 0$$

and

$$\forall x \in (a,b),\ \frac{\partial^2 v}{\partial y^2}(x,c) = \frac{\partial^2 v}{\partial y^2}(x,d) = 0.$$

This type of interpolation is used in practice when (imperatively) $\Omega$ is a *rectangle* and the set of interpolation points is formed by the *vertices of a rectangular mesh*. However, in the smoothing case, that we shall not consider here, there only remains the condition over $\Omega$.

# PART C

# *APPLICATIONS OF DISCRETE $D^m$-SPLINES*

# *INTRODUCTION*

The problems studied in this part of the book mainly arise in oil research and involve disciplines such as Geophysics and Geology. In every case, the question is to construct, from a set of data of the corresponding function, a regular (explicit or possibly parametric) surface, i.e. a surface of class $C^1$ or $C^2$. This set may be made up, either of a *large number* $N$ of point values of the function and, eventually, of its first derivatives, or of an infinity of such values at points continuously distributed on curves or open subsets of $\mathbb{R}^2$. Sometimes these surfaces present discontinuities, e.g. the faulted surfaces in Geophysics.

From the numerical point of view, one cannot use interpolation methods, even if any exist, because they are too expensive. In the case of a finite set of $N$ data, for example, interpolation methods involve processes, such as construction of triangulations or solution of linear systems, that are of order greater than $N$, generally, and $N$ is supposed to be large. For the same reason, one cannot consider interpolating or smoothing $(m, s)$-splines, which both lead to linear systems of order $N + m(m+1)/2$.

The method of discrete $D^m$-splines is worth using on two accounts: flexibility of modelling and cost-in-use. As will be seen, the theory of Chapter VI fits easily to the situation of Chapters VIII–XI. It also fits to the case of parametric surfaces, treated in Chapter XII. On the other hand, the nature of discrete smoothing $D^m$- splines allows us to disconnect the number $M$ of degrees of freedom of the approximant from the number $N$ of data, which enables us to choose $M$ much smaller than $N$.

# *CHAPTER VIII*

# *CONSTRUCTION OF EXPLICIT SURFACES FROM LARGE DATA SETS* †

## *1. FORMULATION OF THE PROBLEM*

We consider the following classical problem:

> given a bounded open subset $\Omega$ of $\mathbb{R}^2$, a set $A$ of $N$ distinct points of $\overline{\Omega}$ and the set of values $\{ f(a) \mid a \in A \}$ of an unknown function $f : \overline{\Omega} \to \mathbb{R}$, construct an approximant $\phi$ of $f$ belonging to $C^k(\overline{\Omega})$, with $k = 1$ or $2$. (1.1)

In geometric language, the problem consists in finding a smooth approximating surface of the explicit surface $z = f(x, y)$ from the set of points $\{ (a, f(a)) \mid a \in A \}$.

We add some constraints to problem (1.1). We suppose that $\Omega$ is an open set with a Lipschitz-continuous boundary (cf. Preliminaries) and that $f$ belongs to $H^{m'}(\Omega)$ for some $m' \geq 2$, assumptions which imply that $f$ is at least of class $C^0$ on $\overline{\Omega}$. Furthermore, we suppose that $N$ is "large" (about several thousands, for example). These conditions are actually imposed in various problems occurring particularly in Geophysics and Geology. For different reasons, of cost specially, we want the approximant $\phi$ to depend on a small number $M$ of degrees of freedom.

Many methods to solve problem (1.1) have been developed over the past decades, giving rise to a vast quantity of literature on scattered data fitting and its applications to the construction of surfaces (cf. R. F. Franke [67], S. K. Lodha and R. F. Franke [95] and references therein). The constraints of the problem, in particular, the assumption on $N$, do not make the use of interpolation methods very realistic, since, of course, $M \geq N$ for any interpolating process. Likewise, in many applications, the data are perturbed by noise or measurement errors. In such cases it is even clearer that smoothing methods, rather than interpolating ones, should be effectively used. However, not every smoothing method is well suited in the present context. Some of them require a computational effort similar to that of an interpolating method, and so they must be discarded. This is the case, for example, of the smoothing $(m, s)$-splines, since their calculation involves the solution of a linear system of order $N+m(m+1)/2$ (cf. Section II–4).

## *2. THE FITTING METHOD*

In order to solve problem (1.1), we present an approximation method based on the computation of smoothing discrete $D^m$-splines, defined in Section VI–3. To this end,

†Cf. D. Apprato, R. Arcangéli and R. Manzanilla [10].

let us first observe that, if we judiciously choose an integer $m$, a finite element space $V_h$ and a smoothing parameter $\varepsilon$, the $V_h$-discrete smoothing $D^m$-spline $\sigma_{\varepsilon h}$ relative to $A$, $(f(a))_{a\in A}$ and $\varepsilon$ constitutes, under some conditions, an accurate approximant of the unknown function $f$ introduced in (1.1) (cf. Theorem VI–3.2). It is also clear from Remark VI–3.1 that the number $M$ of degrees of freedom on which $\sigma_{\varepsilon h}$ depends is just the dimension of $V_h$. Remark VI–3.7 then justifies that, in many cases, $M$ can be small, at least compared with $N$. Therefore, problem (1.1), together with the additional constraints also expressed in the preceding section, can be satisfactorily solved if we simply take $\sigma_{\varepsilon h}$ as the approximant $\phi$ of the function $f$ that we search for. This is, in fact, what we propose.

The results and remarks in Section VI–3 enlighten us how to choose $m$, $V_h$ and $\varepsilon$, as well as the complementary sets and parameters needed to define $\sigma_{\varepsilon h}$ (cf. Section VI–1 and Subsection VI–3.1). Let us begin with the polygonal open set $\widetilde{\Omega}$. Usually, $\Omega$ is a rectangle that contains the data points in $A$ or their convex hull, and hence it suffices to take $\widetilde{\Omega} = \Omega$. In general, $\widetilde{\Omega}$ can be the minimum open rectangle with sides parallel to the coordinates axes such that $\Omega \subset \widetilde{\Omega}$.

Next, one selects a generic Hermite finite element $(K, P_K, \Sigma_K)$ of class $C^{k'}$ (cf. Section 3 for some possible options). The integer $k'$ indicates the order of continuity that will attain the corresponding finite element space (and the approximant of $f$), so $k' \geq k$. Since it is not worth the trouble to get a degree of smoothness greater than $k$, one typically sets $k' = k$.

Once the element to be used is fixed, one makes a triangulation $\widetilde{\mathcal{T}}_h$, bearing in mind the convergence hypothesis (VI–1.8) (implied by (VI–1.6) or (VI–1.7)), (VI–3.7) and, eventually, (VI–3.20). Therefore, as far as possible, the elements of $\widetilde{\mathcal{T}}_h$ should not be too slim, the density of data points should be more or less uniform in all the elements of $\widetilde{\mathcal{T}}_h$, and, finally, the size of these elements should be relatively similar. The geometric meaning of the index $h$ is, as usual, the maximum of the diameters of elements in $\widetilde{\mathcal{T}}_h$.

The construction of a finite element space $\widetilde{V}_h$ on the triangulation $\widetilde{\mathcal{T}}_h$ from the generic element $(K, P_K, \Sigma_K)$ is a straightforward process. Likewise, one also gets from $\widetilde{\mathcal{T}}_h$ and $\widetilde{V}_h$ the open set $\Omega_h$ and the space $V_h$ given by (VI–1.2) and (VI–1.5), respectively. Let us observe that, by construction, $V_h \subset C^k(\overline{\Omega}_h)$.

With respect to the integer $m$, its value is restricted by the condition $m > n/2 = 1$ and hypothesis (VI–1.1), which, in turn, involves the inequality $m \leq k' + 1$. In real problems, one usually takes $k' = 1$ and $m = 2$, if $k = 1$, and $k' = 2$ and $m = 3$, if $k = 2$. Of course, one should verify that $A$ contains a $P_{m-1}$-unisolvent subset (i.e., for $m = 2$, at least three non-aligned points), the condition that guarantees the existence and uniqueness of $\sigma_{\varepsilon h}$. As suggested in Remark VI–3.7, another condition that should be checked is the inequality $N^{m/m'} < M < N$, where $M$ is the dimension of $V_h$ and $m'$ is the order of the Sobolev space defined over $\Omega$ to which $f$ belongs. It is clear that $m'$ is unknown, but, for this purpose, one can take $m' = m + 1$.

The last step before computing the $V_h$-discrete smoothing $D^m$-spline $\sigma_{\varepsilon h}$ is the choice of the smoothing parameter $\varepsilon$. For this, we refer to the methods mentioned in Remark VI–3.2.

We recall that $\sigma_{\varepsilon h}$ is the solution of problems (VI–3.1) and (VI–3.2). Let $a_1, \ldots, a_N$ and $w_1, \ldots, w_M$ be, respectively, the points in $A$ and the basis functions of $V_h$. If $\sigma_{\varepsilon h}$ is expressed as $\sigma_{\varepsilon h} = \sum_{j=1}^{M} \alpha_j w_j$, Remark VI–3.1 shows that, by (VI–3.2), the vector $\alpha = (\alpha_j)_{1 \le j \le M}$ of unknown coefficients is the solution of the linear system

$$(\mathcal{A}^T \mathcal{A} + \varepsilon \mathcal{R})\alpha = \mathcal{A}^T \rho f, \tag{2.1}$$

where $\rho f = \big(f(a_i)\big)_{1 \le i \le N}$, and $\mathcal{A}$ and $\mathcal{R}$ denote the matrices

$$\mathcal{A} = \big(w_j(a_i)\big)_{1 \le i \le N,\, 1 \le j \le M} \quad \text{and} \quad \mathcal{R} = \big((w_j, w_i)_{m,\Omega_h}\big)_{1 \le i,j \le M}.$$

An advantage of the proposed smoothing method consists in the fact that the matrix of the linear system (2.1) is positive definite, symmetric, sparse and, if the unknowns $\alpha_j$ are suitably numbered, also banded. In addition, it is possible to refine *locally* the approximant of $f$ obtained in a first calculation. We do not detail this technique, for which we refer to [10].

## 3. EXAMPLES OF FINITE ELEMENTS

There are several Hermite finite elements which can be selected to apply the fitting method detailed in the preceding section. Among them, we cite the Argyris and Bell triangles of class $C^1$ and the Bogner-Fox-Schmit rectangle of class $C^1$ (cf., for example, P. G. Ciarlet [45]), and the Argyris and Bell triangles of class $C^2$ (cf. A. Ženišek [152]) and the Bogner-Fox-Schmit rectangle of class $C^2$ (cf. D. Apprato, R. Arcangéli and R. Manzanilla [10]). For the above elements, P. Clément's results (VI–1.6) and (VI–1.7) are satisfied.

Let us remember that the elements of class $C^{k'}$ can be used for the computation of discrete $D^m$-splines with $m \le k' + 1$. Thus, satisfactory combinations are $(k', m) = (1, 2)$, $(2, 2)$ and $(2, 3)$. In the strictly formal case $m > 3$, one could consider some triangular finite element of class $C^{m-1}$ (cf. A. Ženišek [152], A. Le Méhauté [90]) or the Bogner-Fox-Schmit rectangle of class $C^{m-1}$ (cf. J. J. Torrens [139]).

While the Argyris, Bell and Bogner-Fox-Schmit finite elements of class $C^1$ are quite standard, their counterparts of class $C^2$ are not well-known. To give a clear account of them, we shall recall their definition as a triple $(K, P, \Sigma)$ (cf. P. G. Ciarlet [45]). Figures 1 to 3 represent graphically the corresponding degrees of freedom. Likewise, for any $l \in \mathbb{N}$ and for any subset $E$ of $\mathbb{R}^2$, $P_l(E)$ (resp. $Q_l(E)$) denotes the space of the restrictions to $E$ of the polynomial functions over $\mathbb{R}^2$ of degree $\le l$ with respect to the set of variables (resp. with respect to each variable).

### *Argyris triangle of class $C^1$*

- $K$ is a triangle with vertices $b_1$, $b_2$ and $b_3$,
- $P = P_5(K)$,
- $\Sigma = \{ v \mapsto \partial^\alpha v(b_i) \mid |\alpha| \le 2,\ 1 \le i \le 3 \} \cup \{ v \mapsto \partial_\nu v(b_{ij}) \mid 1 \le i < j \le 3 \}$, where $\partial_\nu$ denotes the (outer) normal derivative operator, and $b_{ij} = (b_i + b_j)/2$.

***Argyris triangle of class $C^2$***

- $K$ is a triangle with vertices $b_1$, $b_2$ and $b_3$,
- $P = P_9(K)$,
- $\Sigma = \{v \mapsto v(b_0)\} \cup \{v \mapsto \partial^\alpha v(b_i) \mid |\alpha| \leq 4,\ 1 \leq i \leq 3\} \cup \{v \mapsto \partial_\nu v(b_{ij}) \mid 1 \leq i < j \leq 3\} \cup \{v \mapsto \partial_{\nu\nu} v(b_{iij}) \mid 1 \leq i < j \leq 3\} \cup \{v \mapsto \partial_{\nu\nu} v(b_{ijj}) \mid 1 \leq i < j \leq 3\}$, where $b_0 = (b_1 + b_2 + b_3)/3$, $\partial_{\nu\nu}$ denotes the (outer) second normal derivative operator, $b_{iij} = (2b_i + b_j)/3$ and $b_{ijj} = (b_i + 2b_j)/3$.

***Bell triangle of class $C^1$***

- $K$ is a triangle with vertices $b_1$, $b_2$ and $b_3$,
- $P = \{p \in P_5(K) \mid \partial_\nu p \in P_3(K_i'),\ 1 \leq i \leq 3\}$, where $K_1'$, $K_2'$ and $K_3'$ are the sides of $K$,
- $\Sigma = \{v \mapsto \partial^\alpha v(b_i) \mid |\alpha| \leq 2,\ 1 \leq i \leq 3\}$.

***Bell triangle of class $C^2$***

- $K$ is a triangle with vertices $b_1$, $b_2$ and $b_3$,
- $P = \{p \in P_9(K) \mid \partial_\nu p \in P_7(K_i'),\ \partial_{\nu\nu} p \in P_5(K_i'),\ 1 \leq i \leq 3\}$,
- $\Sigma = \{v \mapsto v(b_0)\} \cup \{v \mapsto \partial^\alpha v(b_i) \mid |\alpha| \leq 4,\ 1 \leq i \leq 3\}$.

***Bogner-Fox-Schmit rectangle of class $C^{k'}$***

- $K$ is a rectangle with vertices $b_1$, $b_2$, $b_3$ and $b_4$,
- $P = Q_{2k'+1}(K)$,
- $\Sigma = \{v \mapsto \partial^\alpha v(b_i) \mid |\alpha|_\infty \leq k',\ 1 \leq i \leq 4\}$, where, if $\alpha = (\alpha_1, \alpha_2)$, $|\alpha|_\infty = \max\{\alpha_1, \alpha_2\}$.

The number of degrees of freedom of the Bogner-Fox-Schmit rectangles is smaller than that of the Argyris and Bell triangles of the same class (16 against 21 and 18, respectively, in the $C^1$ case, 36 against 55 and 46, respectively, in the $C^2$ case). It is also clear that the tensor product structure that underlies the Bogner-Fox-Schmit rectangles alleviates many programming tasks and computational requirements. In short, the implementation of these elements is appreciably cheaper than that of Argyris and Bell triangles. Hence, the best we can do is to select systematically the Bogner-Fox-Schmit rectangles as often as possible. This occurs, for example, when one has to apply the fitting method presented in the preceding section. However, for

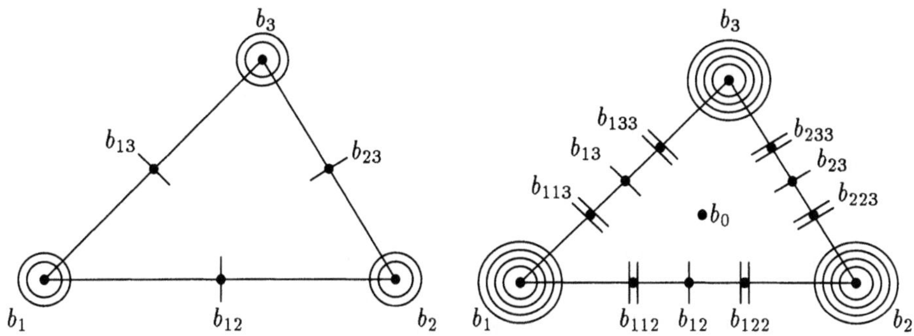


Figure 1: Argyris triangles of classes $C^1$ (left) and $C^2$ (right).

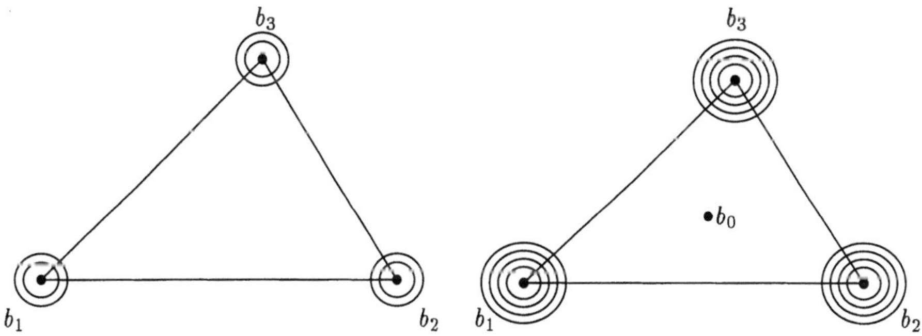


Figure 2: Bell triangles of classes $C^1$ (left) and $C^2$ (right).

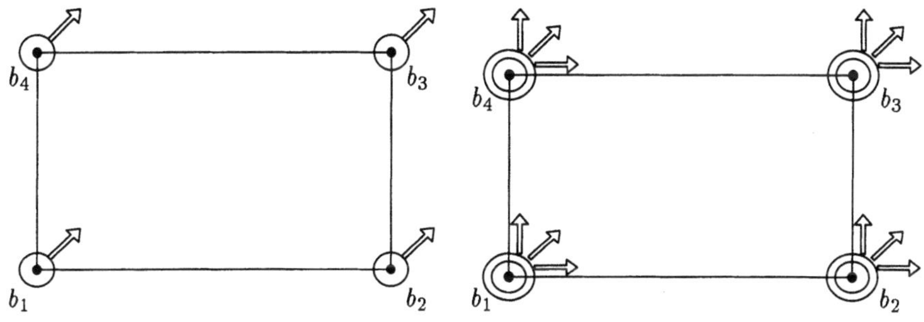


Figure 3: Bogner-Fox-Schmit rectangles of classes $C^1$ (left) and $C^2$ (right).

the computation of the $D^m$-splines which we shall introduce in Chapter IX, as well as in interpolation problems (cf. Remark VI–2.2), the use of triangular elements is almost always unavoidable.

To conclude this section, we note that there also exist other widely known finite elements of class $C^1$, such as the Hsieh-Clough-Tocher, the reduced Hsieh-Clough-Tocher or the Powell-Sabin triangles. On the other hand, the implementation of Argyris and Bell triangles of class $C^2$ seems (at the present time) quite unrealistic and, if one must absolutely use triangular finite elements of class $C^2$, maybe it is better to resort to composite elements such as the Hsieh-Clough-Tocher element of class $C^2$ (with 33 or 27 degrees of freedom; cf. P. Sablonnière [126], M. Laghchim-Lahlou and P. Sablonnière [85]) or the Powell-Sabin element of class $C^2$ (cf. P. Sablonnière [126], M. Laghchim-Lahlou and P. Sablonnière [86]).

## 4. *CONVERGENCE RESULTS*

The solution we propose to solve problem (1.1) is satisfactory inasmuch as one can prove that the approximant $\phi = \sigma_{\varepsilon h}$ converges to $f$, as $N \to +\infty$, in a adequate sense and under suitable hypotheses. It is indeed what follows from the convergence results we have obtained in Chapter VI. Let us recall them.

To be precise, let us use the notations introduced in Subsection VI–3.2. Let $f$ be a function in $H^{m'}(\Omega)$, where $m'$ denotes an integer $\geq m$. We suppose verified hypotheses (II–5.1), (III–3.1), (VI–1.1), (VI–1.2), (VI–1.5), (VI–3.7) and (VI–3.9). If $m' > m$, we also assume that (VI–1.6) is verified with $q = m'$, whereas, if $m' = m$, we suppose that (VI–1.7) holds with $q = m$. Then, as stated in Theorem VI–3.2, the $V_h$-discrete smoothing $D^m$-spline $\sigma^d_{\varepsilon h}$ relative to $A^d$, $\big(f(a)\big)_{a \in A^d}$ and $\varepsilon$ converges to $f$ in $H^m(\Omega)$.

In fact, Remark VI–3.7 shows that, if (VI–3.9) is replaced by the stronger conditions (V–4.1), (VI–3.20) and (VI–3.21), the result of Theorem VI–3.2 still holds. Moreover, one also has the error estimates given in Theorem VI–3.4. From this, we infer that, if $m' > m$, the approximation of $f$ by $\sigma^d_{\varepsilon h}$ is a problem which can be posed in a space $V_h$ whose dimension $M$ is much smaller than the data number $N$, while the convergence of $\sigma^d_{\varepsilon h}$ to $f$ is ensured. This is indeed what we were a priori wishing for. In the applications, we have $m = 2$ or $m = 3$. Thus, to justify this formulation, $f$ should belong to $H^3(\Omega)$, at least, in the case of $D^2$-splines, and to $H^4(\Omega)$ in the case of $D^3$-splines. Examples of Section 6 confirm these theoretical results.

## 5. *AN INTERPOLATION-SMOOTHING MIXED METHOD*

In Geophysics, we sometimes find a variation of problem (1.1). For approximating an unknown function $f$ of class $C^1$ on the closure of a bounded open set $\Omega \subset \mathbb{R}^2$, we have at our disposal two different nonempty data sets: a set $\Sigma_1$ of $N_1$ *hard* data, considered to be exact, and a set $\Sigma_2$ of $N_2$ *noisy* data. Hard data are Lagrange and first order Hermite data coming from measuring, at boreholes, the position and the orientation of the surface (the graph of $f$). The orientation is expressed in terms of two angles, called *dip* and *strike*, from which we deduce values of the first partial derivatives of $f$. Noisy data are Lagrange data of seismic origin. $N_1$ is supposed to be "small" (some units), while $N_2$ is "large" (several thousands, for example).

The elements of $\Sigma_1$ are linear forms of one of the following types

$$\phi : v \mapsto v(a), \tag{5.1a}$$

$$\phi : v \mapsto \partial^{(1,0)}v(a), \tag{5.1b}$$

$$\phi : v \mapsto \partial^{(0,1)}v(a), \tag{5.1c}$$

where $a$ belongs to a finite set $A_1$ of points of $\overline{\Omega}$. We obviously suppose that, for any $a \in A_1$, $\Sigma_1$ contains at least one of the linear forms detailed in (5.1), but not necessarily the three. The elements of $\Sigma_2$ are linear forms of type (5.1a), with $a$ belonging to a finite set $A_2$ of points of $\overline{\Omega}$.

In this context, let us see how to approximate $f$. To simplify notations, we assume that $\Omega$ is polygonal. Let $m$ and $k$ be two integers such that $2 \leq m \leq k+1$ and let $V_h$ be a finite element space, constructed on a triangulation of $\overline{\Omega}$, such that

$$V_h \subset H^m(\Omega) \cap C^k(\overline{\Omega}).$$

We now introduce the affine linear variety

$$\mathcal{K}_h = \{\, v_h \in V_h \mid \rho_1 v_h = \rho_1 f \,\}$$

and the associated vector subspace

$$\mathcal{K}_{0h} = \{\, v_h \in V_h \mid \rho_1 v_h = 0 \,\},$$

where

$$\forall v \in C^1(\overline{\Omega}),\ \rho_1 v = \big(\phi(v)\big)_{\phi \in \Sigma_1}.$$

We shall also write

$$\forall v \in C^0(\overline{\Omega}),\ \rho_2 v = \big(\phi(v)\big)_{\phi \in \Sigma_2}.$$

Likewise, for $i = 1, 2$, we denote by $\langle \cdot\,, \cdot \rangle_{N_i}$ and $\langle \cdot \rangle_{N_i}$ the Euclidean scalar product and the Euclidean norm in $\mathbb{R}^{N_i}$, respectively.

In the sequel, we assume that

$$\mathcal{K}_h \text{ is nonempty and } A_2 \text{ contains a } P_{m-1}\text{-unisolvent subset.} \tag{5.2}$$

Let us observe that the condition $\mathcal{K}_h \neq \emptyset$ is verified if $\Sigma_1 \subset \Sigma_h$, where $\Sigma_h$ stands for the set of degrees of freedom of $V_h$.

For any $\varepsilon > 0$, we denote by $J_\varepsilon$ the functional defined on $H^m(\Omega)$ by

$$J_\varepsilon(v) = \langle \rho_2 v - \rho_2 f \rangle_{N_2}^2 + \varepsilon |v|_{m,\Omega}^2.$$

Then we consider the problem: find $\sigma_{\varepsilon h}$ such that

$$\begin{cases} \sigma_{\varepsilon h} \in \mathcal{K}_h, \\ \forall v_h \in \mathcal{K}_h,\ J_\varepsilon(\sigma_{\varepsilon h}) \leq J_\varepsilon(v_h). \end{cases} \tag{5.3}$$

Taking into account that $V_h$ is a finite-dimensional space, Stampacchia's Theorem (cf. H. Brézis [40, Theorem V.6], P. G. Ciarlet [45, Theorem 1.1.2]) shows that problem (5.3) has a unique solution $\sigma_{\varepsilon h}$, which is also the unique solution of the equation

$$\begin{cases} \sigma_{\varepsilon h} \in \mathcal{K}_h, \\ \forall w_h \in \mathcal{K}_{0h}, \langle \rho_2 \sigma_{\varepsilon h}, \rho_2 w_h \rangle_{N_2} + \varepsilon(\sigma_{\varepsilon h}, w_h)_{m,\Omega} = \langle \rho_2 f, \rho_2 w_h \rangle_{N_2}. \end{cases} \tag{5.4}$$

We propose to take $\sigma_{\varepsilon h}$, which is at the same time an interpolating and a smoothing discrete $D^m$-spline, for approximant of $f$. Of course, we should study again the convergence and approximation error, because the general theory of Part B does not directly apply to this case, but we shall not do that in this work.

Let us finish this section with some comments on the implementation of this interpolation-smoothing method. If $\Sigma_1 \subset \Sigma_h$ is verified, one can directly use equation (5.4). Proceeding as in Remark VI–2.1, one obtains a linear system of order $M - N_1$, where $M$ still denotes the dimension of $V_h$. But this solution presents, in particular, the disadvantage of requiring the use of triangular finite elements, since the number of rectangles to be introduced may be too large (for instance, in the case of a set $A_1$ made up of 9 points, 100 rectangles are generally needed). It is perhaps preferable to modify the statement (5.3) and introduce a *Lagrange multiplier.* This method, which applies under the general condition (5.2), leads to solving the problem

$$\begin{cases} (\sigma_{\varepsilon h}, \lambda) \in \mathcal{K}_h \times \mathbb{R}^{N_1}, \\ \forall v_h \in V_h,\ \langle \rho_2(\sigma_{\varepsilon h} - f), \rho_2 v_h \rangle_{N_2} + \varepsilon(\sigma_{\varepsilon h}, v_h)_{m,\Omega} = \langle \lambda, \rho_1 v_h \rangle_{N_1}. \end{cases} \tag{5.5}$$

The vector $-2\lambda$, where $\lambda$ is the vector appearing in (5.5), is just the Lagrange multiplier of problem (5.3).

**Remark 5.1** – Let $w_1, \ldots, w_M$ be the basis functions of $V_h$ and let us denote by $\phi_1, \ldots, \phi_{N_1}$ and $a_1, \ldots, a_{N_2}$, respectively, the linear forms in $\Sigma_1$ and the points in $A_2$. The solution $\sigma_{\varepsilon h}$ of (5.5) can be expressed as

$$\sigma_{\varepsilon h} = \sum_{j=1}^{M} \alpha_j w_j,$$

with $\alpha_1, \ldots, \alpha_M \in \mathbb{R}$. Assuming that $\mathcal{K}_h \neq \emptyset$, it follows from (5.5) that the vector $(\alpha_j)_{1 \leq j \leq M}$ of unknown coefficients, together with the vector $\tilde{\lambda} = -\lambda$, is the solution of the linear system

$$\begin{pmatrix} \mathcal{A}_2^T \mathcal{A}_2 + \varepsilon \mathcal{R} & \mathcal{A}_1^T \\ \mathcal{A}_1 & 0 \end{pmatrix} \begin{pmatrix} \alpha \\ \tilde{\lambda} \end{pmatrix} = \begin{pmatrix} \mathcal{A}_2^T \rho_2 f \\ \rho_1 f \end{pmatrix}, \tag{5.6}$$

where

$$\mathcal{A}_1 = \big(\phi_i(w_j)\big)_{1\le i\le N_1,\, 1\le j\le M},$$
$$\mathcal{A}_2 = \big(w_j(a_i)\big)_{1\le i\le N_2,\, 1\le j\le M},$$

and

$$\mathcal{R} = \big((w_j, w_i)_{m,\Omega}\big)_{1\le i,j\le M}.$$

When there is not a sufficient number of degrees of freedom in $V_h$ to match the interpolation conditions coherently, the matrix of the above linear system is singular or ill-conditioned. This provides a way to verify *a posteriori* if the hypothesis $\mathcal{K}_h \neq \emptyset$ holds. Let us finally observe that the dimension of the system (5.6) is $M + N_1$. □

## 6. NUMERICAL RESULTS

We next give several examples of solution of problem (1.1) by the smoothing method in Section 2 and the interpolation-smoothing mixed method developed in Section 5. In all cases, we have taken $\Omega = (0,1) \times (0,1)$ and, given a function $f$ defined on $\overline{\Omega}$, we have found an approximant of $f$ of class $C^1$ (i.e. $k = 1$). For the tests, we have selected the well known Franke's function (cf. R. F. Franke [68])

$$\begin{aligned} f(x,y) = {} & 0.75\exp\big(-0.25(9x-2)^2 - 0.25(9y-2)^2\big) \\ & + 0.75\exp\big(-(9x+1)^2/49 - (9y+1)/10\big) \\ & + 0.5\exp\big(-0.25(9x-7)^2 - 0.25(9y-3)^2\big) \\ & - 0.2\exp\big(-(9x-4)^2 - (9y-7)^2\big) \end{aligned} \tag{6.1}$$

and Nielson's function (cf. G. M. Nielson [110])

$$f(x,y) = \frac{y}{2}\cos\big(4(x^2+y-1)^4\big). \tag{6.2}$$

The graphs and contour plots of these functions are represented in Figures 8 and 13.

To apply the smoothing method, we have considered three sets $A$ of data points. Each of them consists of $N$ points randomly distributed on $\overline{\Omega}$, with $N = 400, 800$ and $1600$. These sets are displayed in Figures 4, 5 and 6. Likewise, we have set $\widetilde{\Omega} = \Omega$ and we have taken several triangulations $\widetilde{\mathcal{T}}_h$ of $\overline{\widetilde{\Omega}}$, made up of $n_{el} \times n_{el}$ equal squares, where $n_{el}$ ranges from 3 to 9. The finite element space $\widetilde{V}_h$ is constructed on $\widetilde{\mathcal{T}}_h$ from the Bogner-Fox-Schmit rectangle of class $C^1$ (i.e. $k' = 1$). Since $\widetilde{\Omega} = \Omega$, we have $\Omega_h = \Omega$, $\mathcal{T}_h = \widetilde{\mathcal{T}}_h$ and $V_h = \widetilde{V}_h$. The dimension $M$ of $V_h$ is shown in Table 1. Let us observe that $M = 4(n_{el}+1)^2$.

We have set $m = 2$ and $\varepsilon = 10^{-5}$. Then, for both test functions and for any combination of $N$ and $n_{el}$ such that $N^{m/m'} < M < N$, with $m' = m+1$, we have computed the corresponding $V_h$-discrete smoothing $D^2$-spline $\sigma_{\varepsilon h}$ relative to $A$, $\big(f(a)\big)_{a\in A}$ and $\varepsilon$. Some of these splines are represented in Figures 9–11 and 14–16.

| $n_{el}$ | 3 | 4 | 5 | 6 | 7 | 8 | 9 |
|---|---|---|---|---|---|---|---|
| $M$ | 64 | 100 | 144 | 196 | 256 | 324 | 400 |

Table 1: Dimension $M$ of the finite element space $V_h$, constructed from the Bogner-Fox-Schmit rectangle of class $C^1$ on a triangulation having $n_{el} \times n_{el}$ elements.

The interpolation-smoothing mixed method has been implemented in the form detailed in Remark 5.1. We have also taken $m = 2$ and $\varepsilon = 10^{-5}$ and, for simplicity, we have considered only Lagrange data. The interpolation and smoothing conditions have been imposed on two sets $A_1$ and $A_2$ having, respectively, 80 and 720 points (cf. Figure 7). In fact, $A_1 \cup A_2$ is just the set $A$ for $N = 800$. Figures 12 and 17 show some discrete interpolation-smoothing $D^2$-splines, also denoted by $\sigma_{\varepsilon h}$.

To get a quantitative measure of the degree of approximation provided by every discrete $D^2$-spline $\sigma_{\varepsilon h}$, we have computed the relative error $r(f)$ in the Euclidean norm, given by

$$r(f) = \left( \frac{\sum_{i=1}^{2500} \left(f(b_i) - \sigma_{\varepsilon h}(b_i)\right)^2}{\sum_{i=1}^{2500} f(b_i)^2} \right)^{1/2}, \tag{6.3}$$

where $\{b_1, \ldots, b_{2500}\}$ is a fixed set of points regularly distributed on $\overline{\Omega}$. Figure 18 shows the decimal logarithm of $r(f)$ for every discrete $D^2$-spline computed, as described before, by the smoothing and the interpolation-smoothing methods. The plots of this figure clearly suggest, coherently with the theoretical results, that the convergence of the smoothing method is achieved if both $N$ and $M$ tend to $+\infty$, although at different rates. With respect to the interpolation-smoothing method, for low values of $n_{el}$, the spline lacks enough flexibility to match the interpolation conditions while retaining good global approximation properties. This initial stiffness disappears as $n_{el}$ increases, so that the interpolation-smoothing spline yields almost the same errors as the smoothing spline for the same number of data points.

The value $\varepsilon = 10^{-5}$ used in all the preceding calculations has been suggested by the GCV method, applied as explained in Remark VI–3.3. For most of the considered data sets, we have observed that the GCV function $\overline{\mathcal{V}}$ follows the same pattern: $\overline{\mathcal{V}}$ attains a global minimum at a point $\varepsilon^*$ relatively close to $10^{-6}$, it has quite small slopes on the left of a point $\tilde{\varepsilon} \in (\varepsilon^*, 10^{-4})$, and it rapidly grows on the right of $\tilde{\varepsilon}$. Most of these characteristics can be seen in Figure 19, which displays, for $n_{el} = 6$, the decimal logarithm of the function $\overline{\mathcal{V}}$ corresponding to every data set. We have realized that, at least in our examples, it suffices to fix the value of $\varepsilon$ between two rough estimates of $\varepsilon^*$ and $\tilde{\varepsilon}$. This is why we have chosen $\varepsilon = 10^{-5}$.

We conclude this section showing in Figure 20 the behaviour of the relative error $r(f)$, for the fixed value $n_{el} = 6$, when the smoothing parameter $\varepsilon$ varies. From this figure we conclude that, for most data sets and, at least, for $n_{el} = 6$, the value $\varepsilon = 10^{-5}$ is close to being optimal from the standpoint of the relative error.

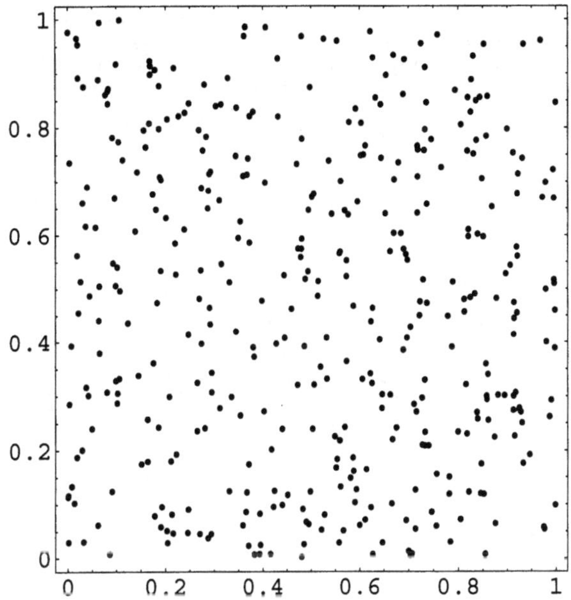


Figure 4: Data set $A$ for the computation of smoothing splines. Case $N = 400$.

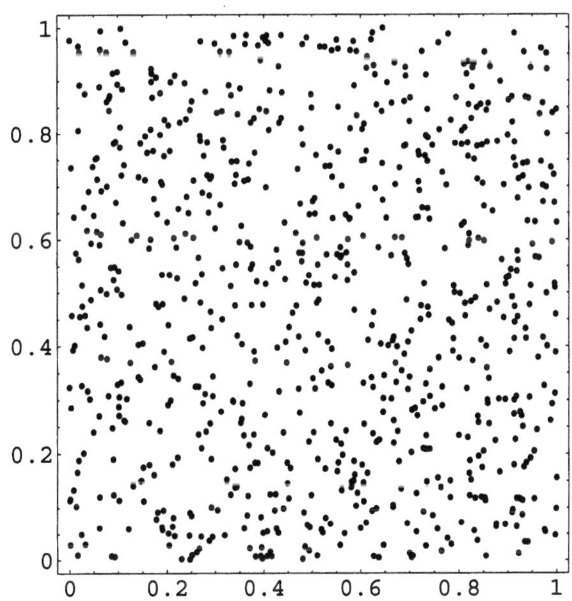


Figure 5: Data set $A$ for the computation of smoothing splines. Case $N = 800$.

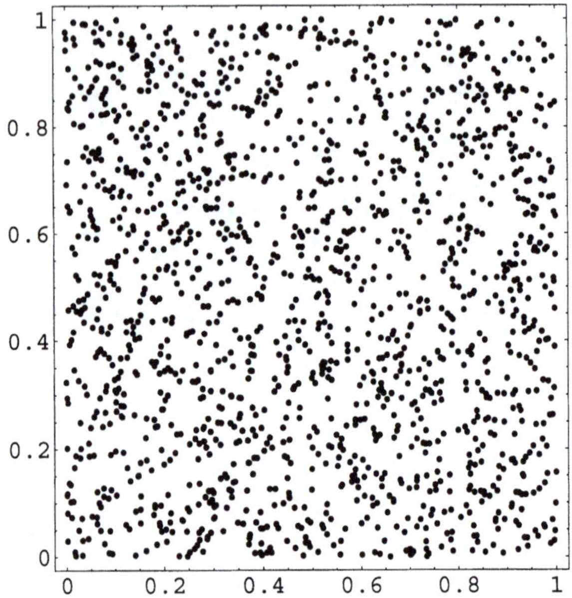


Figure 6: Data set $A$ for the computation of smoothing splines. Case $N = 1600$.

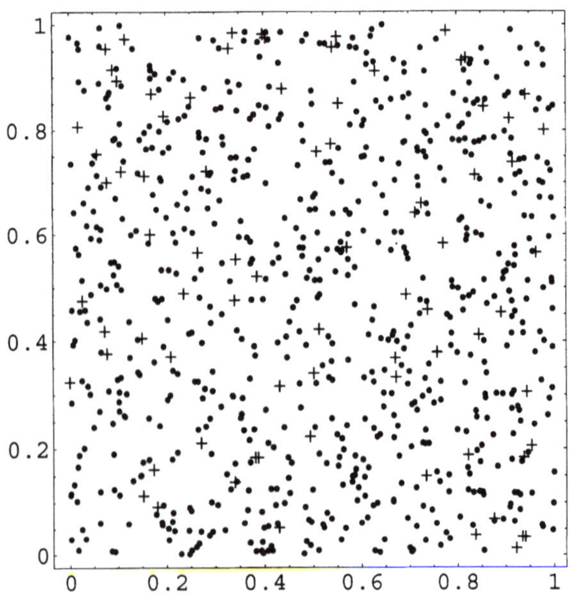


Figure 7: Data sets $A_1$ (crosses) and $A_2$ (dots) for the computation of interpolation-smoothing splines.

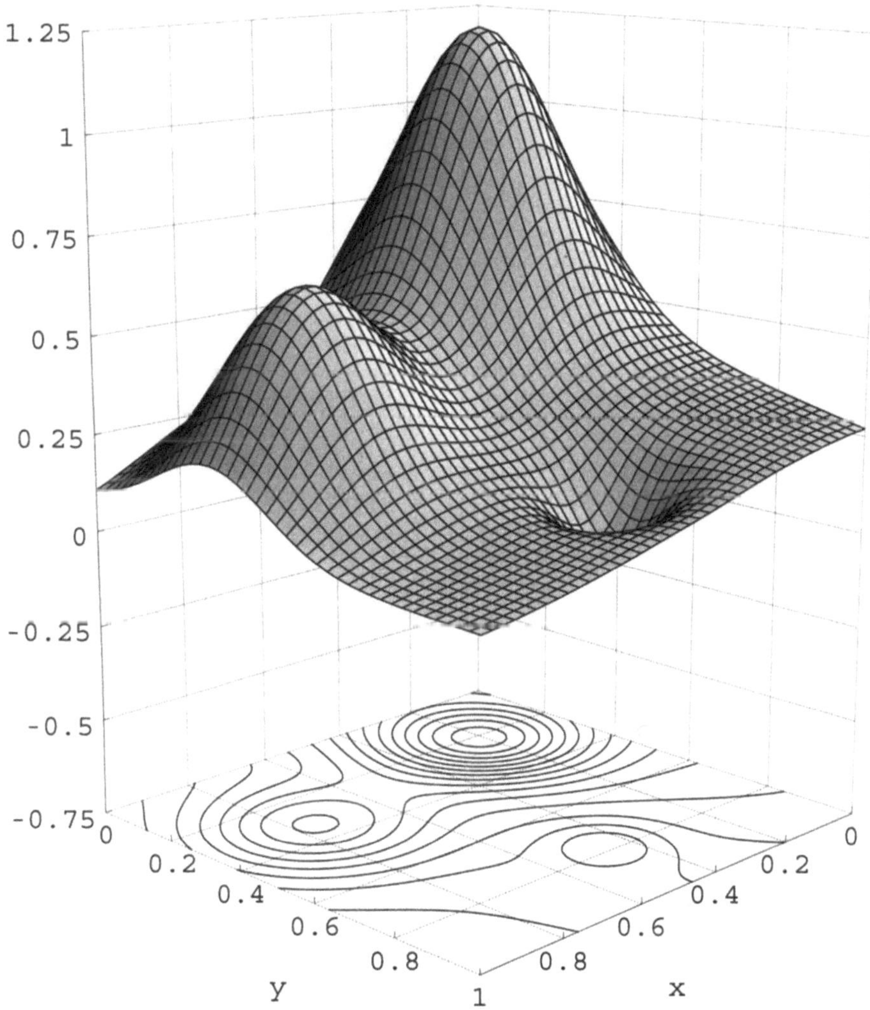


Figure 8: Graph and contour plot of Franke's function.

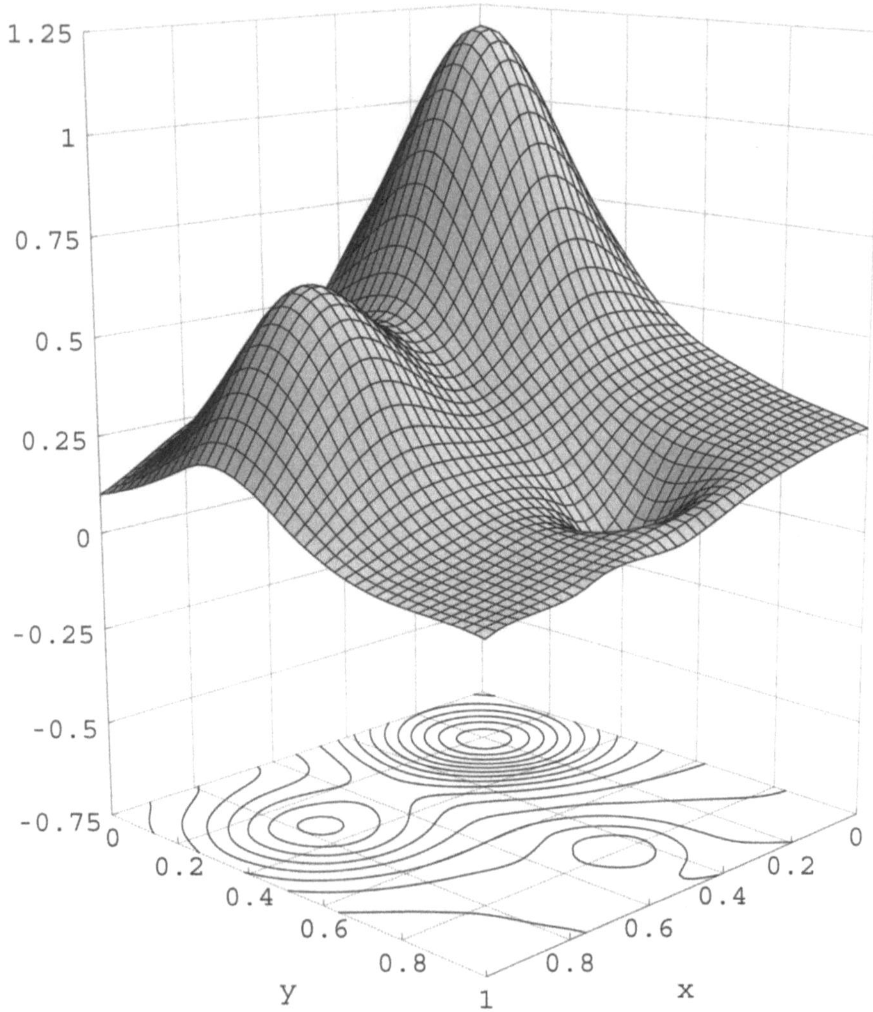


Figure 9: Fitting of Franke's function. Graph and contour plot of the discrete smoothing $D^2$-spline of class $C^1$ corresponding to $N = 400$, $n_{el} = 4$ and $\varepsilon = 10^{-5}$. Relative error: $r(f) = 0.00616472$.

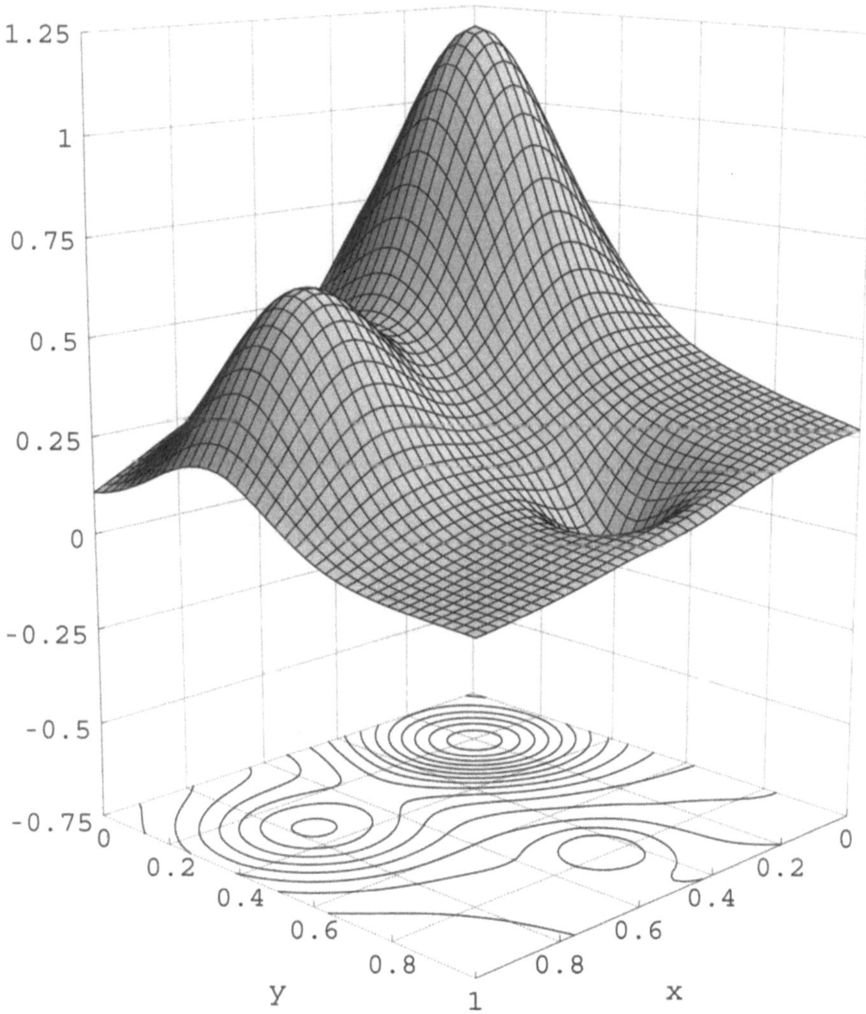


Figure 10: Fitting of Franke's function. Graph and contour plot of the discrete smoothing $D^2$-spline of class $C^1$ corresponding to $N = 800$, $n_{el} = 5$ and $\varepsilon = 10^{-5}$. Relative error: $r(f) = 0.00294882$.

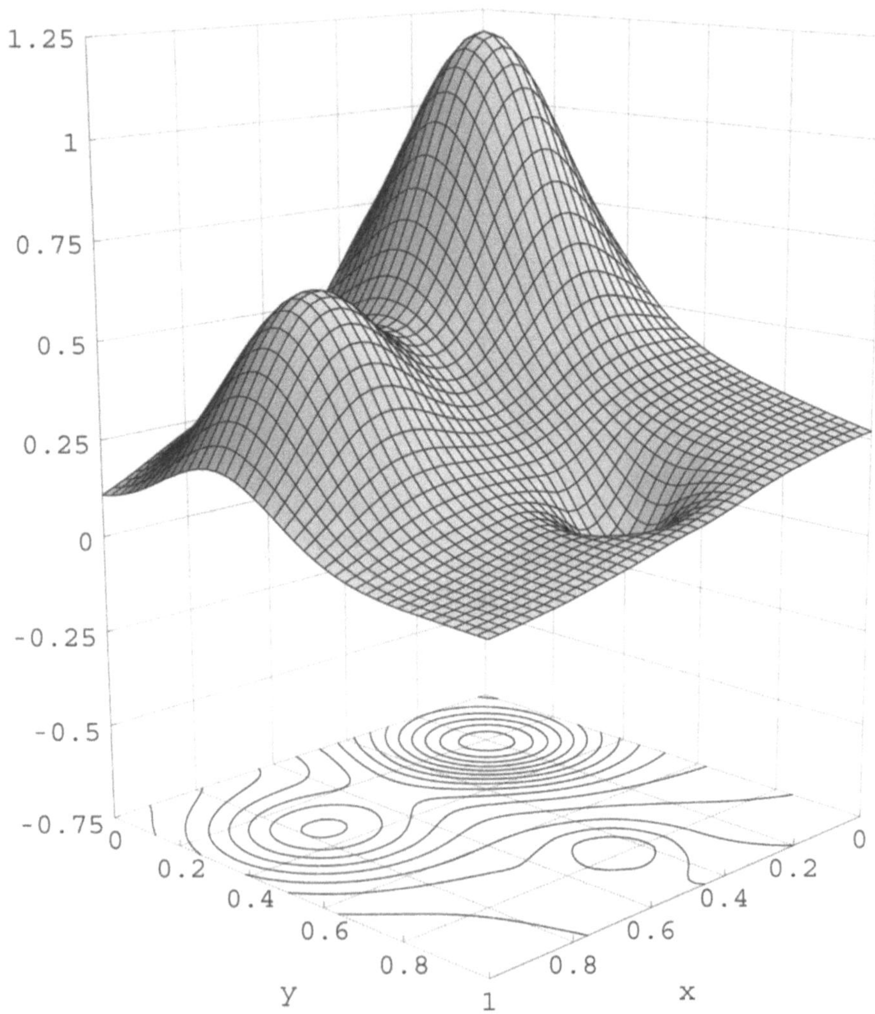


Figure 11: Fitting of Franke's function. Graph and contour plot of the discrete smoothing $D^2$-spline of class $C^1$ corresponding to $N = 1600$, $n_{el} = 6$ and $\varepsilon = 10^{-5}$. Relative error: $r(f) = 0.00176478$.

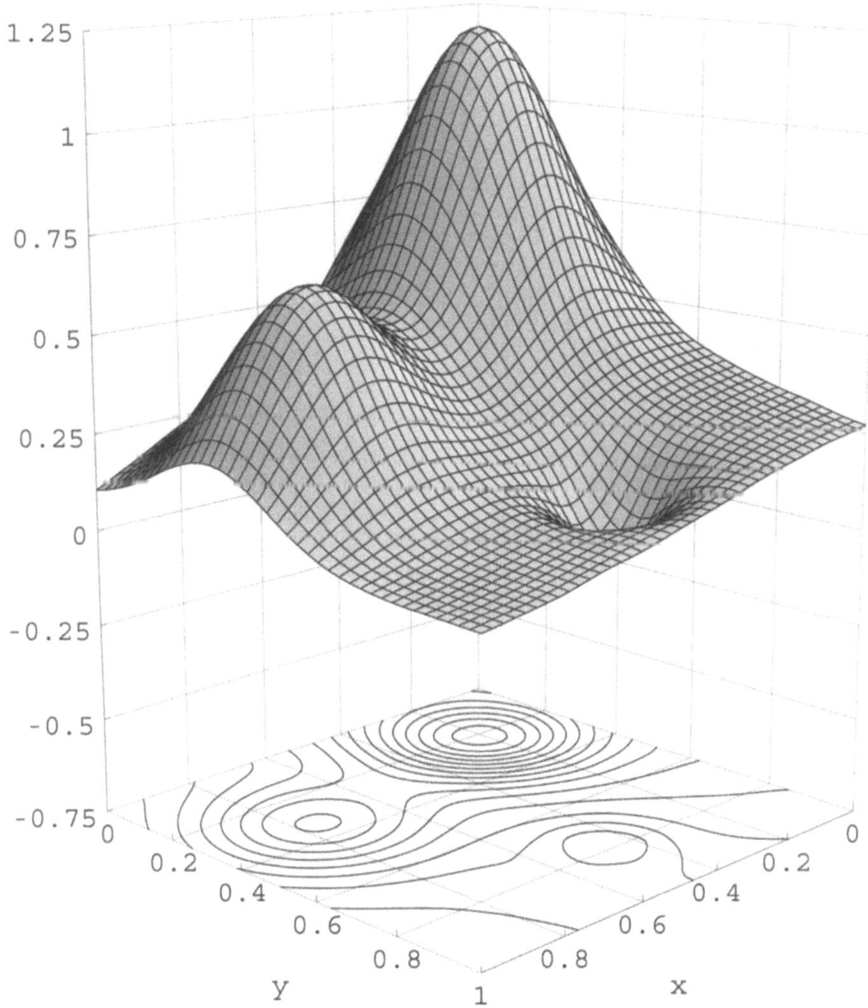


Figure 12: Fitting of Franke's function. Graph and contour plot of the discrete interpolation-smoothing $D^2$-spline of class $C^1$ corresponding to $n_{el} = 6$ and $\varepsilon = 10^{-5}$. Relative error: $r(f) = 0.00245605$.

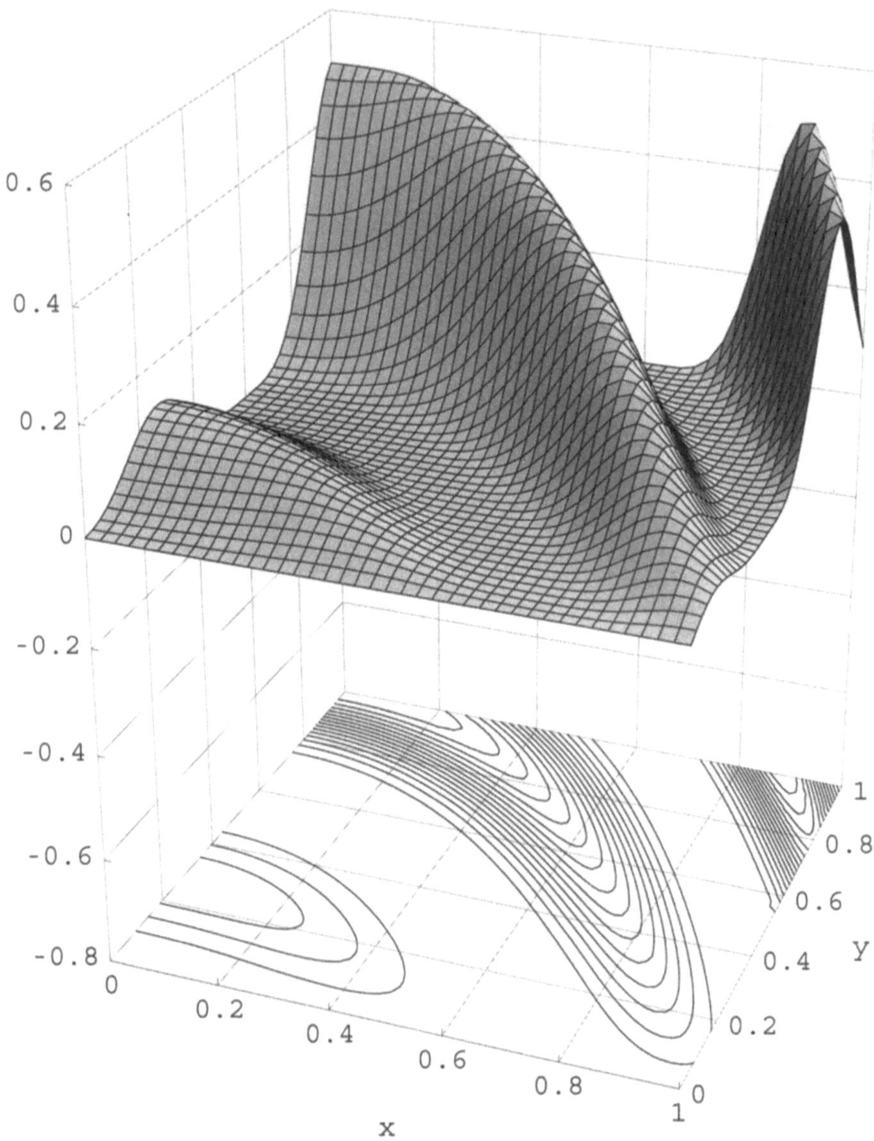


Figure 13: Graph and contour plot of Nielson's function.

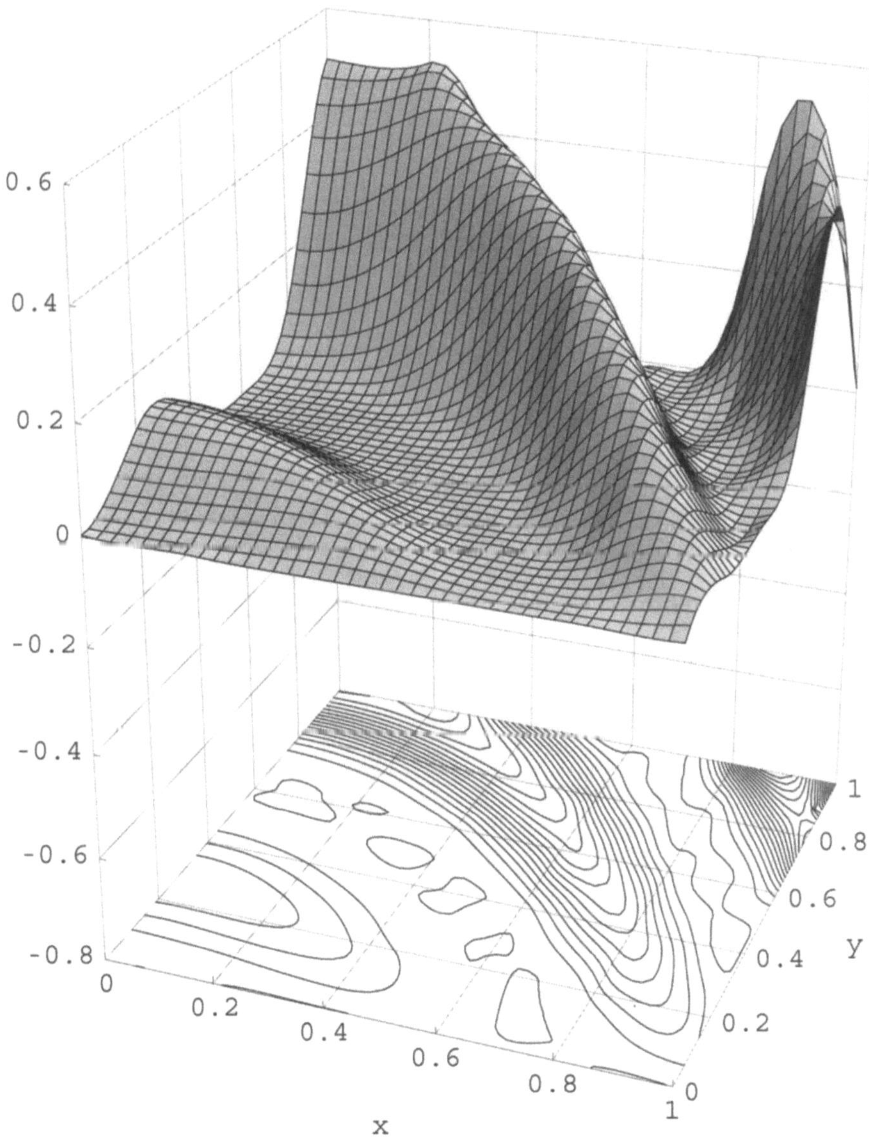


Figure 14: Fitting of Nielson's function. Graph and contour plot of the discrete smoothing $D^2$-spline of class $C^1$ corresponding to $N = 400$, $n_{el} = 5$ and $\varepsilon = 10^{-5}$. Relative error: $r(f) = 0.0533174$.

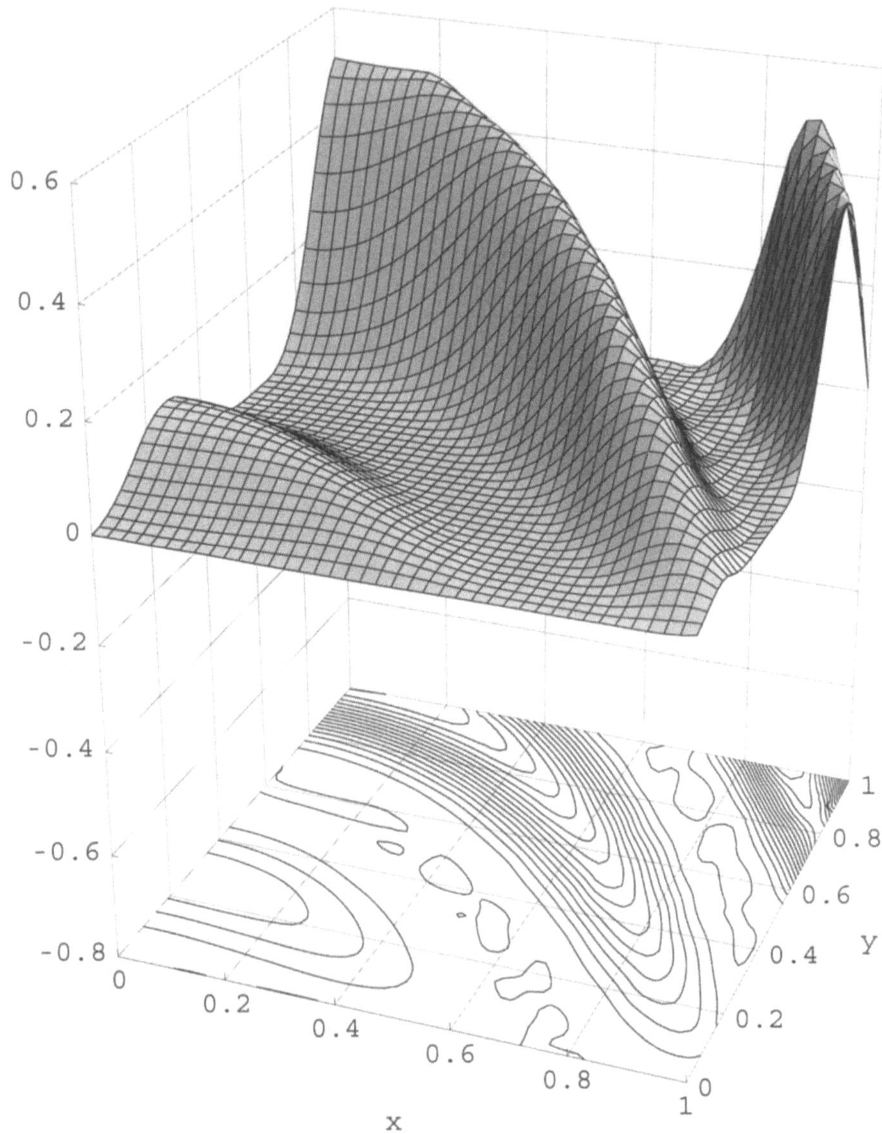


Figure 15: Fitting of Nielson's function. Graph and contour plot of the discrete smoothing $D^2$-spline of class $C^1$ corresponding to $N = 800$, $n_{el} = 6$ and $\varepsilon = 10^{-5}$. Relative error: $r(f) = 0.0220282$.

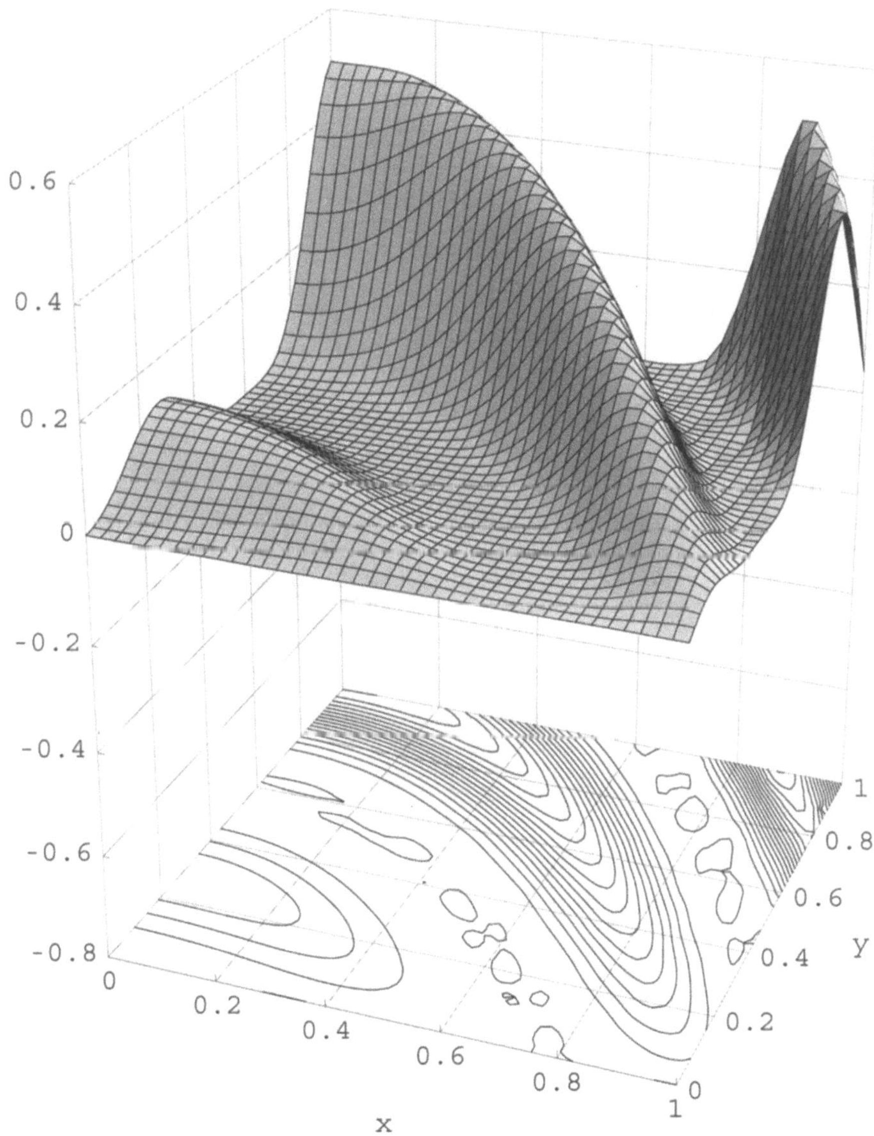


Figure 16: Fitting of Nielson's function. Graph and contour plot of the discrete smoothing $D^2$-spline of class $C^1$ corresponding to $N = 1600$, $n_{el} = 7$ and $\varepsilon = 10^{-5}$. Relative error: $r(f) = 0.0111137$.

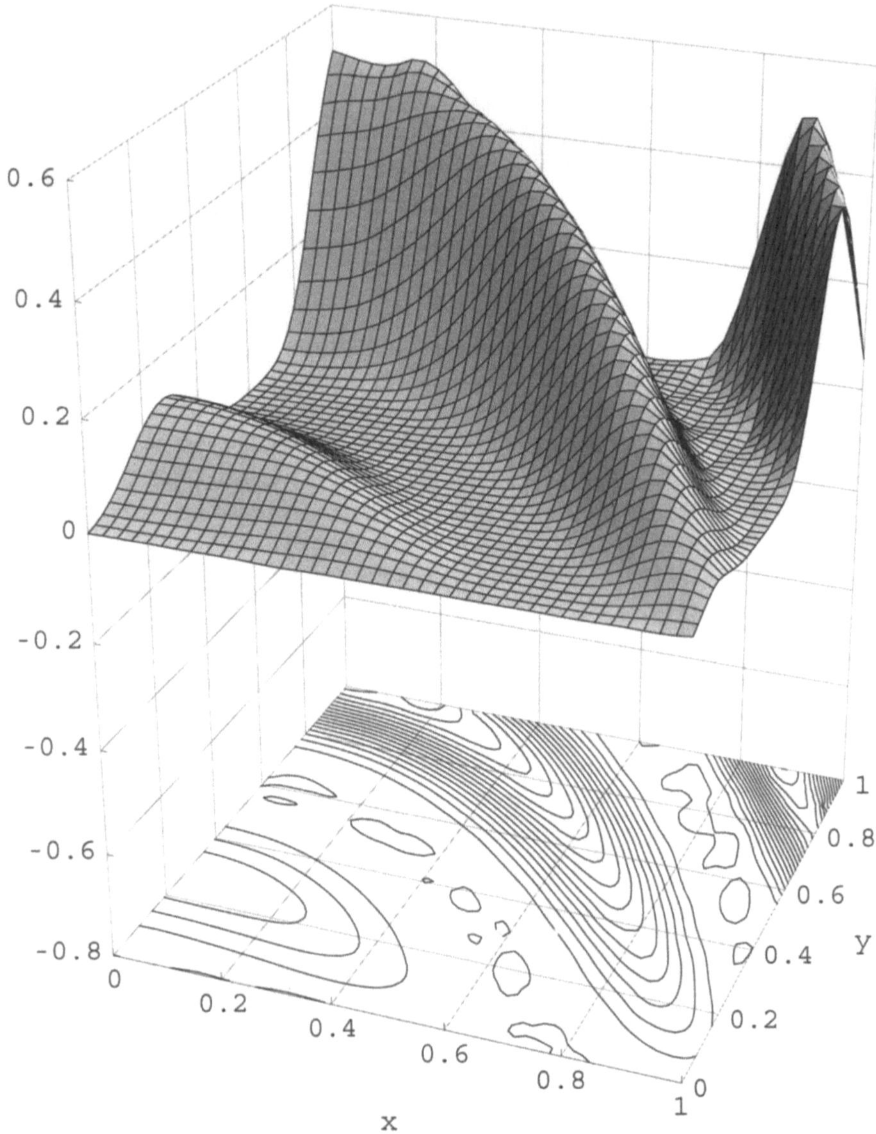


Figure 17: Fitting of Nielson's function. Graph and contour plot of the discrete interpolation-smoothing $D^2$-spline of class $C^1$ corresponding to $n_{el} = 7$ and $\varepsilon = 10^{-5}$. Relative error: $r(f) = 0.0153479$.

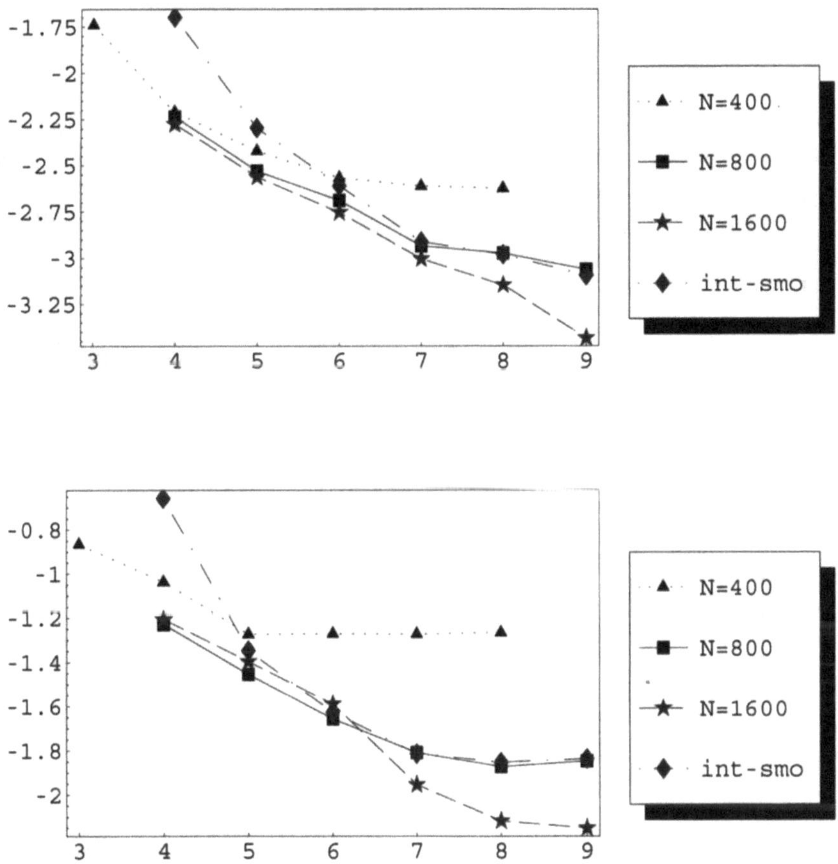


Figure 18: Plot of $\log_{10} r(f)$ versus $n_{el}$, for $\varepsilon = 10^{-5}$. Top: Franke's function, bottom: Nielson's function. The results for the smoothing splines are labelled by `N=400`, `N=800` and `N=1600`, depending on the number of data points. The label `int-smo` refers to the interpolation-smoothing splines.

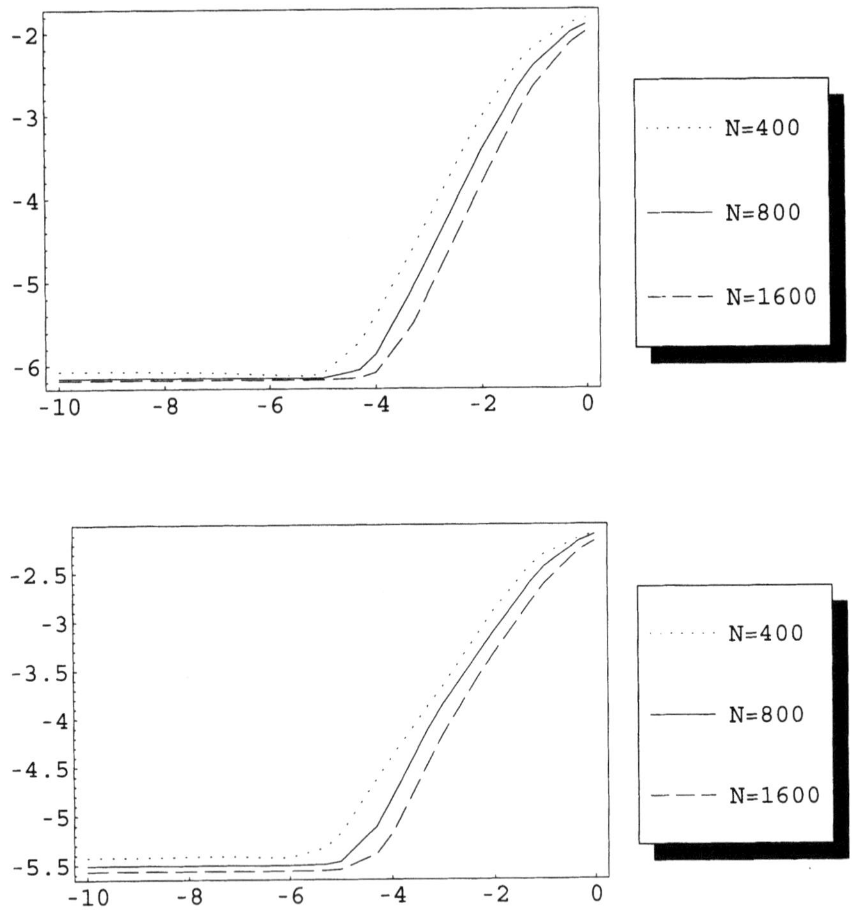


Figure 19: Plot of $\log_{10} \overline{V}(\varepsilon)$ versus $\log_{10} \varepsilon$, for $n_{el} = 6$. Top: Franke's function, bottom: Nielson's function. The results for the smoothing splines are labelled by N=400, N=800 and N=1600, depending on the number of data points.

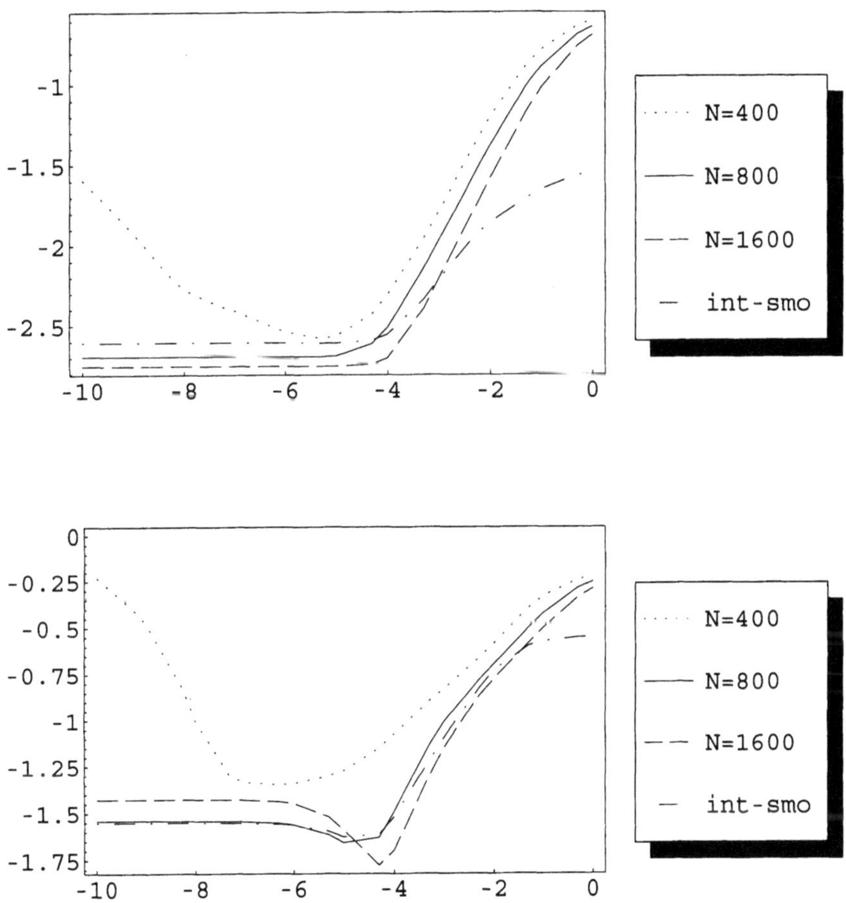


Figure 20: Plot of $\log_{10} r(f)$ versus $\log_{10} \varepsilon$, for $n_{el} = 6$. Top: Franke's function, bottom: Nielson's function. The results for the smoothing splines are labelled by `N=400`, `N=800` and `N=1600`, depending on the number of data points. The label `int-smo` refers to the interpolation-smoothing splines.

# CHAPTER IX

# APPROXIMATION OF FAULTED EXPLICIT SURFACES †

## 1. FORMULATION OF THE PROBLEM

In certain domains of Applied Mathematics, it is necessary to approximate "non-regular functions" in the sense of the following definition: let $\Omega$ be a bounded open subset of $\mathbb{R}^n$ and $f$ a mapping from $\Omega$ into $\mathbb{R}$; if there exists a nonempty subset $F$ of $\overline{\Omega}$ and an integer $m > n/2$ such that (the restriction to $\Omega \setminus \overline{F}$ of) $f$ belongs to the Sobolev space $H^m(\Omega \setminus \overline{F})$ whereas $f$ does not belong to $H^m(\Omega)$, then we say that $f$ is *non-regular of order* $m$ *over* $\Omega$. Such functions or some of their derivatives often present discontinuities.

We consider the following problem: *from Lagrange or first order Hermite data of a non-regular function $f$, construct an approximant of $f$ over $\Omega$, of class $C^1$ or $C^2$ over* $\Omega' = \Omega \setminus \overline{F}$. The set $F$, which we call the *discontinuity set of $f$*, is supposedly known. When this set is unknown, one has further difficulties to face (cf. Section 7). Of course, $F$ cannot be any set. The definition of $F$ is specified in Section 2.

In this way, we model the problem of *representation of faults* in geophysical sciences or oil engineering, where vertical faults correspond to discontinuities of the function $f$ and direct oblique faults (creases) are associated with discontinuities of some of its first partial derivatives. This problem is, in fact, the origin of the expression *approximation of faulted explicit surfaces.* Data sets usually contain position data (points of the surface) and orientation data (two angles, called *dip* and *strike*, that determine the slope of the tangent plane to the surface and which are converted to first order Hermite data). The case of inverse oblique faults (overlapping folds) will not be considered. It can be set as a parametric surface problem and reduced to the approximation of $\mathbb{R}^3$-valued functions (see the study made later in Chapter XII, specially Remark XII–3.3). The problem of representation of (direct) faults in Geophysics can be formulated in terms of bivariate functions. We shall show in Section 2 that, to treat the vertical or the oblique case, one has only to consider functions $f$ belonging to $H^2(\Omega')$ or $H^2(\Omega') \cap C^0(\overline{\Omega})$. However, in order to deal possibly with the case of curves or trivariate functions such as, for instance, the porosity of an oilfield, we prefer to study the general problem ($n$-variate functions presenting discontinuities of some derivatives of order $< \min\{m, m - n/2 + 1\}$).

The fault problem is not restricted to geophysical applications. At much smaller (microscopic) scales, material scientists and solid state physicists have to deal with atomic or molecular arrangements where different kinds of dislocations and defects of the crys-

†Cf. R. Arcangéli, R. Manzanilla and J. J. Torrens [17] and M. C. López de Silanes, M. C. Parra, M. Pasadas and J. J. Torrens [102].

tal lattice determine the properties of technological interest (cf. R. Phillips [117]). In this setting, non-regular functions are needed to model microstructural properties of materials. There are many other fields where the fault problem may appear in a natural way, such as computer graphics, computer vision, signal processing, medical imaging, or, in general, those areas which analyze objects or phenomena represented, somehow, by non-regular functions.

The numerical approximation of functions presenting discontinuities has received increasing attention, due to its many applications. In addition to the articles on which this chapter is based (cf. R. Arcangéli, R. Manzanilla and J. J. Torrens [17] and M. C. López de Silanes, M. C. Parra, M. Pasadas and J. J. Torrens [102]), let us quote, among others, the works of R. F. Franke and G. M. Nielson [69], P.-J. Laurent [89], R. Manzanilla [106], P. Klein [84], V. A. Vasilenko and A. I. Rozhenko [148], C. Serres [134], E. Arge and M. S. Floater [20], J. Springer [135], C. Tarrou [137], and R. Besenghi and G. Allasia [30].

The usual approximation methods (by means of polynomials or standard splines) have on the approximated function $f$ a smoothing effect which tends to erase the discontinuities of $f$ or of its derivatives, whereas we want to reproduce them precisely. In addition, unwanted oscillations may appear near the discontinuity sets due to abrupt changes in the data values. Therefore, the fitting of non-regular functions requires the development of specific methods which take due account of the discontinuities. In [84] and [89], for example, this is achieved by incorporating into the approximation space special functions tied to the discontinuity sets. But implementing these methods is not easy to do in the $n$-dimensional case, unless $n = 1$.

Due to their local structure, finite element spaces are suitable for the construction of discontinuous functions. This is why many of the fitting methods presented in the above references follow a finite element approach. In our problem, we should use Hermite-type finite element spaces, since we need approximants of class $C^1$ or $C^2$ outside the discontinuity set. Unfortunately, we cannot do this directly, because we usually have no suitable Hermite data set at our disposal. In the particular case where, at any point $a$ of a set $A$, the data $f(a)$, $\frac{\partial f}{\partial x}(a)$ and $\frac{\partial f}{\partial y}(a)$ are available, we can interpolate $f$ using the reduced Hsieh-Clough-Tocher triangle (cf. P. G. Ciarlet [45]) or the Powell-Sabin triangle (cf. M. J. D. Powell and M. A. Sabin [119]). We can also use other $C^1$ or $C^2$ finite elements and triangulations not necessarily tied to the interpolation points, but then we have to estimate somehow the unknown degrees of freedom (point values of $f$ and its partial derivatives). For this purpose, we can consider, for example, the "plaquettes splines" method (cf. H. Akima [3], A. Le Méhauté [90], J. J. Torrens [142]).

We are going to give an answer to the problem of approximation of non-regular functions by adapting the theory of $D^m$-splines over a bounded domain of $\mathbb{R}^n$ and discrete $D^m$-splines given in Chapters V and VI. In Section 2, we shall define the discontinuity set $F$ as well as the spaces of functions on $\Omega'$ that we shall use hereafter. In Section 3, we shall introduce the $D^m$-splines over $\Omega'$, which are $D^m$-splines whose definition integrates the particular nature of the open subset $\Omega'$. In Section 4, we shall study the discrete $D^m$-splines over $\Omega'$, which constitute one of the tools we propose to use for the numerical approximation of the non-regular functions introduced in

Section 2 (in order to simplify, we shall treat only the case of discontinuous functions without derivative discontinuities). Sections 5 and 6 will be devoted to convergence results of discrete smoothing $D^m$-splines, obtained in two different frameworks. In Section 7 we shall present, in the bivariate case, two approximation methods for the fitting of non-regular functions, based on the computation of discrete smoothing $D^m$-splines. The chapter ends with Section 8, where we shall give some numerical and graphical examples.

## 2. SPACES OF FUNCTIONS ON $\Omega'$

### 2.1. DEFINITION OF THE DISCONTINUITY SET F

Let $\Omega$ be an open subset of $\mathbb{R}^n$ with a Lipschitz-continuous boundary (cf. Preliminaries). We say that a nonempty subset $F$ of $\overline{\Omega}$ is a *discontinuity set* if there exists a finite family $\{R_1, \ldots, R_I\}$ of open subsets in $\Omega$, with a Lipschitz-continuous boundary, satisfying the following conditions:

$$\forall i, j = 1, \ldots, I,\ i \neq j,\ R_i \cap R_j = \emptyset; \tag{2.1a}$$

$$\bigcup_{i=1}^{I} \overline{R}_i = \overline{\Omega}; \tag{2.1b}$$

$$F \text{ is a subset of } \partial R, \text{ where } R = \bigcup_{i=1}^{I} R_i; \tag{2.1c}$$

$$\left| \begin{array}{l} F \text{ is contained in the interior of } F \cup \partial\Omega \text{ (in the trace topology of } \partial R \\ \text{induced by that of } \mathbb{R}^n); \end{array} \right. \tag{2.1d}$$

$$\text{the interior of } \overline{F} \cap \Omega \text{ (in the trace topology of } \partial R) \text{ is contained in } F; \tag{2.1e}$$

$$\overline{F} \cap \partial\Omega \text{ is contained in } F. \tag{2.1f}$$

Then we also say that *the family* $\{R_1, \ldots, R_I\}$ *represents* $F$ *in* $\Omega$ and we write

$$\Omega' = \Omega \setminus \overline{F}$$

(cf. Figure 1). Notice that $\Omega'$ is an open set in $\mathbb{R}^n$.

This definition generalizes the definition originally introduced by R. Manzanilla [106] and corresponds, more or less, to that suggested by J. J. Torrens [139, 141]. It allows us to model discontinuity sets which are quite general, such as those which may occur in Geophysics to represent faults: nonconnected sets, branching out into connected components, or touching the boundary of $\Omega$. Strictly speaking, conditions (2.1d), (2.1e) and (2.1f) are not essential, but they permit the simplification of the study of the functions on $\Omega'$, defined below in Subsection 2.2, and introducing multilateral interpolation conditions on $F$ (cf. Section 3).

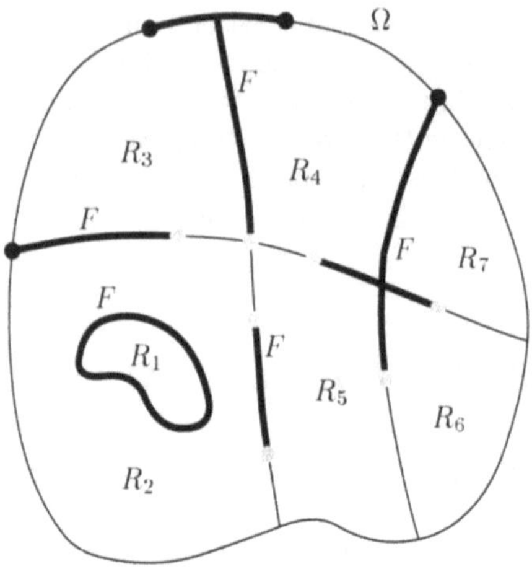


Figure 1: Example of family $\{R_1, \ldots, R_I\}$, where $I = 7$, representing the discontinuity set $F$ in the open set $\Omega$ (grey dots do not belong to $F$).

## 2.2. SPACES $C^k_F(\Omega')$, $H^m(\Omega')$ AND $H^m(\Omega') \cap C^r(\overline{\Omega})$

For any $k \in \mathbb{N}$, let $C^k_F(\Omega')$ be the space defined by

$$C^k_F(\Omega') = \{ v \in C^k(\Omega') \mid \forall i = 1, \ldots, I, \ v|_{R_I} \in C^k(\overline{R}_i) \}. \tag{2.2}$$

For any $v \in C^k_F(\Omega')$ and for any $i = 1, \ldots, I$, we write $v_i = v|_{R_i}$.

***Theorem* 2.1** – *The space $C^k_F(\Omega')$, endowed with the norm*

$$\|v\|_{C^k_F(\Omega')} = \max_{1 \le i \le I} \|v_i\|_{C^k(\overline{R}_i)}, \tag{2.3}$$

*is a Banach space. Moreover, the space $C^k_F(\Omega')$ and the norm (2.3) are independent of the choice of the family $\{R_1, \ldots, R_I\}$ which represents $F$ in $\Omega$.*

**Proof** – The result is obvious when $F = \partial R$. Let us suppose that $F \neq \partial R$.

1) Let $(v^{(l)})_{l \in \mathbb{N}}$ be a Cauchy sequence in $C^k_F(\Omega')$. For any $l \in \mathbb{N}$ and for any $i = 1, \ldots, I$, let us write $v_i^{(l)} = v^{(l)}|_{R_i}$. It is clear, by definition of the norm (2.3), that every sequence $(v_i^{(l)})_{l \in \mathbb{N}}$ admits a limit $w_i \in C^k(\overline{R}_i)$. Let us show that there exists a function $v \in C^k(\Omega')$ such that, for any $i = 1, \ldots, I$, $v_i = w_i$. Any point of $\Omega'$ either belongs to $R$ or to $\partial R$. If $x \in R_j$, with $1 \le j \le I$, we write $v(x) = w_j(x)$. If $x \in \partial R$, then, for any $j_1, j_2 = 1, \ldots, I$, $j_1 \neq j_2$, such that $x \in \partial R_{j_1} \cap \partial R_{j_2}$, we have, for any $l \in \mathbb{N}$, $v_{j_1}^{(l)}(x) = v_{j_2}^{(l)}(x)$, hence $w_{j_1}(x) = w_{j_2}(x)$ and we take the common value for $v(x)$.

2) Let $\{O_1, \ldots, O_{\tilde{I}}\}$ be another family representing $F$ in $\Omega$. Let $v \in C^k_F(\Omega')$. For any $i = 1, \ldots, \tilde{I}$, any $j = 1, \ldots, I$ and any $\alpha \in \mathbb{N}^n$, with $|\alpha| \le k$, the function $\partial^\alpha v$

is continuous on $\overline{O}_i \cap \overline{R}_j$. For any $i = 1, \dots, \tilde{I}$, any $j_1, j_2 = 1, \dots, I$, $j_1 \neq j_2$, any $\alpha \in \mathbb{N}^n$, $|\alpha| \leq k$, and any $x \in O_i \cap \partial R_{j_1} \cap \partial R_{j_2}$, one has obviously $\partial^\alpha v_{j_1}(x) = \partial^\alpha v_{j_2}(x) = \partial^\alpha v(x)$. By passing to the limit, we deduce that this relation is still valid for any $x \in \overline{O}_i \cap \overline{R}_{j_1} \cap \overline{R}_{j_2}$. The result of this is that, for any $i = 1, \dots, \tilde{I}$ and any $\alpha \in \mathbb{N}^n$, $|\alpha| \leq k$, $\partial^\alpha v$ is continuous on $\bigcup_{j=1}^{I} (\overline{O}_i \cap \overline{R}_j) = \overline{O}_i$. Therefore, we conclude that $C_F^k(\Omega')$ is independent of the choice of the family which represents $F$ in $\Omega$.

3) Let $\{R_1, \dots, R_I\}$ and $\{O_1, \dots, O_{\tilde{I}}\}$ be two families representing $F$ in $\Omega$, and let $\| \cdot \|^R_{C_F^k(\Omega')}$ and $\| \cdot \|^O_{C_F^k(\Omega')}$ be the corresponding norms (2.3). For any $j = 1, \dots, I$, every point $x \in \overline{R}_j$ belongs to a set $\overline{O}_j$. Therefore

$$\forall v \in C_F^k(\Omega'),\ \forall j = 1, \dots, I,\ \forall \alpha \in \mathbb{N}^n,\ |\alpha| \leq k,\ \forall x \in \overline{R}_j,\ |\partial^\alpha v(x)| \leq \|v\|^0_{C_F^k(\Omega')}.$$

Hence,

$$\forall x \in C_F^k(\Omega'),\ \|v\|^R_{C_F^k(\Omega')} \leq \|v\|^0_{C_F^k(\Omega')}.$$

From this, we deduce that the norm (2.3) is independent of the family that represents $F$ in $\Omega$. □

**Remark 2.1** – For any $x \in F$, let us write $I_x = \{ i \in \{1, \dots, I\} \mid x \in \partial R_i \}$. It follows from (2.1d) that, for any $x \in F$, there exists $\delta > 0$ such that, for any $i \in I_x$, $B(x, \delta) \cap R_i$ is a connected component of $\Omega \cap B(x, \delta) \setminus F$. Thus, for any $v \in C_F^k(\Omega')$, there exist exactly $\operatorname{card} I_x$ limit values of $v$ at the point $x$, namely $v_i(x)$ for every $i \in I_x$.

Likewise, when $F \neq \partial R$, we deduce from (2.1) and point 2) of the proof of Theorem 2.1 that, for any $x \in \overline{\Omega} \setminus (\Omega' \cup F) = (\overline{F} \setminus F) \cup (\partial\Omega \setminus F)$ and for any $i, j = 1, \dots, I$, $i \neq j$, such that $x \in \partial R_i \cap \partial R_j$,

$$\forall v \in C_F^k(\Omega'),\ v_i(x) = v_j(x).$$

So, it is possible to extend continuously any function $v \in C_F^k(\Omega')$ at every point $x \in \overline{\Omega} \setminus (\Omega' \cup F)$, in a unique way, by defining

$$v(x) = \lim_{\substack{y \to x \\ y \in \Omega'}} v(y).$$

These results are also valid for the derivatives of order $\leq k$ of the functions of $C_F^k(\Omega')$.

In short, we can consider that the elements of $C_F^k(\Omega')$ and their derivatives of order $\leq k$ are functions defined and continuous on $\overline{\Omega} \setminus F$ which possibly present discontinuities with a finite jump on $F$. □

Before using the usual Sobolev space $H^l(\Omega')$, for any $l \in \mathbb{N}$, let us point out that this space is defined over an open set in $\mathbb{R}^n$ which *has not* a Lipschitz-continuous boundary (in the sense of the definition given in Preliminaries), because it is not lying locally on only one side of its boundary.

***Theorem* 2.2** – *For any $m, k \in \mathbb{N}$, with $m > n/2 + k$, $H^m(\Omega') \overset{c}{\subset} C_F^k(\Omega')$.*

**Proof** – For any $x \in \Omega'$, there exists $\delta > 0$ such that $B(x,\delta) \subset \Omega'$. Then, from Sobolev's Continuous Imbedding Theorem (cf. (4) in Preliminaries), we derive that $H^m(B(x,\delta)) \hookrightarrow C^k(\overline{B(x,\delta)})$, and therefore $H^m(\Omega') \subset C^k(\Omega')$. On the other hand, since the open sets $R_i$ have a Lipschitz-continuous boundary, we deduce from Rellich-Kondrašov Compact Imbedding Theorem (cf. (3) ibidem) that

$$\forall i = 1,\dots,I,\ H^m(R_i) \overset{c}{\subset} C^k(\overline{R}_i). \tag{2.4}$$

Thus, $H^m(\Omega') \hookrightarrow C^k_F(\Omega')$. The proof ends observing that, by (2.4), any sequence weakly convergent in $H^m(\Omega')$ is strongly convergent in $C^k(\overline{R_i})$ for every $i = 1,\dots,I$, and also, taking (2.3) into account, in $C^k_F(\Omega')$. Notice that it follows from Theorem 2.1 that the continuity of the canonical injection does not depend on the family which represents $F$ in $\Omega$. □

***Theorem* 2.3** – *For any $l, l' \in \mathbb{N}$, with $l > l'$, $H^l(\Omega') \overset{c}{\subset} H^{l'}(\Omega')$.*

**Proof** – This result is a particular case of the Rellich-Kondrašov Theorem, just as stated, for instance, in R. A. Adams [1], because $\Omega'$ is an open set having the cone property. One cannot directly apply here the Compact Imbedding Theorem (2) in Preliminaries, where the open set is supposed to possess a Lipschitz-continuous boundary. □

***Theorem* 2.4** – *For any $m, r \in \mathbb{N}$, with $m > n/2+r$, the subspace $H^m(\Omega') \cap C^r(\overline{\Omega})$ is closed in $H^m(\Omega')$.*

**Proof** – Let $(u^{(l)})_{l\in\mathbb{N}}$ be a sequence of elements of $H^m(\Omega') \cap C^r(\overline{\Omega})$ which is a Cauchy sequence in $H^m(\Omega')$. This sequence converges to an element $u \in H^m(\Omega')$ and, by Theorem 2.2, $u \in C^r_F(\Omega')$. Now, $C^r(\overline{\Omega})$ is a closed subspace of $C^r_F(\Omega')$, since

$$\forall v \in C^r(\overline{\Omega}),\ \|v\|_{C^r(\overline{\Omega})} = \|v\|_{C^r_F(\Omega')}.$$

Therefore, $u \in C^r(\overline{\Omega})$ and the Theorem follows. □

**Remark 2.2** – When $m > n/2 + r$, the space $H^m(\Omega') \cap C^r(\overline{\Omega})$ consists of functions which may admit discontinuities of derivatives of order $r+1$. For $n \geq 2$, we have $r + 1 \leq m - 1$, so the elements of $H^m(\Omega') \cap C^r(\overline{\Omega})$ are really non-regular functions in the sense of the definition we adopted. On the other hand, for $n = 1$, it is possible that $r + 1 = m$, but in this case $H^m(\Omega') \cap C^{m-1}(\overline{\Omega}) = H^m(\Omega)$. This is the reason we are interested, as a general rule, in the approximation of functions presenting discontinuities of derivatives of order $< \min\{m, m - n/2 + 1\}$. □

## *3. $D^m$-SPLINES OVER $\Omega'$*

Let $\Omega$ be an open subset of $\mathbb{R}^n$ with a Lipschitz-continuous boundary (cf. Preliminaries). Let $F \subset \overline{\Omega}$ and $\{R_1,\dots,R_I\}$ be, respectively, a discontinuity set and a family of open sets that represents $F$ in $\Omega$ (cf. Subsection 2.1). We write $\Omega' = \Omega \setminus \overline{F}$.

Let $A$ be a finite set of distinct points of $\overline{\Omega}$ and let $\Sigma$ be a set of linear forms of the type

$$\phi : v \mapsto \partial^\alpha v(a), \tag{3.1}$$

with $a \in A \setminus F$, $\alpha \in \mathbb{N}^n$, $|\alpha| \leq 1$, or of the type

$$\phi : v \mapsto \partial^\alpha (v|_{R_i})(a), \tag{3.2}$$

with $a \in A \cap \partial R_i \cap F$, $1 \leq i \leq I$, $\alpha \in \mathbb{N}^n$, $|\alpha| \leq 1$. Of course, we suppose that every point of $A$ is associated, at least, with one element of $\Sigma$. Moreover, we denote by $\mu$ the maximal order of the derivatives occurring in the definition of the elements of $\Sigma$, so $\mu = 0$ or 1. From now on, we suppose that

$$m > \frac{n}{2} + \mu. \tag{3.3}$$

According to Remark 2.1, the elements of $\Sigma$ are continuous linear forms on $C_F^\mu(\Omega')$.

**Remark 3.1** – Every point of $A \setminus F$ is a node corresponding, at most, to $n+1$ elements of $\Sigma$, whereas every point $a$ of $A \cap F$ can be associated, at most, with $n+1$ elements of $\Sigma$ for each open set $R_i$ such that $a \in \partial R_i$. □

Now, we write $N = \operatorname{card} \Sigma$. We denote by $\mathbb{R}^N$ the Euclidean space of dimension $N$, and by $\langle \cdot \rangle$ and $\langle \cdot , \cdot \rangle$ the norm and the scalar product in $\mathbb{R}^N$. As the set $\Sigma$ is supposedly ordered, we introduce the linear continuous operator $\rho : C_F^\mu(\Omega') \to \mathbb{R}^N$, defined by

$$\rho v = \big(\phi(v)\big)_{\phi \in \Sigma}.$$

From now on, we shall suppose that $\Sigma$ contains a $\widetilde{P}_{m-1}(\Omega')$-unisolvent subset, i.e. that

$$\operatorname{Ker} \rho \cap \widetilde{P}_{m-1}(\Omega') = \{0\}, \tag{3.4}$$

where $\widetilde{P}_{m-1}(\Omega')$ denotes the space of functions on $\Omega'$ which are polynomial functions of degree $\leq m-1$, with respect to the set of variables, over each connected component of $\Omega'$. Hereafter, every time (3.4) is assumed, we shall implicitly suppose that $\Sigma$ is defined by (3.1) and (3.2), with $|\alpha| \leq 1$ and $\mu = 0$ or 1.

It follows from (3.3) and Theorem 2.2 that $\rho$ is also a linear continuous operator from $H^m(\Omega')$ into $\mathbb{R}^N$. So, we can consider the mapping $[\![ \cdot ]\!]_{m,\Omega'}$, defined on $H^m(\Omega')$ by

$$[\![ v ]\!]_{m,\Omega'} = \left( \langle \rho v \rangle^2 + |v|_{m,\Omega'}^2 \right)^{1/2}. \tag{3.5}$$

***Proposition* 3.1** – *Suppose that hypotheses* (3.3) *and* (3.4) *are verified. Then, the mapping* $[\![ \cdot ]\!]_{m,\Omega'}$ *is a Hilbertian norm on* $H^m(\Omega')$, *equivalent to the norm* $\| \cdot \|_{m,\Omega'}$.

**Proof** – The fact that $[\![ \cdot ]\!]_{m,\Omega'}$ is a norm on $H^m(\Omega')$ comes from (3.4) and it is clear that it derives from a scalar product. In order to prove the equivalence of norms, one reasons by compactness, as in the proof of Theorem 2.7.1 of J. Nečas [109], by using Theorem 2.3. □

Now, we give the definition of interpolating $D^m$-splines over $\Omega'$. Let $\beta \in \mathbb{R}^N$ and consider the affine linear variety

$$\mathcal{K} = \{ v \in H^m(\Omega') \mid \rho v = \beta \}$$

as well as the associated vector subspace

$$\mathcal{K}_0 = \{ v \in H^m(\Omega') \mid \rho v = 0 \}.$$

Then, we call *interpolating $D^m$-spline over $\Omega'$ relative to $\rho$ and $\beta$* any solution, if any exists, of the problem: find $\sigma$ such that

$$\begin{cases} \sigma \in \mathcal{K}, \\ \forall v \in \mathcal{K}, \ |\sigma|_{m,\Omega'} \le |v|_{m,\Omega'}. \end{cases} \tag{3.6}$$

***Theorem* 3.1** – *Suppose that hypotheses* (3.3) *and* (3.4) *are verified. Then, problem* (3.6) *has a unique solution $\sigma$ characterized by*

$$\begin{cases} \sigma \in \mathcal{K}, \\ \forall w \in \mathcal{K}_0, \ (\sigma, w)_{m,\Omega'} = 0. \end{cases} \tag{3.7}$$

**Proof** –

1) First, let us show that $\mathcal{K}$ is nonempty. Since $A$ is finite, there exists $\delta > 0$ such that

(i) for any $a \in A$, $B(a,\delta) \cap A = \{a\}$;

(ii) for any $i = 1, \dots, I$ and any $a \in A \cap F \cap \partial R_i$, $B(a,\delta) \cap R_i$ is a connected component of $\Omega \cap B(a,\delta) \setminus F$ (cf. Remark 2.1).

Let $\psi : \mathbb{R}^n \to \mathbb{R}$ be the function defined by

$$\psi(x) = \begin{cases} \exp\left(1 - \frac{1}{1-|x|^2}\right), & |x| < 1, \\ 0, & |x| \ge 1. \end{cases}$$

We number $\phi_1, \dots, \phi_N$ the elements of $\Sigma$ and we set $\beta = (\beta_1, \dots, \beta_N)$, where, for any $j = 1, \dots, N$, $\beta_j$ corresponds to $\phi_j$. For any $j = 1, \dots, N$, if $\phi_j$ is of type (3.1), we define the function $\varphi_j : \mathbb{R}^n \to \mathbb{R}$ by

$$\varphi_j(x) = (x-a)^\alpha \psi\Big(\frac{x-a}{\delta}\Big),$$

where $\alpha \in \mathbb{N}^n$ and $a \in A$ are associated with $\phi_j$. Likewise, if $\phi_j$ is of type (3.2), we define $\varphi_j$ by

$$\varphi_j(x) = (x-a)^\alpha \psi\Big(\frac{x-a}{\delta}\Big) \chi_{R_i}(x),$$

where $\chi_{R_i}$ is the characteristic function of $R_i$. Then, we easily see that the function

$$u_\beta = \sum_{j=1}^{N} \beta_j \varphi_j \tag{3.8}$$

belongs to $\mathcal{K}$.

2) Any possible solution of (3.6) is an element of minimal norm $[\![ \cdot ]\!]_{m,\Omega'}$ in the set $\mathcal{K}$. Now, $\mathcal{K}$ is convex, closed in $H^m(\Omega')$, since $\rho$ is continuous, and nonempty. Thus, problem (3.6) admits a unique solution $\sigma$, namely the element of minimal norm $[\![ \cdot ]\!]_{m,\Omega'}$ in $\mathcal{K}$. We know that this element is characterized by the relations

$$\begin{cases} \sigma \in \mathcal{K}, \\ \forall v \in \mathcal{K},\ [\![ -\sigma, v-\sigma ]\!]_{m,\Omega'} \leq 0, \end{cases}$$

where $[\![ \cdot\, , \cdot ]\!]_{m,\Omega'}$ denotes the scalar product in $H^m(\Omega')$ associated with the norm $[\![ \cdot ]\!]_{m,\Omega'}$. From this, one deduces (3.7). □

***Proposition* 3.2** – *There exists one and only one pair $(\sigma, \lambda) \in H^m(\Omega') \times \mathbb{R}^N$ which is the solution of*

$$\begin{cases} \sigma \in \mathcal{K}, \\ \forall v \in H^m(\Omega'),\ (\sigma, v)_{m,\Omega'} = \langle \lambda, \rho v \rangle, \end{cases} \tag{3.9}$$

*and $\sigma$ is just the solution of problem* (3.6).

**Proof** – If $(\sigma, \lambda)$ is a solution of (3.9), then $\sigma \in \mathcal{K}$ and, for any $w \in \mathcal{K}_0$, $(\sigma, w)_{m,\Omega'} = 0$, hence $\sigma$ is the solution of (3.6) and $\sigma$ is unique. Moreover, if $(\sigma, \lambda')$ and $(\sigma, \lambda")$ are two solutions of (3.9), we have, for any $v \in H^m(\Omega')$, $\langle \lambda' - \lambda", \rho v \rangle = 0$, which implies $\lambda' = \lambda"$. Thus, there exists, at most, one solution of (3.9).

On the other hand, for any $v \in H^m(\Omega')$, the function $w = v - u_{\rho v}$ belongs to $\mathcal{K}_0$, where $u_{\rho v}$ is the function introduced in (3.8) for $\beta = \rho v$. Let $\sigma$ be the solution of (3.6) and let $\lambda$ be the vector in $\mathbb{R}^N$ of components $(\sigma, \varphi_j)_{m,\Omega'}$, for $j = 1, \ldots, N$, with $\varphi_j$ defined as in the previous proof. Then, by (3.7), the pair $(\sigma, \lambda)$ is a solution of (3.9). □

We next introduce the smoothing $D^m$-splines over $\Omega'$. For any $\varepsilon > 0$, we put

$$\forall v \in H^m(\Omega'),\ J_\varepsilon(v) = \langle \rho v - \beta \rangle^2 + \varepsilon |v|^2_{m,\Omega'}. \tag{3.10}$$

Then we consider the minimization problem: find $\sigma_\varepsilon$ such that

$$\begin{cases} \sigma_\varepsilon \in H^m(\Omega'), \\ \forall v \in H^m(\Omega'),\ J_\varepsilon(\sigma_\varepsilon) \leq J_\varepsilon(v). \end{cases} \tag{3.11}$$

Any solution $\sigma_\varepsilon$, if any exists, is called *smoothing $D^m$-spline over $\Omega'$ relative to $\rho$, $\beta$ and $\varepsilon$.*

***Theorem* 3.2** – *Suppose that hypotheses* (3.3) *and* (3.4) *are verified. Then, problem* (3.11) *has a unique solution $\sigma_\varepsilon$ characterized by*

$$\begin{cases} \sigma_\varepsilon \in H^m(\Omega'), \\ \forall v \in H^m(\Omega'),\ \langle \rho\sigma_\varepsilon, \rho v \rangle + \varepsilon(\sigma_\varepsilon, \rho v)_{m,\Omega'} = \langle \beta, \rho v \rangle. \end{cases} \tag{3.12}$$

**Proof** – Replacing $\Omega$ by $\Omega'$ and taking Proposition 3.1 into account, one resumes the proof of Theorem V–3.1. □

**Remark 3.2** – Reasoning as in the proof of Theorem V–2.1, one can show that the interpolating and smoothing $D^m$-splines over $\Omega'$ relative to $\rho$ belong to the space $C^{2m-n-1-\mu}(\Omega')$. Therefore, in particular, $\sigma$ and $\sigma_\varepsilon$ belong to $C^1(\Omega')$ if $m = 2$, $n = 2$ and $\mu = 0$, and to $C^2(\Omega')$ if $m = 3$, $n = 2$ and $\mu = 1$. □

**Remark 3.3** – Suppose, in this section, that $\Sigma$ is a set of linear forms either of type $v \mapsto v(a)$, with $a \in \overline{\Omega}$, or of types (3.1) or (3.2), with $|\alpha| = 1$. Replace $H^m(\Omega')$ by the space $V = H^m(\Omega') \cap C^0(\overline{\Omega})$. Then, taking Theorem 2.4 into account, one defines in the same way the $V$-interpolating (resp. $V$-smoothing) $D^m$-spline relative to $\rho$ and $\beta$ (resp. $\rho$, $\beta$ and $\varepsilon$). □

## 4. DISCRETE $D^m$-SPLINES

In order to simplify the exposition, we suppose from now on that $\Omega$ is a *polyhedral* subset of $\mathbb{R}^n$ (this assumption is verified in the applications). On the other hand, we suppose that the closure of the discontinuity set $F$ is a finite union of (closed) faces of polyhedrons in $\mathbb{R}^n$, that $m$ is any positive integer and we denote by $k$ an integer equal to 1 or 2. We write $\Omega' = \Omega \setminus \overline{F}$, we keep the notations $A$, $\Sigma$, $\mu$, $N$ and $\rho$ of Section 3 and we suppose that (3.4) is verified.

Let $\mathbb{H}$ be a bounded subset of $(0, +\infty)$ such that $0 \in \overline{\mathbb{H}}$. For any $h \in \mathbb{H}$, suppose we are given

- a triangulation $\mathcal{T}_h$ of $\overline{\Omega}$ by means of $n$-simplices $K$ with diameters $h_K \le h$ and pairwise disjoint interiors $\overset{\circ}{K}$, such that

$$\forall K \in \mathcal{T}_h, \ \overset{\circ}{K} \cap F = \emptyset, \tag{4.1}$$

$$\left| \begin{array}{l} \text{any face of a } n\text{-simplex } K \in \mathcal{T}_h \text{ is either the face of another } n\text{-} \\ \text{simplex in } \mathcal{T}_h, \text{ or a part of } \partial\Omega, \text{ or a part of } \overline{F} \end{array} \right. \tag{4.2}$$

  (cf. Figure 2),

- a finite element space $V_h$, constructed on $\mathcal{T}_h$, such that

$$V_h \text{ is a finite-dimensional subspace of } H^m(\Omega') \cap C^k_{\overline{F}}(\Omega'). \tag{4.3}$$

**Remark 4.1** – Let $k'$ be the class of the generic finite element of the space $V_h$. Then, hypothesis (4.3) implies that $k \le k'$ and that $m \le k' + 1$, so that the inclusions $V_h \subset C^k_{\overline{F}}(\Omega')$ and $V_h \subset H^m(\Omega')$, respectively, are obtained. When dealing with real problems, one takes $k' = k$, for reasons of cost.

For the problem of approximating surfaces from Lagrange or first order Hermite data, we have $n = 2$ and so, taking account of hypothesis (5.2), needed in study of the convergence, the usual choices are $k' = 1$ and $m = 2$, if $k = 1$, and $k' = 2$ and $m = 3$,

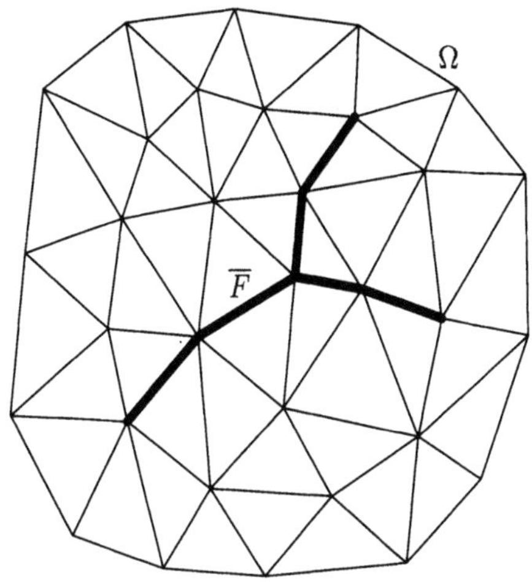


Figure 2: Example of triangulation of the set $\overline{\Omega}$.

if $k = 2$ (see Section VIII-2). Notice that we must use *triangular* finite elements, because $F$ may be any polygonal set. For examples of generic finite elements, the reader is referred to Section VIII-3. □

**Remark 4.2** – Let us detail how to obtain a finite element space $V_h$ satisfying (4.3). In a first step, one follows the usual process in the Finite Element Method, without taking $F$ into account, in order to construct a finite element space $V_h^*$ such that $V_h^* \subset H^m(\Omega) \cap C^k(\overline{\Omega})$. Let $w_1^*, \ldots, w_{M^*}^*$ be the basis functions of $V_h^*$. For $i = 1, \ldots, M^*$, let $b_i \in \overline{\Omega}$ be the node with which $w_i^*$ is associated and let $\gamma_i$ be the number of connected components of $(\operatorname{supp} w_i^*) \setminus \overline{F}$, whose respective closures are denoted by $U_i^1, \ldots, U_i^{\gamma_i}$ (cf. Figure 3). It is obvious that $\gamma_i > 1$ only if $b_i$ belongs to $F$ (it may happen, however, that $\gamma_i = 1$ for some nodes $b_i \in \partial\Omega \cap F$).

Now, let $W = \{ w_i^* \chi_{U_i^j} \mid i = 1, \ldots, M^*,\ j = 1, \ldots, \gamma_i \}$, where $\chi_{U_i^j}$ is the characteristic function of $U_i^j$. It is clear that $W$ is a finite family of linearly independent functions of $H^m(\Omega') \cap C_F^k(\Omega')$. Then, the space $V_h$ is just the linear space spanned by $W$. Let us observe that the sets $U_i^j$ are the supports of the functions in $W$ and so the supports of the basis functions of $V_h$. □

According to (4.3) and since $\mu \le k$, we can define on $V_h$, for any $h \in \mathbb{H}$, the mapping $[\![ \cdot ]\!]_{m,\Omega'}$, introduced in (3.5). It follows from (3.4) that $[\![ \cdot ]\!]_{m,\Omega'}$ is a norm on $V_h$. Of course, endowed with this norm, $V_h$ is a Hilbert space, because it is finite dimensional.

Let $\beta \in \mathbb{R}^N$. For any $h \in \mathbb{H}$, we define the vector space

$$\mathcal{K}_{0h} = \{ v_h \in V_h \mid \rho v_h = 0 \}$$

and the affine linear variety

$$\mathcal{K}_h = \{ v_h \in V_h \mid \rho v_h = \beta \}.$$

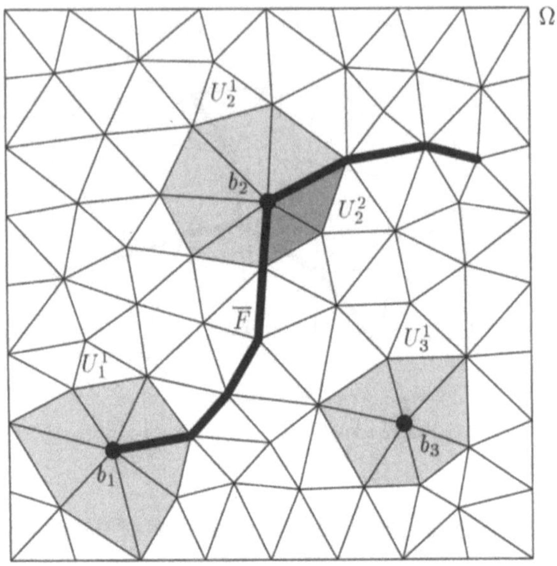


Figure 3: Example of sets $U_i^j$ associated with three nodes $b_i$ (cf. Remark 4.2). Let us observe that $\gamma_1 = \gamma_3 = 1$, whereas $\gamma_2 = 2$.

Then we consider the problem: find $\sigma_h$ solution of

$$\begin{cases} \sigma_h \in \mathcal{K}_h, \\ \forall v_h \in \mathcal{K}_h, \ |\sigma_h|_{m,\Omega'} \le |v_h|_{m,\Omega'}. \end{cases} \tag{4.4}$$

Any solution $\sigma_h$ of (4.4), if any exists, is called *$V_h$-discrete interpolating $D^m$-spline relative to $\rho$ and $\beta$.*

**Theorem 4.1** – *Suppose that hypotheses* (3.4), (4.1), (4.2) *and* (4.3) *are verified. Moreover, suppose that*

$$\forall h \in \mathbb{H}, \ \Sigma \subset \Sigma_h, \tag{4.5}$$

*where $\Sigma_h$ denotes the set of degrees of freedom of $V_h$. Then, for any $h \in \mathbb{H}$, problem* (4.4) *has a unique solution $\sigma_h$, characterized by*

$$\begin{cases} \sigma_h \in \mathcal{K}_h, \\ \forall w_h \in \mathcal{K}_{0h}, \ (\sigma_h, w_h)_{m,\Omega'} = 0. \end{cases} \tag{4.6}$$

**Proof** – Let us show that $\mathcal{K}_h$ is nonempty. Let us denote by $\phi_1, \ldots, \phi_N$ the elements of $\Sigma$. Let $M$ be the dimension of $V_h$ and let $w_1, \ldots, w_M$ be the basis functions of $V_h$, numbered so that, for any $j = 1, \ldots, N$, $\phi_j(w_j) = 1$ (which, by (4.5), we are able to do). Then, the function $v_h = \sum_{j=1}^{N} \beta_j w_j$ belongs to $\mathcal{K}_h$. Thus, the closed convex nonempty subset $\mathcal{K}_h$ of $V_h$ has a unique element $\sigma_h$ of minimal norm $[\![ \cdot ]\!]_{m,\Omega'}$, characterized by (4.6). □

When (3.3) is verified, it is clear that (4.4) (resp. (4.6)) constitutes a discretization of (3.6) (resp. (3.7)).

**Remark 4.3** – With the notations of the previous proof and under hypothesis (4.5), the solution $\sigma_h$ of (4.4) can be written as

$$\sigma_h = \sum_{j=1}^{N} \beta_j w_j + \sum_{j=N+1}^{M} \alpha_j w_j,$$

with $\alpha_j \in \mathbb{R}$, for $j = N+1, \dots, M$. Reasoning as in Remark VI–2.1, we see that the unknown coefficients $\alpha_j$ are the solution of the linear system

$$\sum_{j=N+1}^{M} (w_j, w_i)_{m,\Omega'} \alpha_j = -\sum_{j=1}^{N} \beta_j (w_j, w_i)_{m,\Omega'}, \quad N+1 \le i \le M,$$

whose matrix is *regular*. □

For any $\varepsilon > 0$ and any $h \subset \mathbb{H}$, we now consider the problem: find $\sigma_{\varepsilon h}$ verifying

$$\begin{cases} \sigma_{\varepsilon h} \in V_h, \\ \forall v_h \in V_h, \ J_\varepsilon(\sigma_{\varepsilon h}) \le J_\varepsilon(v_h), \end{cases} \tag{4.7}$$

where $J_\varepsilon$ denotes the functional introduced in (3.10) (let us observe that, independently of any condition on $m$, such as (3.3), $J_\varepsilon$ is defined, in fact, on $H^m(\Omega') \cap C^\mu_F(\Omega')$ and hence on $V_h$).

***Theorem 4.2*** – *Under hypotheses* (3.4), (4.1), (4.2) *and* (4.3), *for any* $h \subset \mathbb{H}$, *problem* (4.7) *has a unique solution* $\sigma_{\varepsilon h}$, *called* $V_h$-discrete smoothing $D^m$-spline relative to $\rho, \beta$ and $\varepsilon$, *which is also the unique solution of the problem: find* $\sigma_{\varepsilon h}$ *such that*

$$\begin{cases} \sigma_{\varepsilon h} \in V_h, \\ \forall v_h \in V_h, \ \langle \rho\sigma_{\varepsilon h}, \rho v_h \rangle + \varepsilon(\sigma_{\varepsilon h}, v_h)_{m,\Omega'} = \langle \beta, \rho v_h \rangle. \end{cases} \tag{4.8}$$

**Proof** – Taking into account that $V_h$, endowed with the norm $[\![ \cdot ]\!]_{m,\Omega'}$, is a Hilbert space, the proof is similar to that of Theorem V–3.1. □

When (3.3) is verified, (4.7) (resp. (4.8)) is clearly a discretization of (3.11) (resp. (3.12)).

**Remark 4.4** – Let us write $\sigma_{\varepsilon h} = \sum_{j=1}^{M} \alpha_j w_j$, where $w_1, \dots, w_M$ denote the basis functions of $V_h$, and let us introduce the matrices

$$\mathcal{A} = \big(\phi_i(w_j)\big)_{1 \le i \le N,\, 1 \le j \le M}$$

and

$$\mathcal{R} = \big((w_j, w_i)_{m,\Omega'}\big)_{1 \le i,j \le M}.$$

Then, reasoning as in Remark VI–3.1, we see that (4.8) is equivalent to the problem: find $\alpha = (\alpha_j)_{1\le j\le M}$ solution of

$$\begin{cases} \alpha \in \mathbb{R}^M, \\ (\mathcal{A}^T\mathcal{A} + \varepsilon\mathcal{R})\alpha = \mathcal{A}^T\beta, \end{cases}$$

where $\mathcal{A}^T$ denotes the transpose matrix of $\mathcal{A}$. □

**Remark 4.5** – In real problems, the number $N$ of data points may be large. That is the reason why, in that case, one specifically uses the discrete smoothing $D^m$-splines, which can be defined in a space $V_h$ of dimension $M$ much smaller than $N$ (cf. Remark VI–3.7). □

**Remark 4.6** – When $n = 2$, under condition (4.3), the $V_h$-discrete $D^m$-splines are suitable tools for the representation of vertical faults in Geophysics (cf. the numerical and graphical results of Section 8). We recall that functions in the space $H^m(\Omega') \cap C^k_F(\Omega')$, and so in $V_h$, may present a jump discontinuity at any point of $F$ (cf. Remark 2.1).

To represent surfaces which present (direct) oblique faults, we must change in two directions the assumptions made at the beginning of this section. On the one hand, we must replace (4.3) by the following condition:

$$V_h \text{ is a finite-dimensional subspace of } H^m(\Omega') \cap C^0(\overline{\Omega}) \cap C^k_F(\Omega'),$$

in order to ensure that the functions in $V_h$ are continuous functions on $\overline{\Omega}$ whose first order partial derivatives may admit discontinuities on $F$ (cf. Remark 2.2). On the other hand, the triangulation $\mathcal{T}_h$ on which $V_h$ is constructed must also contain curved elements, so that the fault line $F$ is contained in the union of the curved sides of those elements.

The need of curved finite elements is justified by the following result, established by C. Tarrou (cf. [137, Part II, Lemma 2]):

*Let $\Omega$, $\Omega_1$ and $\Omega_2$ be three domains in $\mathbb{R}^2$ such that $\overline{\Omega} = \overline{\Omega}_1 \cup \overline{\Omega}_2$ and $F = \overline{\Omega}_1 \cap \overline{\Omega}_2$ is a polygonal curve with vertices $b_1, \dots, b_\nu$, supposedly numbered in a monotone way on $F$. Let $f$ be a function in $C^0(\overline{\Omega})$ and assume that, for $j = 1, 2$, $f_j \in C^1(\overline{\Omega}_j)$, where $f_j = f|_{\Omega_j}$. Then, at every vertex $b_i$, with $2 \le i \le \nu - 1$, such that $b_{i-1}$, $b_i$ and $b_{i+1}$ are not collinear, we have*

$$\nabla f_1(b_i) = \nabla f_2(b_i),$$

*where, for $i = 2, \dots, \nu - 1$ and $j = 1, 2$, $\nabla f_j(b_i)$ denotes the value of the gradient vector of $f_j$ at the point $b_i$.*

As a consequence of this result, the gradient vectors of continuous functions which are at least of class $C^1$ on both sides of a polygonal curve $F$ cannot be discontinuous along $F$, as would be required when modelling oblique faults in Geophysics. This fact suggests, as pointed out by C. Tarrou, that the fault lines associated with oblique faults should be represented by means of, at least, $G^1$-continuous curves (i.e. curves for which the unit tangent vector is defined and continuous at any point). To this end, C. Tarrou uses composite quadratic Bézier curves. Then, to construct the finite element space $V_h$, he develops a triangular finite element with only one curved side, based on a Clough-Tocher split (cf. [137]). Of course, one could use, instead, isoparametric Hermite finite elements (cf. P. G. Ciarlet [45]). Anyway, the analysis of the convergence that we shall present in Section 5 is no longer valid and has to be re-elaborated. In particular, it is necessary to have an extension of Clément's result (see later the hypothesis (5.5)), like that obtained by C. Bernardi for the isoparametric case (cf. [29]). We shall not make this study in our book. □

## 5. *GLOBAL CONVERGENCE*

This section is devoted to the study of the convergence of the $V_h$-discrete smoothing $D^m$-splines, in view of its major interest. We shall not consider the case of (continuous) $D^m$-splines over $\Omega'$ or that of $V_h$-discrete interpolating $D^m$-splines, for which we could obtain convergence results similar to those of Theorems V–4.1, V–4.2 and VI–2.2.

Hereafter, we shall keep the notations $\Omega$, $F$, $\Omega'$ and $k$ of Section 4 and we shall suppose we are given

- two bounded sets $\mathbb{D}$ and $\mathbb{H}$ in $(0, +\infty)$ such that $0 \in \overline{\mathbb{D}} \cap \overline{\mathbb{H}}$,
- an integer $\mu$ equal to 0 or 1,
- for any $d \in \mathbb{D}$, a set $A^d$ of points in $\overline{\Omega}$ and a set $\Sigma^d$ of $N = N(d)$ linear forms on $C_F^\mu(\Omega')$ of types (3.1) or (3.2), with $|\alpha| \le \mu$, associated with the points of $A^d$, verifying
$$\sup_{x \in \Omega'} \delta(x, A_L^d) = d, \tag{5.1}$$
where $\delta(\,\cdot\,,\,\cdot\,)$ denotes the Euclidean distance in $\mathbb{R}^n$ and $A_L^d$ is the set of Lagrange nodes of $\Sigma^d$ (i.e. the set of points in $A^d$ for which there exists an element in $\Sigma^d$ of types (3.1) or (3.2) with $|\alpha| = 0$),
- for any $d \in \mathbb{D}$, the operator $\rho^d \in \mathcal{L}\big(C_F^\mu(\Omega'), \mathbb{R}^N\big)$ defined by
$$\rho^d v = \big(\phi(v)\big)_{\phi \in \Sigma^d},$$
- for any $h \in \mathbb{H}$, a triangulation $\mathcal{T}_h$ of $\overline{\Omega}$ by means of $n$-simplices $K$ with diameters $h_K \le h$, verifying (4.1) and (4.2),
- for any $h \in \mathbb{H}$, a finite element space $V_h$, constructed on $\mathcal{T}_h$, satisfying (4.3), where $m$ is now an integer such that
$$m > \frac{n}{2}. \tag{5.2}$$

Moreover, we suppose that the families $(A^d)_{d \in \mathbb{D}}$ and $(\mathcal{T}_h)_{h \in \mathbb{H}}$ are linked by the relation
$$\exists C > 0,\ \forall (d,h) \in \mathbb{D} \times \mathbb{H},\ \forall K \in \mathcal{T}_h,\ \frac{\operatorname{card}(A^d \cap K)}{\operatorname{meas} K} \le C d^{-n} \tag{5.3}$$

(cf. hypothesis (VI–3.7) and also the comment following this assumption).

Let $m'$ be an integer such that

$$m' > \max\{\frac{n}{2} + \mu, m - 1\}. \tag{5.4}$$

We suppose that the family $(V_h)_{h\in\mathbb{H}}$ and the integer $m'$ are such that the following result is verified:

> There exists a constant $C > 0$ and, for any $h \in \mathbb{H}$, a linear operator $\widetilde{\Pi}_h : L^2(\Omega') \to V_h$ such that, for any $l = 0, \dots, m'$, one has
>
> $$\forall v \in H^{m'}(\Omega'), \ \left(\sum_{K\in\mathcal{T}_h} |v - \widetilde{\Pi}_h v|^2_{l,K}\right)^{1/2} \leq C h^{m'-l} |v|_{m',\Omega'}.$$
>
> Moreover,
>
> $$\forall v \in H^{m'}(\Omega'), \ \lim_{h\to 0}\left(\sum_{K\in\mathcal{T}_h} |v - \widetilde{\Pi}_h v|^2_{m',K}\right)^{1/2} = 0. \tag{5.5}$$

**Remark 5.1** – Let us observe that (5.5) is, essentially, P. Clément's result (VI–1.7), replacing $q$, $\widetilde{\Omega}$, $\widetilde{\mathcal{T}}_h$ and $\widetilde{V}_h$ by $m'$, $\Omega'$, $\mathcal{T}_h$ and $V_h$, respectively. Therefore, this result implicitly assumes that the generic finite element $(K, P_K, \Sigma_K)$ of the family $(V_h)_{h\in\mathbb{H}}$ verifies the condition $P_{m'}(K) \subset P_K \subset H^{m'}(K)$, and also, as pointed out in Remark VI–1.1, that the family $(\mathcal{T}_h)_{h\in\mathbb{H}}$ is regular and that the basis functions of any space of the family $(V_h)_{h\in\mathbb{H}}$ satisfy a uniformity property analogous to (VI–1.9). □

***Lemma* 5.1** – *Suppose that hypotheses* (5.3), (5.4) *and* (5.5) *are verified. Then, there exists* $C > 0$ *such that*

$$\forall v \in H^{m'}(\Omega'), \ \forall d \in \mathbb{D}, \ \forall h \in \mathbb{H}, \ \langle \rho^d(\widetilde{\Pi}_h v - v)\rangle^2 \leq C\frac{h^{2(m'-\mu)}}{d^n}|v|^2_{m',\Omega'}.$$

**Proof** – We adapt the proof of Lemma VI–3.1, taking account of the fact that here the definition of $\rho^d$ involves derivatives of order $\leq \mu$. □

We also need the following result.

***Proposition* 5.1** – *Let* $\{R_1, \dots, R_I\}$ *be any family of open sets that represents* $F$ *in* $\Omega$. *Let* $B_0 = \{b_{01}, \dots, b_{0\widetilde{\mathfrak{M}}}\}$ *be a* $\widetilde{P}_{m-1}(\Omega')$*-unisolvent subset of points of* $R = \bigcup_{i=1}^{I} R_i$. *For any* $r > 0$, *we denote by* $\mathcal{B}_r$ *the family of all subsets* $B = \{b_1, \dots, b_{\widetilde{\mathfrak{M}}}\}$ *of points of* $\Omega'$ *which satisfy the condition*

$$\forall j = 1, \dots, \widetilde{\mathfrak{M}}, \ |b_j - b_{0j}| \leq r.$$

*Then, there exists $r_0 > 0$ such that the mapping $[\![ \cdot ]\!]_{B,m,\Omega'}$, defined for any subset $B = \{b_1, \ldots, b_{\widetilde{\mathfrak{M}}}\}$ of points of $\Omega'$ by*

$$\forall v \in H^m(\Omega'),\ [\![v]\!]_{B,m,\Omega'} = \left( \sum_{j=1}^{\widetilde{\mathfrak{M}}} |v(b_j)|^2 + |v|^2_{m,\Omega'} \right)^{1/2},$$

*is, for any $B \in \mathcal{B}_{r_0}$ a norm on $H^m(\Omega')$, uniformly equivalent over $\mathcal{B}_{r_0}$ to the norm $\| \cdot \|_{m,\Omega'}$.*

**Proof** – We adapt the proof of Proposition I–2.3, replacing $X^{m,s}$ (resp. $H^{m+s}(\Omega)$) by $H^m(\Omega')$ and using, in particular, Theorem 2.2. □

Finally, we suppose, as in Subsection VI–3.2, that $\varepsilon$ and $h$ are functions of $d$ that, in this case, verify the relation

$$\frac{h^{2(m'-\mu)}}{d^n \varepsilon} = o(1),\ d \to 0, \tag{5.6}$$

and also (III–3.1), i.e.

$$\varepsilon = o(d^{-n}),\ d \to 0.$$

It follows from (III–3.1), (5.4) and (5.6) that $d \to 0$ implies $h \to 0$. From now on, to simplify the notations, we shall write $\varepsilon$ and $h$ instead of $\varepsilon(d)$ and $h(d)$.

We get the following convergence theorem for discrete smoothing $D^m$-splines.

**Theorem 5.1** – *Suppose that hypotheses* (III–3.1), (4.1), (4.2), (4.3), (5.1), (5.2), (5.3), (5.4), (5.5) *and* (5.6) *are verified. Let $f$ be a given function in $H^{m'}(\Omega')$. For any $d \in \mathbb{D}$, we denote by $\sigma^d_{\varepsilon h}$ the $V_h$-discrete smoothing $D^m$-spline relative to $\rho^d$, $\rho^d f$ and $\varepsilon$. Then*

$$\lim_{d \to 0} \|\sigma^d_{\varepsilon h} - f\|_{m,\Omega'} = 0.$$

**Proof** – We adapt the proof of Theorem VI–3.2, using, in particular, Theorems 2.2 and 2.3, Lemma 5.1 and Proposition 5.1 (cf. R. Arcangéli, R. Manzanilla and J. J. Torrens [17]). □

**Remark 5.2** – Even when $\mu = 1$ (i.e. even when Hermite conditions are involved), *the convergence requires only the condition $m > n/2$*, and not the condition $m > n/2 + 1$, which is implied by the inclusion $H^m(\Omega') \subset C^1_F(\Omega')$. Theorem 5.1 actually shows that it suffices that $V_h \subset C^1_F(\Omega')$. This result is very important in practice for approximating surfaces from Lagrange or first order Hermite data. It justifies the fact that we can use finite element spaces whose generic finite element is precisely of class $C^1$. □

**Remark 5.3** – To study the discrete $D^m$-splines in Sections 4 and 5, we have supposed that the open subset $\Omega$ is polyhedral. There is no theoretical difficulty in treating the general case, i.e. $\Omega$ being an open set with a Lipschitz-continuous boundary (cf. Preliminaries), by reasoning as in Chapter VI. We achieve the same convergence result. □

# 6. LOCAL CONVERGENCE

Let $\Omega$ be, for simplicity, a polyhedral subset of $\mathbb{R}^n$ and let $F \subset \overline{\Omega}$ be a discontinuity set (in the sense of Subsection 2.1). Likewise, let $\Omega' = \Omega \setminus \overline{F}$.

Assuming that $F$ is polyhedral, it is a consequence of Theorem 5.1 that the discrete smoothing $D^m$-splines introduced in Section 4 can be useful tools to fit a non-regular function $f$ over $\Omega'$ from Lagrange and, if available, Hermite data. Obviously, these splines provide an approximation of $f$ on the whole set $\Omega'$. In some situations, however, it may be sufficient (or even necessary) to fit $f$ only on a set $\omega_h \subset \Omega$ such that $\overline{\omega}_h \cap \overline{F} = \emptyset$ (the sense of the index $h$ will be clear later). Since $f$ is regular on $\omega_h$, one can approximate $f$ by "usual" discrete smoothing $D^m$-splines (like those considered in Chapter VI or, for $n = 2$, in Chapter VIII). Of course, if one really needs a global approximation of $f$ on $\Omega'$, one can extend the "local" spline defined on $\omega_h$ to a function defined on $\Omega'$, as we shall see later in Section 7.

The main advantage of this approach is that one does not need to worry about the geometry of $F$, since all the computations take place on $\omega_h$. The particular form of $F$ is only relevant when an extension to $\Omega'$ is required. In this way, we can obtain convergence results that are valid for arbitrary geometries of the discontinuity set $F$. These results are, in fact, the main object of this section.

Suppose we are given

- three nonnegative integers $m$, $m'$ and $\mu$ such that $m > n/2$, $\mu = 0$ or $1$ and $m' > \max\{n/2 + \mu, m-1\}$,
- two bounded sets $\mathbb{D}$ and $\mathbb{H}$ in $(0, +\infty)$ such that $0 \in \overline{\mathbb{D}} \cap \overline{\mathbb{H}}$,
- for any $d \in \mathbb{D}$, a set $A^d$ of points in $\overline{\Omega}$ and a set $\Sigma^d$ of $N = N(d)$ linear forms on $C^{\mu}_{F}(\Omega')$ of types (3.1) or (3.2), with $|\alpha| \le \mu$, associated with the points of $A^d$, that verify (5.1),
- for any $d \in \mathbb{D}$, the operator $\rho^d \in \mathcal{L}\big(C^{\mu}_{F}(\Omega'), \mathbb{R}^N\big)$ defined by

$$\rho^d v = \big(\phi(v)\big)_{\phi \in \Sigma^d},$$

- for any $h \in \mathbb{H}$, a triangulation $\mathcal{T}_h$ of $\overline{\Omega}$ by means of $n$-simplices $K$ with diameters $h_K \le h$,
- a family $(\omega_h)_{h \in \mathbb{H}}$ of open subsets of $\Omega$ with a Lipschitz-continuous boundary (cf. Preliminaries) such that

$$\text{for any } h \in \mathbb{H},\ \overline{\omega}_h \text{ is a union of elements of } \mathcal{T}_h, \tag{6.1}$$

$$\forall h \in \mathbb{H},\ \overline{F} \cap \overline{\omega}_h = \emptyset, \tag{6.2}$$

$$\lim_{h \to 0} \sup_{x \in \Omega \setminus \omega_h} \delta(x, \overline{F}) = 0, \tag{6.3}$$

• a family $(\widehat{V}_h)_{h\in\mathbb{H}}$ of finite element spaces such that, for any $h \in \mathbb{H}$, $\widehat{V}_h$ is constructed on the triangulation $\widehat{\mathcal{T}}_h = \{ K \in \mathcal{T}_h \mid K \subset \overline{\omega}_h \}$ of $\overline{\omega}_h$ and that

$$\widehat{V}_h \text{ is a finite-dimensional subspace of } H^m(\omega_h). \tag{6.4}$$

We also assume that the following result holds:

There exists a constant $C > 0$ and, for any $h \in \mathbb{H}$, a linear operator $\widetilde{\Pi}_h : L^2(\omega_h) \to \widehat{V}_h$ such that, for any $l = 0, \dots, m'$, one has

$$\forall v \in H^{m'}(\omega_h),\ \Big( \sum_{K\in\widehat{\mathcal{T}}_h} |v - \widetilde{\Pi}_h v|^2_{l,K} \Big)^{1/2} \le C h^{m'-l} |v|_{m',\omega_h}.$$

Moreover,

$$\forall v \in H^{m'}(\omega_h),\ \lim_{h\to 0} \Big( \sum_{K\in\widehat{\mathcal{T}}_h} |v - \widetilde{\Pi}_h v|^2_{m',K} \Big)^{1/2} = 0. \tag{6.5}$$

**Remark 6.1** – Like the relation (5.5), the preceding hypothesis is another variant of P. Clément's result (VI–1.7). In this case, we have replaced $q$, $\widetilde{\Omega}$, $\widetilde{\mathcal{T}}_h$ and $\widetilde{V}_h$ by $m'$, $\omega_h$, $\widehat{\mathcal{T}}_h$ and $\widehat{V}_h$, respectively. Let us observe that $\widetilde{\Pi}_h$ is now defined on a space that also depends on $h$. Of course, it is implicitly assumed that the generic finite element $(K, P_K, \Sigma_K)$ of the family $(\widehat{V}_h)_{h\in\mathbb{H}}$ verifies the condition $P_{m'}(K) \subset P_K \subset H^{m'}(K)$, that the family $(\widehat{\mathcal{T}}_h)_{h\in\mathbb{H}}$ is regular and that the basis functions of any space of the family $(\widehat{V}_h)_{h\in\mathbb{H}}$ satisfy a uniformity property similar to (VI–1.9). □

**Remark 6.2** – We have implicitly supposed that, for any $h \in \mathbb{H}$, $\omega_h$ is connected. However, it is readily seen that all the results in this section are still valid if, for any $h \in \mathbb{H}$, any two connected components of $\omega_h$ have disjoint closures and every connected component of $\omega_h$ is an open set with a Lipschitz-continuous boundary (cf. [102]). □

For any $h \in \mathbb{H}$, it follows from (5.1) that, for any $d \in \mathbb{D}$ sufficiently small, the set $A^d_L \cap \overline{\omega}_h$ contains a $P_{m-1}$-unisolvent subset (we recall that $A^d_L$ is the set of Lagrange nodes of $\Sigma^d$). Hereafter, for simplicity, we shall assume that this fact is verified for any $(h, d) \in \mathbb{H} \times \mathbb{D}$. Likewise, for any $(h, d) \in \mathbb{H} \times \mathbb{D}$, let $\Sigma^d_h$ be the set of linear forms in $\Sigma^d$ which are associated with the points in $A^d \cap \overline{\omega}_h$, and let us denote by $\rho^d_h \in \mathcal{L}(H^m(\omega_h), \mathbb{R}^{N^d_h})$, with $N^d_h = \operatorname{card} \Sigma^d_h$, the operator defined by

$$\rho^d_h v = \big(\phi(v)\big)_{\phi\in\Sigma^d_h},$$

whose continuity follows from (3) in Preliminaries.

For any $(h, d) \in \mathbb{H} \times \mathbb{D}$, we denote by $\langle \cdot \rangle$ (respectively, by $\langle \cdot , \cdot \rangle$) the Euclidean norm (respectively, the Euclidean scalar product) in $\mathbb{R}^{N^d_h}$. We also suppose that the families $(A^d)_{d\in\mathbb{D}}$ and $(\mathcal{T}_h)_{h\in\mathbb{H}}$ satisfy the relation (5.3). In these conditions, we have the following auxiliary results.

***Lemma* 6.1** – *Suppose that* (5.3), (6.1), (6.4) *and* (6.5) *hold. Then, there exists* $C > 0$ *such that*

$$\forall (h,d) \in \mathbb{H} \times \mathbb{D},\ \forall v \in H^{m'}(\omega_h),\ \langle \rho_h^d(\widetilde{\Pi}_h v - v)\rangle^2 \leq C\,\frac{h^{2(m'-\mu)}}{d^n}|v|^2_{m',\omega_h}.$$

**Proof** – Cf. the proof of Lemma 5.1. □

***Lemma* 6.2** – *Suppose that* (6.2) *and* (6.3) *hold. Then,*

*(i) For any open set* $\omega \subset \Omega$ *with* $\overline{\omega} \cap \overline{F} = \emptyset$, *there exists* $h^* \in \mathbb{H}$ *such that*

$$\forall h \in \mathbb{H},\ h \leq h^*,\ \omega \subset \omega_h.$$

*(ii)* $\lim_{h \to 0} \text{meas}(\overline{\Omega \setminus \omega_h}) = 0.$

**Proof** –

1) Let $\omega$ be an open subset of $\Omega$ such that $\overline{\omega} \cap \overline{F} = \emptyset$. Since $\overline{\omega}$ and $\overline{F}$ are closed sets, it is clear that $\delta(\overline{\omega}, \overline{F}) > 0$. In addition, for any $h \in \mathbb{H}$, we have

$$\delta(\overline{\omega}, \overline{F}) \leq \delta(\overline{\omega}, \Omega \setminus \omega_h) + \sup_{x \in \Omega \setminus \omega_h} \delta(x, \overline{F}).$$

By (6.3), there exists $h^* \in \mathbb{H}$ such that

$$\forall h \in \mathbb{H},\ h \leq h^*,\ \sup_{x \in \Omega \setminus \omega_h} \delta(x, \overline{F}) \leq \frac{1}{2}\,\delta(\overline{\omega}, \overline{F}).$$

Thus,

$$\forall h \in \mathbb{H},\ h \leq h^*,\ \delta(\overline{\omega}, \Omega \setminus \omega_h) > 0,$$

which implies that

$$\forall h \in \mathbb{H},\ h \leq h^*,\ \overline{\omega} \cap (\Omega \setminus \omega_h) = \emptyset.$$

Hence,

$$\forall h \in \mathbb{H},\ h \leq h^*,\ \omega \subset \omega_h.$$

Therefore, point (i) is satisfied.

2) For any $h \in \mathbb{H}$, let $\mathcal{F}_h = \overline{\Omega \setminus \omega_h}$. Let us first point out that, from (i) and (6.2), we have

$$\forall h_0 \in \mathbb{H},\ \exists h^* \in \mathbb{H},\ \forall h \in \mathbb{H},\ h \leq h^*,\ \mathcal{F}_h \subset \mathcal{F}_{h_0}. \tag{6.6}$$

Now, we shall prove that (ii) holds, arguing by contradiction. Suppose that

$$\lim_{h \to 0} \text{meas}(\mathcal{F}_h) \neq 0.$$

Then, there exists a real number $\beta > 0$ and a sequence $(\mathcal{F}_{h_l})_{l \in \mathbb{N}}$, with $\lim_{l \to +\infty} h_l = 0$, extracted from the family $(\mathcal{F}_h)_{h \in \mathbb{H}}$, such that

$$\forall l \in \mathbb{N},\ \text{meas}(\mathcal{F}_{h_l}) > \beta. \tag{6.7}$$

By (6.6), it can be assumed without loss of generality that

$$\mathcal{F}_{h_1} \supset \mathcal{F}_{h_2} \supset \cdots \supset \mathcal{F}_{h_l} \supset \mathcal{F}_{h_{l+1}} \supset \cdots .$$

Since $\operatorname{meas}(\mathcal{F}_{h_1}) < +\infty$, it follows that

$$\operatorname{meas}\Big(\bigcap_{l\in\mathbb{N}} \mathcal{F}_{h_l}\Big) = \lim_{l\to+\infty} \operatorname{meas}(\mathcal{F}_{h_l}). \tag{6.8}$$

Let us show that

$$\bigcap_{l\in\mathbb{N}} \mathcal{F}_{h_l} = \overline{F}. \tag{6.9}$$

It follows from (6.2) that, for any $l \in \mathbb{N}$, $\overline{F} \subset \mathcal{F}_{h_l}$. Hence,

$$\overline{F} \subset \bigcap_{l\in\mathbb{N}} \mathcal{F}_{h_l}.$$

Reciprocally, if $x \in \bigcap_{l\in\mathbb{N}} \mathcal{F}_{h_l}$, it is obvious that

$$\forall l \in \mathbb{N},\ \delta(x,\overline{F}) \leq \sup_{y\in\mathcal{F}_{h_l}} \delta(y,\overline{F}).$$

We deduce from (6.3) that $\delta(x,\overline{F}) = 0$ and hence $x \in \overline{F}$. Therefore,

$$\bigcap_{l\in\mathbb{N}} \mathcal{F}_{h_l} \subset \overline{F}.$$

Since $\operatorname{meas}(\overline{F}) = 0$, we conclude from (6.8) and (6.9) that $\lim_{l\to+\infty} \operatorname{meas}(\mathcal{F}_{h_l}) = 0$, in contradiction with (6.7). Thus, the result holds. □

**Lemma 6.3** – *Suppose that* (5.1) *holds. Let $\omega$ be an open subset of $\Omega$ with a Lipschitz-continuous boundary, and let $(g^d)_{d\in\mathbb{D}}$ be a family of $H^m(\omega)$ such that*

$$\sum_{a\in A_L^d\cap\overline{\omega}} |g^d(a)|^2 = o(d^{-n}),\ d \to 0. \tag{6.10}$$

*Then, for any $r > 0$ and for any $b \in \omega$, there exists a family $(b^d)_{d\in\mathbb{D}} \subset \overline{\omega} \cap \overline{B}(b,r)$ such that*

$$g^d(b^d) = o(1),\ d \to 0. \tag{6.11}$$

**Proof** – Let $r > 0$ and $b \in \omega$. Since $\omega \cap B(b,r)$ is an open set, there exists $\tilde{r} > 0$ such that $B(b,\tilde{r}) \subset \omega \cap B(b,r)$. Let $\omega_0 = B(b,\tilde{r})$.

For any $d \in \mathbb{D}$, with $d < \tilde{r}$, it follows from (5.1) that $A_L^d \cap \overline{\omega}_0$ is a nonempty finite set. Hence, there exists at least one point $b^d \in A_L^d \cap \overline{\omega}_0$ such that

$$|g^d(b^d)| = \min_{a\in A_L^d\cap\overline{\omega}_0} |g^d(a)|.$$

Likewise, for any $d \in \mathbb{D}$, with $d \geq \tilde{r}$, let $b^d$ be any element of $\overline{\omega}_0$. Obviously, the family $(b^d)_{d \in \mathbb{D}}$ is included in $\overline{\omega} \cap \overline{B}(b, r)$. Let us see that (6.11) also holds. It follows from (5.1) that

$$\forall d \in \mathbb{D},\ d < \tilde{r},\ \overline{B}(b, \tilde{r} - d) \subset \bigcup_{a \in A_L^d \cap \overline{\omega}_0} \overline{B}(a, d).$$

Taking the corresponding Lebesgue measures and simplifying, we get

$$\forall d \in \mathbb{D},\ d < \tilde{r},\ (\tilde{r} - d)^n \leq d^n \operatorname{card}(A_L^d \cap \overline{\omega}_0).$$

Let $d_0 \in (0, \tilde{r})$. We deduce that

$$\forall d \in \mathbb{D},\ d \leq d_0,\ (\tilde{r} - d_0)^n \leq d^n \operatorname{card}(A_L^d \cap \overline{\omega}_0).$$

Therefore,

$$\forall d \in \mathbb{D},\ d \leq d_0,\ |g^d(b^d)|^2 = \min_{a \in A_L^d \cap \overline{\omega}_0} |g^d(a)|^2 \leq \frac{1}{\operatorname{card}(A_L^d \cap \overline{\omega}_0)} \sum_{a \in A_L^d \cap \overline{\omega}_0} |g^d(a)|^2$$

$$\leq \frac{1}{(\tilde{r} - d_0)^n} d^n \sum_{a \in A_L^d \cap \overline{\omega}} |g^d(a)|^2.$$

This relation and (6.10) finally imply (6.11). □

Let $f$ be a given function in $H^{m'}(\Omega')$. For any $d \in \mathbb{D}$, for any $h \in \mathbb{H}$ and for any $\varepsilon > 0$, we denote by $\hat{\sigma}_{\varepsilon h}^d$ the $\widehat{V}_h$*-discrete smoothing* $D^m$*-spline relative to* $\rho_h^d$, $\rho_h^d f$ *and* $\varepsilon$, which is the unique solution of the following problem: find $\hat{\sigma}_{\varepsilon h}^d$ such that

$$\begin{cases} \hat{\sigma}_{\varepsilon h}^d \in \widehat{V}_h, \\ \forall v_h \in \widehat{V}_h,\ J_{\varepsilon h}^d(\hat{\sigma}_{\varepsilon h}^d) \leq J_{\varepsilon h}^d(v_h), \end{cases} \tag{6.12}$$

where

$$J_{\varepsilon h}^d(v_h) = \langle \rho_h^d v_h - \rho_h^d f \rangle^2 + \varepsilon |v_h|_{m,\omega_h}^2.$$

**Remark 6.3** – The existence and uniqueness of a solution of problem (6.12) is a consequence of the $P_{m-1}$-unisolvency of a subset of $A_L^d \cap \overline{\omega}_h$. It suffices to reason as in the proof of Theorem V–3.1, with $\widehat{V}_h$ instead of $H^m(\Omega)$. □

As in Section 5, we suppose that $\varepsilon$ and $h$ are functions of $d$ that verify the relations (III–3.1) and (5.6). We recall that these hypotheses imply that $h \to 0$ as $d \to 0$. As usual, we shall write $\varepsilon$ and $h$ instead of $\varepsilon(d)$ and $h(d)$.

For any $v \in H^m(\Omega')$ (respectively, $v \in H^m(\omega_h)$) and for any $E \subset \Omega'$ (respectively, $E \subset \omega_h$), we shall write $v$ instead of $v|_E$.

***Theorem* 6.1** – *Let $\omega$ be a nonempty open subset of $\Omega$ such that $\overline{\omega} \cap \overline{F} = \emptyset$. Suppose that hypotheses* (III–3.1), (5.1), (5.3), (5.6), (6.1), (6.2), (6.3), (6.4) *and* (6.5) *hold. Then, we have*

$$\lim_{d \to 0} \|\hat{\sigma}_{\varepsilon h}^d - f\|_{m,\omega} = 0. \tag{6.13}$$

**Proof –**

1) We shall first assume that $\omega$ is an open set with a Lipschitz-continuous boundary (cf. Preliminaries).

By (6.2) and point (i) of Lemma 6.2, there exists $h^* \in \mathbb{H}$ such that

$$\forall h \in \mathbb{H},\ h \leq h^*,\ \omega \subset \omega_h \subset \Omega'. \tag{6.14}$$

Hence, for any $d \in \mathbb{D}$ sufficiently small, $\hat{\sigma}^d_{\varepsilon h} - f$ is a function defined on $\omega$. Let us show that there exist two positive real numbers $C$ and $d_0$ such that

$$\forall d \in \mathbb{D},\ d \leq d_0,\ \|\hat{\sigma}^d_{\varepsilon h}\|_{m,\omega} \leq C. \tag{6.15}$$

From (6.5) and point (ii) of Lemma 6.2, it is clear that

$$|\widetilde{\Pi}_h f|_{m,\omega_h} = |f|_{m,\Omega'} + o(1),\ h \to 0. \tag{6.16}$$

Likewise, from (6.12) with $v_h = \widetilde{\Pi}_h f$, we have

$$\forall d \in \mathbb{D},\ \langle \rho^d_h(\hat{\sigma}^d_{\varepsilon h} - f)\rangle^2 \leq \varepsilon \left( \frac{1}{\varepsilon} \langle \rho^d_h(\widetilde{\Pi}_h f - f)\rangle^2 + |\widetilde{\Pi}_h f|^2_{m,\omega_h} \right)$$

and

$$\forall d \in \mathbb{D},\ |\hat{\sigma}^d_{\varepsilon h}|^2_{m,\omega_h} \leq \frac{1}{\varepsilon} \langle \rho^d_h(\widetilde{\Pi}_h f - f)\rangle^2 + |\widetilde{\Pi}_h f|^2_{m,\omega_h}.$$

From Lemma 6.1, (5.6) and (6.16), we deduce that

$$\langle \rho^d_h(\hat{\sigma}^d_{\varepsilon h} - f)\rangle^2 = O(\varepsilon),\ d \to 0, \tag{6.17}$$

and

$$|\hat{\sigma}^d_{\varepsilon h}|_{m,\omega_h} \leq |f|_{m,\Omega'} + o(1),\ d \to 0. \tag{6.18}$$

This last inequality, together with (6.14), implies that

$$|\hat{\sigma}^d_{\varepsilon h}|_{m,\omega} \leq |f|_{m,\Omega'} + o(1),\ d \to 0. \tag{6.19}$$

Now, let $B_0 = \{b_{01}, \ldots, b_{0\mathfrak{M}}\}$ be a $P_{m-1}$-unisolvent subset of $\omega$ and let $r_0$ be the constant provided by Proposition V–1.2 (using $\omega$ instead of $\Omega$). From (III–3.1) and (6.17), it follows that

$$\sum_{a \in A^d_L \cap \overline{\omega}} |(\hat{\sigma}^d_{\varepsilon h} - f)(a)|^2 = o(d^{-n}),\ d \to 0. \tag{6.20}$$

Applying Lemma 6.3, we deduce that, for $j = 1, \ldots, \mathfrak{M}$, there exists a family $(b^d_j)_{d \in \mathbb{D}} \subset \overline{\omega} \cap \overline{B}(b_{0j}, r_0)$ such that

$$(\hat{\sigma}^d_{\varepsilon h} - f)(b^d_j) = o(1),\ d \to 0.$$

For all $d \in \mathbb{D}$, let $B^d = \{b_1^d, \ldots, b_{\mathfrak{M}}^d\}$. By (6.19), there exists $C^* > 0$ such that, for any $d \in \mathbb{D}$ sufficiently small,

$$[\![\hat{\sigma}_{\varepsilon h}^d - f]\!]_{B^d,m,\omega} \leq C^*.$$

Therefore, by Proposition V–1.2, (6.15) holds.

2) Since the family $(\hat{\sigma}_{\varepsilon h}^d)_{\substack{d\in\mathbb{D}\\ d\leq d_0}}$ is bounded in $H^m(\omega)$, there exists a sequence $\left(\hat{\sigma}_{\varepsilon_j h_j}^{d_j}\right)_{j\in\mathbb{N}}$, extracted from the family $(\hat{\sigma}_{\varepsilon h}^d)_{\substack{d\in\mathbb{D}\\ d\leq d_0}}$, with $\varepsilon_j = \varepsilon(d_j)$ and $h_j = h(d_j)$, for any $j \in \mathbb{N}$, and

$$\lim_{j\to+\infty} d_j = \lim_{j\to+\infty} d_j^n \varepsilon_j = \lim_{j\to+\infty} \frac{h_j^{2(m'-\mu)}}{d_j^n \varepsilon_j} = 0$$

(and so $\lim_{j\to+\infty} h_j = 0$), and there also exists an element $f^* \in H^m(\omega)$ such that

$$\hat{\sigma}_{\varepsilon_j h_j}^{d_j} \to f^*, \quad \text{weakly in } H^m(\omega). \tag{6.21}$$

We shall prove that $f^* = f|_\omega$, arguing by contradiction. Suppose that $f^* \neq f|_\omega$. Hence, there exist $\beta > 0$, $b \in \omega$ and $r > 0$ such that $B(b,r) \subset \omega$ and

$$\forall x \in \overline{B}(b,r), \ |f^*(x) - f(x)| > \beta.$$

It follows from (3) in Preliminaries and (6.21) that the sequence $\left(\hat{\sigma}_{\varepsilon_j h_j}^{d_j}\right)_{j\in\mathbb{N}}$ converges uniformly on $\overline{\omega}$ to $f^*$. Therefore, there exists $j_0 \in \mathbb{N}$ such that

$$\forall j \in \mathbb{N},\ j \geq j_0,\ \forall x \in \overline{B}(b,r),\ |\hat{\sigma}_{\varepsilon_j h_j}^{d_j}(x) - f^*(x)| \leq \frac{\beta}{2},$$

and then

$$\begin{aligned}&\forall j \in \mathbb{N},\ j \geq j_0,\ \forall x \in \overline{B}(b,r),\\ &\quad |\hat{\sigma}_{\varepsilon_j h_j}^{d_j}(x) - f(x)| \geq |f^*(x) - f(x)| - |\hat{\sigma}_{\varepsilon_j h_j}^{d_j}(x) - f^*(x)| > \frac{\beta}{2}.\end{aligned} \tag{6.22}$$

Now, by (6.20) and Lemma 6.3, there exists a family $(b^d)_{d\in\mathbb{D}} \subset \overline{B}(b,r)$ such that

$$(\hat{\sigma}_{\varepsilon h}^d - f)(b^d) = o(1),\ d \to 0.$$

In particular,

$$\lim_{j\to+\infty} (\hat{\sigma}_{\varepsilon_j h_j}^{d_j} - f)(b^{d_j}) = 0,$$

in contradiction with (6.22). Therefore, $f^* = f|_\omega$.

3) Next, we shall show that

$$\lim_{j\to+\infty} \|\hat{\sigma}_{\varepsilon_j h_j}^{d_j} - f\|_{m,\omega} = 0. \tag{6.23}$$

Since $(\hat{\sigma}^{d_j}_{\varepsilon_j h_j})$ is weakly convergent to $f$ in $H^m(\omega)$ and this space is compactly imbedded into $H^{m-1}(\omega)$ (cf. (2) in Preliminaries), it follows that

$$\lim_{j\to+\infty}\|\hat{\sigma}^{d_j}_{\varepsilon_j h_j} - f\|_{m-1,\omega} = 0.$$

Hence, (6.23) holds if and only if

$$\lim_{j\to+\infty}|\hat{\sigma}^{d_j}_{\varepsilon_j h_j} - f|_{m,\omega} = 0. \tag{6.24}$$

To prove this last equality, we shall argue again by contradiction. Suppose that $\lim_{j\to+\infty}|\hat{\sigma}^{d_j}_{\varepsilon_j h_j} - f|_{m,\omega} \neq 0$. Then, there exists a real number $\beta > 0$ and a subsequence $(\hat{\sigma}^{d_{j_l}}_{\varepsilon_{j_l} h_{j_l}})_{l\in\mathbb{N}}$, extracted from $(\hat{\sigma}^{d_{j_l}}_{\varepsilon_{j_l} h_{j_l}})_{j\in\mathbb{N}}$, such that

$$\forall l \in \mathbb{N},\ |\hat{\sigma}^{d_{j_l}}_{\varepsilon_{j_l} h_{j_l}} - f|_{m,\omega} > \beta. \tag{6.25}$$

Let $\alpha$ be a real positive number such that

$$2\alpha^2 + 2\alpha|f|_{m,\Omega'} < \beta^2. \tag{6.26}$$

By point (ii) of Lemma 6.2, there exists $\tilde{h} \in \mathbb{H}$ such that

$$|f|_{m,\Omega'\setminus\omega_{\tilde{h}}} \leq \alpha. \tag{6.27}$$

Let $\omega_\alpha = \omega_{\tilde{h}}$. We observe that, by (6.2), $\omega_\alpha$ is an open subset of $\Omega$ with a Lipschitz-continuous boundary such that $\overline{\omega}_\alpha \cap \overline{F} = \emptyset$. Then, by point (i) of Lemma 6.2, there exists $h_\alpha \in \mathbb{H}$ such that

$$\forall h \in \mathbb{H},\ h \leq h_\alpha,\ \omega_\alpha \subset \omega_h.$$

Since $\lim_{j\to+\infty} h_j = 0$, there exists $l_\alpha \in \mathbb{N}$ such that, for all $l \geq l_\alpha$, $h_{j_l} \leq \min\{h^*, h_\alpha\}$, where $h^*$ is the constant given in (6.14). We can assume that $l_\alpha = 0$. Thus, for any $l \in \mathbb{N}$, $\omega_{h_{j_l}}$ contains the sets $\omega_\alpha$ and $\omega$. In particular,

$$(\hat{\sigma}^{d_{j_l}}_{\varepsilon_{j_l} h_{j_l}})_{l\in\mathbb{N}} \subset H^m(\omega_\alpha).$$

Reasoning as in points 1) and 2), we deduce that there exists a subsequence of $(\hat{\sigma}^{d_{j_l}}_{\varepsilon_{j_l} h_{j_l}})_{l\in\mathbb{N}}$ that is weakly convergent to $f$ in $H^m(\omega_\alpha)$. Since there is no danger of confusion, in order to simplify the notation, we denote such a subsequence by $(\hat{\sigma}^{d_{j_l}}_{\varepsilon_{j_l} h_{j_l}})_{l\in\mathbb{N}}$, like the sequence from which it is extracted.

Now, for all $l \in \mathbb{N}$, we have

$$\begin{aligned} |\hat{\sigma}^{d_{j_l}}_{\varepsilon_{j_l} h_{j_l}} - f|^2_{m,\omega} &\leq |\hat{\sigma}^{d_{j_l}}_{\varepsilon_{j_l} h_{j_l}} - f|^2_{m,\omega_{h_{j_l}}} \\ &\leq |\hat{\sigma}^{d_{j_l}}_{\varepsilon_{j_l} h_{j_l}}|^2_{m,\omega_{h_{j_l}}} + |f|^2_{m,\omega_{h_{j_l}}} - 2(\hat{\sigma}^{d_{j_l}}_{\varepsilon_{j_l} h_{j_l}}, f)_{m,\omega_\alpha} - 2(\hat{\sigma}^{d_{j_l}}_{\varepsilon_{j_l} h_{j_l}}, f)_{m,\omega_{h_{j_l}}\setminus\omega_\alpha}. \end{aligned}$$

Using (6.18), we obtain

$$|\hat{\sigma}^{d_{j_l}}_{\varepsilon_{j_l}h_{j_l}} - f|^2_{m,\omega} \leq 2|f|^2_{m,\Omega'} - 2\big(\hat{\sigma}^{d_{j_l}}_{\varepsilon_{j_l}h_{j_l}}, f\big)_{m,\omega_\alpha} + 2|f|_{m,\Omega'}|f|_{m,\Omega'\setminus\omega_\alpha} + o(1),\ l \to +\infty.$$

Taking limits in this inequality, we derive from (6.26) and (6.27) that

$$\limsup_{l\to+\infty}|\hat{\sigma}^{d_{j_l}}_{\varepsilon_{j_l}h_{j_l}} - f|^2_{m,\omega} \leq 2|f|^2_{m,\Omega'\setminus\omega_\alpha} + 2|f|_{m,\Omega'}|f|_{m,\Omega'\setminus\omega_\alpha} < \beta^2,$$

in contradiction with (6.25). Therefore, (6.24) (and then (6.23)) holds.

4) We are now in a position to prove that (6.13) holds (assuming, let us remember, that $\omega$ is an open set with a Lipschitz-continuous boundary). We argue once more by contradiction. Suppose that $\lim_{d\to 0}\|\hat{\sigma}^d_{\varepsilon h} - f\|_{m,\omega} \neq 0$. This means that there exists a real number $\beta > 0$ and three sequences $(d_j)_{j\in\mathbb{N}}$, $(h_j)_{j\in\mathbb{N}}$ and $(\varepsilon_j)_{j\in\mathbb{N}}$, with $\varepsilon_j = \varepsilon(d_j)$ and $h_j = h(d_j)$, for any $j \in \mathbb{N}$, such that

$$\lim_{j\to+\infty} d_j = \lim_{j\to+\infty} d_j^n\varepsilon_j = \lim_{j\to+\infty} \frac{h_j^{2(m'-\mu)}}{d_j^n\varepsilon_j} = 0$$

and

$$\forall j \in \mathbb{N},\ \|\hat{\sigma}^{d_j}_{\varepsilon_j h_j} - f\|_{m,\omega} > \beta. \tag{6.28}$$

But it follows from (6.15) that the sequence $\big(\hat{\sigma}^{d_j}_{\varepsilon_j h_j}\big)_{j\in\mathbb{N}}$ is bounded in $H^m(\omega)$. A similar argument to that of points 2) and 3) shows that there exists a subsequence of $\big(\hat{\sigma}^{d_j}_{\varepsilon_j h_j}\big)_{j\in\mathbb{N}}$ that converges to $f$ in $H^m(\omega)$. Hence, we get a contradiction with (6.28). In consequence, (6.13) holds.

5) To complete the proof, it remains to show that (6.13) is verified if $\omega$ has not a Lipschitz-continuous boundary. But, in such a case, by (6.2) and point (i) of Lemma 6.2, there exists an open subset $\omega^*$ of $\Omega$ with a Lipschitz-continuous boundary such that $\overline{\omega}^* \cap \overline{F} = \emptyset$ and $\omega \subset \omega^*$. Hence, for any $d \in \mathbb{D}$ sufficiently small, we have

$$\|\hat{\sigma}^d_{\varepsilon h} - f\|_{m,\omega} \leq \|\hat{\sigma}^d_{\varepsilon h} - f\|_{m,\omega^*}.$$

In addition, the reasoning of points 1)–4) proves that

$$\lim_{d\to 0}\|\hat{\sigma}^d_{\varepsilon h} - f\|_{m,\omega^*} = 0.$$

Therefore, (6.13) holds. □

***Corollary* 6.1** – *Suppose that hypotheses* (III–3.1), (5.1), (5.3), (5.6), (6.1), (6.2), (6.3), (6.4) *and* (6.5) *hold. Then,*

$$\lim_{d\to 0}\tilde{\sigma}^d_{\varepsilon h} = f \quad in\ H^m_{\mathrm{loc}}(\Omega'),$$

*where $\tilde{\sigma}^d_{\varepsilon h}$ is any extension of $\hat{\sigma}^d_{\varepsilon h}$ that belongs to $H^m(\Omega')$.*

**Proof** – Let $\omega$ be any nonempty open set such that $\overline{\omega} \subset \Omega'$. Since $\omega \subset \Omega$ and $\overline{\omega} \cap \overline{F} = \emptyset$, it follows from Theorem 6.1 that

$$\lim_{d\to 0}\|\tilde{\sigma}^d_{\varepsilon h} - f\|_{m,\omega} = 0. \quad \square$$

# 7. APPROXIMATION OF EXPLICIT SURFACES WITH VERTICAL FAULTS

Suppose we are given an open set $\Omega$ in $\mathbb{R}^2$ with a Lipschitz-continuous boundary and let $F$ be a discontinuity set on $\overline{\Omega}$. We write, as usual, $\Omega' = \Omega \setminus \overline{F}$. Likewise, let $f$ be an unknown non-regular function over $\Omega$. We assume that $f$ belongs to $H^{m'}(\Omega)$ for some $m' \geq 3$ and that, in fact, $f$ has a jump discontinuity at any point of $F$. In this way, the graph of $f$ is an explicit surface that, in geological language, presents a vertical fault on the fault line $F$. In this context, we consider the approximation problem introduced in Section 1, which is commonly stated as follows:

> given two finite subsets $A_L$ and $A_H$ of $\overline{\Omega} \setminus \overline{F}$, the set of values $\{ f(a) \mid a \in A_L \}$ and the set of gradient vectors $\{ \nabla f(a) \mid a \in A_H \}$, construct an approximant $\phi$ of $f$ belonging to $C^k_F(\Omega')$, with $k = 1$ or 2. (7.1)

**Remark 7.1** – It is clear that $A_L$ and $A_H$ are not necessarily disjoint sets. We always assume that $A_L$ is a nonempty set, but $A_H$ can eventually be empty, in which case there is no Hermite data. For the sake of simplicity, $A_L$ and $A_H$ do not include points of $F$ and we have not considered more general first order Hermite data sets (as already done in Section 3). Finally, if there are only Lagrange data, it suffices to assume that $m' \geq 2$. □

The theory developed in the preceding sections suggests two methods to solve this problem. We shall discuss them under different assumptions.

From now on, we shall denote by $\Sigma$ the set of linear forms on $C^1_F(\Omega')$ defined by

$$\Sigma = \{ v \mapsto v(a) \mid a \in A_L \} \cup \{ v \mapsto \partial^{(1,0)} v(a) \mid a \in A_H \} \\ \cup \{ v \mapsto \partial^{(0,1)} v(a) \mid a \in A_H \}. \tag{7.2}$$

Likewise, we shall write $N = \operatorname{card} \Sigma$ and, assuming that $\Sigma$ is ordered, we shall denote by $\rho$ the linear continuous operator from $C^1_F(\Omega')$ onto $\mathbb{R}^N$ given by $\rho v = \big(\phi(v)\big)_{\phi \in \Sigma}$.

## CASE 1: F IS A KNOWN POLYGONAL SET

We suppose that $F$ is a finite union of polygonal lines and that $F$ is part of the input data of the problem.

### Method 1

The first approximation method is directly based on the theory in Sections 4 and 5. In this method, one first makes a triangulation $\mathcal{T}_h$ of $\overline{\Omega}$, by means of triangles of diameter $\leq h$, that satisfies (4.1). Next, one fixes an integer $m$ and a generic Hermite finite element of class $C^{k'}$, with $k \leq k'$ and $2 \leq m \leq k' + 1$, and then, taking Remark 4.2 into account, one constructs on $\mathcal{T}_h$ a finite element space $V_h$ such that (4.2) and (4.3) hold. Since the triangular finite elements of class $C^2$ are difficult to implement, one typically has $k = 1$ and so one selects $k' = 1$ and $m = 2$ (this will be the case for the numerical examples in Section 8, where we have computed discrete $D^2$-splines using the Bell triangle of class $C^1$). Finally, one chooses a positive number $\varepsilon$, verifies that $\Sigma$ contains a $\widetilde{P}_{m-1}(\Omega')$-unisolvent subset and finds the $V_h$-discrete smoothing $D^m$-spline $\sigma_{\varepsilon h}$ relative to $\rho$, $\rho f$ and $\varepsilon$. The computation of $\sigma_{\varepsilon h}$ is detailed in Remark 4.4.

Theorem 5.1 justifies that one can take $\sigma_{\varepsilon h}$ as the solution $\phi$ of problem (7.1), since, under suitable hypotheses, $\sigma_{\varepsilon h} \to f$ in $H^m(\Omega')$ as $\operatorname{card} A_L \to +\infty$. Of course, the convergence hypotheses needed in that theorem should be taken into account during the process of construction of $V_h$ and the choice of $\varepsilon$. To this purpose, one can apply many of the ideas expressed in Section VIII–2, mainly based on Theorem VI–3.2, since the hypotheses of this theorem are quite similar to those of Theorem 5.1.

### *Method 2*

The theory in Section 6 suggests another way to solve problem (7.1). Now, in a first step, one proceeds as explained previously to make a triangulation $\mathcal{T}_h$ of $\overline{\Omega}$, to construct a finite element space $V_h$ and to fix the integer $m$, so that (4.1), (4.2) and (4.3) hold. Next, one finds an open subset $\omega_h$ of $\Omega$ which verifies (6.1) and (6.2). This can be easily done, for example, by taking $\omega_h$ so that $\overline{\omega}_h$ is the union of those elements in the triangulation $\mathcal{T}_h$ that do not touch $\overline{F}$, i.e. $\omega_h$ is the interior of $\bigcup_{K\in\widehat{\mathcal{T}}_h} K$, with $\widehat{\mathcal{T}}_h = \{ K \in \mathcal{T}_h \mid K \cap \overline{F} = \emptyset \}$. Of course, if $\omega_h$ were not connected or it had not a Lipschitz-continuous boundary, one should refine the triangulation $\mathcal{T}_h$ or work as follows on every connected component of $\omega_h$.

Let $\widehat{V}_h$ be the space of restrictions to $\omega_h$ of functions in $V_h$, which satisfies (6.4). Let $\Sigma_h$ be a set of linear forms defined as $\Sigma$ in (7.2), but replacing $A_L$ and $A_H$, respectively, by $A_L \cap \overline{\omega}_h$ and $A_H \cap \overline{\omega}_h$. Likewise, let $\rho_h$ be the linear continuous operator from $C^1(\overline{\omega}_h)$ onto $\mathbb{R}^{N_h}$, with $N_h = \operatorname{card} \Sigma_h$, given by $\rho_h v = \big(\phi(v)\big)_{\phi\in\Sigma_h}$. Once a positive number $\varepsilon$ is chosen and it is verified that $\Sigma_h$ contains a $P_{m-1}$-unisolvent subset (i.e. $\operatorname{Ker} \rho_h \cap P_{m-1}(\omega_h) = \{0\}$), one computes the $\widehat{V}_h$-discrete smoothing $D^m$-spline $\hat{\sigma}_{\varepsilon h}$ relative to $\rho_h$, $\rho_h f$ and $\varepsilon$. Finally, one finds an extension $\tilde{\sigma}_{\varepsilon h}$ of $\hat{\sigma}_{\varepsilon h}$ that belongs to $H^m(\Omega')$.

Under the conditions of Corollary 6.1, $\tilde{\sigma}_{\varepsilon h}$ tends to $f$ in $H^m_{\text{loc}}(\Omega')$ as $\operatorname{card} A_L \to +\infty$. Thus, we can consider that $\phi = \tilde{\sigma}_{\varepsilon h}$ is a suitable solution of problem (7.1). Let us observe, in particular, that, by (6.3), $\Omega \setminus \overline{\omega}_h$ should be a quite "small" open set in order to get good approximations of $f$.

There exist different ways to extend $\hat{\sigma}_{\varepsilon h}$ from $\omega_h$ to $\Omega'$. To this end, we propose to solve the following problem: find $\tilde{\sigma}_{\varepsilon h}$ such that

$$\begin{cases} \tilde{\sigma}_{\varepsilon h} \in \mathcal{K} = \{ v_h \in V_h \mid v_h|_{\overline{\omega}_h} = \hat{\sigma}_{\varepsilon h} \}, \\ \forall v_h \in \mathcal{K},\ |\tilde{\sigma}_{\varepsilon h}|_{m,\Omega'} \le |v_h|_{m,\Omega'}. \end{cases} \tag{7.3}$$

It is clear that this problem is equivalent to the problem: find $\tilde{\sigma}_{\varepsilon h}$ such that

$$\begin{cases} \tilde{\sigma}_{\varepsilon h} \in \mathcal{K}, \\ \forall v_h \in \mathcal{K},\ \|\tilde{\sigma}_{\varepsilon h}\|^{\omega_h}_{m,\Omega'} \le \|v_h\|^{\omega_h}_{m,\Omega'}, \end{cases} \tag{7.4}$$

where

$$\forall v_h \in V_h,\ \|v_h\|_{m,\Omega'}^{\omega_h} = \left(\int_{\omega_h} |v_h(x)|^2\,dx + |v_h|_{m,\Omega'}^2\right)^{1/2}.$$

Then, it is readily seen that $\|\cdot\|_{m,\Omega'}^{\omega_h}$ is a Hilbertian norm in $V_h$ and that $\mathcal{K}$ is nonempty, convex and closed. In consequence, problems (7.3) and (7.4) have a unique solution, which is the orthogonal projection of the zero element of $V_h$ onto the set $\mathcal{K}$. We then deduce that $\tilde{\sigma}_{\varepsilon h}$ is also characterized by

$$\begin{cases} \tilde{\sigma}_{\varepsilon h} \in \mathcal{K}, \\ \forall v_h \in \mathcal{K}_0,\ (\tilde{\sigma}_{\varepsilon h}, v_h)_{m,\Omega'} = 0, \end{cases} \tag{7.5}$$

where $\mathcal{K}_0 = \left\{ v_h \in V_h \mid v_h|_{\overline{\omega}_h} = 0 \right\}$.

Let us see how to compute $\tilde{\sigma}_{\varepsilon h}$. We denote by $\phi_1, \ldots, \phi_{N_h}$ the elements of $\Sigma_h$ and by $w_1, \ldots, w_{\widehat{M}}, w_{\widehat{M}+1}, \ldots, w_M$, the basis functions of $V_h$, numbered in such a way that $\{w_1|_{\overline{\omega}_h}, \ldots, w_{\widehat{M}}|_{\overline{\omega}_h}\}$ is a basis of $\widehat{V}_h$ (i.e. $w_1, \ldots, w_M$ are the basis functions of $V_h$ attached to the nodes that belong to $\overline{\omega}_h$). We remark that, in turn, $\{w_{\widehat{M}+1}, \ldots, w_M\}$ is a basis of $\mathcal{K}_0$. We have

$$\tilde{\sigma}_{\varepsilon h} = \sum_{j=1}^{\widehat{M}} \alpha_j w_j + \sum_{j=1}^{M-\widehat{M}} \beta_j w_{\widehat{M}+j},$$

with $\alpha_j, \beta_j \in \mathbb{R}$. It is clear that $\hat{\sigma}_{\varepsilon h} = \sum_{j=1}^{\widehat{M}} \alpha_j w_j|_{\overline{\omega}_h}$. Then, taking account of (7.5) and the variational characterization of $\hat{\sigma}_{\varepsilon h}$ (similar to (4.8)), we see that the vectors $\alpha = (\alpha_j)_{1\le j\le \widehat{M}}$ and $\beta = (\beta_j)_{1\le j\le M-\widehat{M}}$ of unknown coefficients are the solution of the linear system

$$\begin{pmatrix} \widehat{\mathcal{A}}^T\widehat{\mathcal{A}} + \varepsilon\mathcal{R}_0 & 0 \\ \mathcal{R}_1 & \mathcal{R}_2 \end{pmatrix} \begin{pmatrix} \alpha \\ \beta \end{pmatrix} = \begin{pmatrix} \widehat{\mathcal{A}}^T \rho_h f \\ 0 \end{pmatrix}, \tag{7.6}$$

where

$$\widehat{\mathcal{A}} = \big(\phi_i(w_j)\big)_{1\le i\le N_h,\,1\le j\le \widehat{M}}, \qquad \mathcal{R}_1 = \big((w_j, w_{\widehat{M}+i})_{m,\Omega'}\big)_{1\le i\le M-\widehat{M},\,1\le j\le \widehat{M}},$$
$$\mathcal{R}_0 = \big((w_j, w_i)_{m,\omega_h}\big)_{1\le i,j\le \widehat{M}}, \qquad \mathcal{R}_2 = \big((w_{\widehat{M}+j}, w_{\widehat{M}+i})_{m,\Omega'}\big)_{1\le i,j\le M-\widehat{M}}.$$

Of course, $\alpha$ and $\beta$ can also be obtained by solving, in order, the systems

$$(\widehat{\mathcal{A}}^T\widehat{\mathcal{A}} + \varepsilon\mathcal{R}_0)\alpha = \widehat{\mathcal{A}}^T\rho_h f \quad \text{and} \quad \mathcal{R}_2\beta = -\mathcal{R}_1\alpha.$$

Let us observe that $\widehat{\mathcal{A}}^T\widehat{\mathcal{A}} + \varepsilon\mathcal{R}_0$ and $\mathcal{R}_2$ are symmetric positive definite matrices ($\mathcal{R}_2$ is the Gram matrix of a basis of $\mathcal{K}_0$ in the space $(V_h, \|\cdot\|_{m,\Omega'}^{\omega_h})$).

### *CASE 2: F IS A KNOWN NON-POLYGONAL SET*

We assume that $F$ is still part of the input data of the problem, but now $F$ contains curved arcs.

Since $F$ is known, one can find an accurate approximant $F_h$ of $F$ made up only of polygonal lines. Of course, the points in $A_L \cup A_H$ should keep the same relative position with respect to $F$ and $F_h$. Points lying on the "wrong" side of $F_h$ may distort the fittings and lead to a bad approximation of $f$ near the discontinuity set.

Once $F_h$ has been fixed, the approximation methods described for Case 1 can be directly applied, replacing $F$ and $\Omega'$ by $F_h$ and $\Omega'_h = \Omega \setminus \overline{F}_h$, respectively. For both methods, the resulting solutions, $\sigma_{\varepsilon h}$ and $\tilde{\sigma}_{\varepsilon h}$, belong to $C^k_{\mathbb{F}}(\Omega'_h)$, which is a space "close" to $C^k_{\mathbb{F}}(\Omega')$ (the space cited in problem (7.1)) if $F_h$ is a sufficiently precise approximant of $F$.

Let us finally remark that Theorem 5.1 cannot be applied in the present situation and so the convergence of $\sigma_{\varepsilon h}$ to $f$ can only be conjectured. However, Theorem 6.1 may still hold, ensuring some kind of convergence of $\tilde{\sigma}_{\varepsilon h}$ to $f$.

### *CASE 3: F IS UNKNOWN*

In this last case we suppose that the only information relative to $f$ at our disposal is the set of values on $A_L$ and the set of gradient vectors on $A_H$. This is the usual situation in many real applications, where one knows position and, eventually, orientation data of a surface, but one does not know the location of the fault lines or even whether or not the surface is faulted.

It is clear that any fitting method now requires a preliminary step, which consists in finding the approximate location of $F$. To this purpose, some *discontinuity* (or singularity or fault) *detection methods* have been recently developed from different points of view. Let us cite, among others, the works of G. Allasia, R. Besenghi and A. de Rossi [4], T. Gutzmer and A. Iske [77], D. Girard and P.-J. Laurent [70], P.-J. Laurent [89], D. Lee [93], S. Mallat and W. L. Hwang [105], M. C. Parra [112], M. C. López de Silanes, M. C. Parra and J. J. Torrens [103, 113] and M. Rossini [123, 124].

From the output of any discontinuity detection method, one can usually derive a thin compact subset $\mathbb{F}$ of $\overline{\Omega}$ where $F$ is supposed to be contained and also an approximant $F_h$ of $F$ made up of polygonal lines. It is advisable to remove from $A_L \cup A_H$ the points belonging to $\mathbb{F}$, since one cannot always be sure that they lie on the correct side of $F_h$. With the set $F_h$ in hand, one proceeds as in Case 2 to finally get an approximant of $f$.

## *8. NUMERICAL RESULTS*

In this section, we shall solve problem (7.1) for different non-regular functions, simulating the three cases considered in Section 7, to which we shall refer as Case 1, Case 2 and Case 3. For every case, we shall use one or both of the two approximation methods, i.e. Method 1 and Method 2, described there.

We keep the notations in Section 7. In all the following examples we assume that

- $k = 1$ and $m = 2$;

- $\Omega$ is the open set $(0,1) \times (0,1)$;

- $A_L$ is a subset of $A_{1600}$, where $A_{1600}$ is the set of 1600 randomly distributed points on $\overline{\Omega}$ represented in Figure VIII–6, and the set $A_H$ is empty (i.e. there is no Hermite data); therefore, the operator $\rho$ is given by $\rho v = \big(v(a)\big)_{a \in A_L}$;

- the finite element space $V_h$ is constructed from the Bell triangle of class $C^1$ (cf. Section VIII–3);

- the value of $\varepsilon$ is fixed with the help of the GCV method (cf. Remark VI–3.3 and also the comments in Section VIII–6);

- the relative error is given by (VIII–6.3) (with $\tilde{\sigma}_{\varepsilon h}$ instead of $\sigma_{\varepsilon h}$ when Method 2 is applied).

*Example 1*

Let $F$ be the discontinuity set whose closure is the polygonal line of vertices $(0.2, 0.2)$, $(0.35, 0.225)$, $(0.42, 0.3)$, $(0.48, 0.4)$, $(0.49, 0.53)$, $(0.5, 0.65)$, $(0.65, 0.725)$, $(0.8, 0.75)$ and $(0.9, 0.725)$, and let $f : \overline{\Omega} \to \mathbb{R}$ be the non-regular function defined by

$$f(x,y) = \frac{2}{(2x-3)^2 + (2y-3)^2} + g(x,y),$$

where $g$ denotes the function given by

$$g(x,y) = \begin{cases} \exp\left(2 + \dfrac{0.35^2}{(x-0.55)^2 - 0.35^2} + \dfrac{0.3^2}{(y-0.5)^2 - 0.3^2}\right), & (x,y) \in \mathcal{O}, \\ 0, & \text{otherwise,} \end{cases}$$

$\mathcal{O}$ being the open subset of the open rectangle $(0.2, 0.9) \times (0.2, 0.8)$ that lies above $\overline{F}$. It is clear that $f$ presents a finite jump discontinuity at any point of $F$. The set $F$ and the graph of $f$ are shown in Figures 4 and 5.

Assuming that $F$ is known, we are placed under the assumptions in Case 1, so we can directly apply Method 1 to fit $f$. To this end, we make the triangulation $\mathcal{T}_h$ depicted in Figure 4, we construct on $\mathcal{T}_h$ the finite element space $V_h$, whose dimension is 522, we set $A_L = A_{1600}$ and finally we compute the $V_h$-discrete smoothing $D^2$-spline $\sigma_{\varepsilon h}$ relative to $\rho$, $\rho f$ and $\varepsilon = 10^{-5}$, represented in Figure 6.

In this context, it is not worth applying Method 2, since, for any subset $\omega_h$ of $\Omega$ verifying (6.1) and (6.2), the set $\overline{\Omega} \setminus \overline{\omega}_h$ would be relatively "big" and would contain a large number of data points.

*Example 2*

Let us now consider the discontinuity set $F = \{ (x, \psi(x)) \mid 0 \le x < 0.9 \}$, with $\psi(x) = 0.5 + 0.2 \sin 5\pi x/3$, and let $f$ be given by

$$f(x,y) = \begin{cases} 0.5 \cos^4 g(x,y), & g(x,y) \le \pi/2,\ y > \psi(x), \\ 0.25(1-x)^2 \cos^2 g(x,y), & g(x,y) \le \pi/2,\ y \le \psi(x), \\ 0, & \text{otherwise}, \end{cases}$$

with

$$g(x,y) = \frac{\pi}{1.3} \left( (x-0.2)^2 + (y-0.7)^2 \right).$$

See in Figures 7 and 9 the set $F$ and the graph of $f$.

We suppose again that $F$ is known. Since $F$ is a sinusoidal curve, we are now in Case 2. In this situation, we first find a polygonal approximant $F_h$ of $F$ made up of nine segments (concretely, $\overline{F}_h$ is the polygonal line of vertices $(i/10, \psi(i/10))$, for $i = 0, \ldots, 9$). Then we make a triangulation $\mathcal{T}_h$ of $\overline{\Omega'_h}$ and we construct the finite element space $V_h$. The set $F_h$ and the triangulation $\mathcal{T}_h$ are shown in Figure 8. The dimension of $V_h$ is 492.

The curves $F$ and $F_h$ delimit a small nonconnected open set $\mathcal{W}$. Points in $\mathcal{W}$ cannot be used as data points, since they lie below $F_h$ but above $F$ or vice versa. Therefore, to apply Method 1, we take $A_L = A_{1600} \cap (\overline{\Omega} \setminus \mathcal{W})$. This set contains 1595 points. In these conditions, for $\varepsilon = 10^{-6}$, Method 1 yields the $V_h$-discrete smoothing $D^2$-spline $\sigma_{\varepsilon h}$ represented in Figure 10.

If we directly take $A_L = A_{1600}$, the GCV method also suggests a value of $\varepsilon$ close to $10^{-6}$, which, in this case, is quite far from optimal. In fact, the corresponding $V_h$-discrete smoothing $D^2$-spline, not shown here, exhibits wide oscillations near the fault line. The failure of the GCV method forces a "trial and error" search, after which we get the value $\varepsilon = 10^{-3}$. The oscillations have almost disappeared (cf. Figure 11). It is clear, however, that the quality of the approximation is worse than that obtained with the preceding choice of $A_L$.

For the same reasons as in Example 1, it is not worth applying Method 2.

*Example 3*

In this example we take the discontinuity set $F = \{ (x, \psi(x)) \mid 0.15 < x < 0.85 \}$, with $\psi(x) = 1.4 - 5.8x + 10.7x^2 - 5.7x^3$, and the non-regular function $f$ given by

$$f(x,y) = g_1(x,y)(g_2(x,y) + 1.1 - x),$$

where $g_1$ denotes Franke's function (cf. (VIII–6.1)) and $g_2$ is defined by

$$g_2(x,y) = \begin{cases} -\exp\left(2 + \dfrac{0.35^2}{(x-0.5)^2 - 0.35^2} + \dfrac{0.3^2}{(y-0.6)^2 - 0.3^2}\right), & (x,y) \in \mathcal{O}, \\ 0, & \text{otherwise}, \end{cases}$$

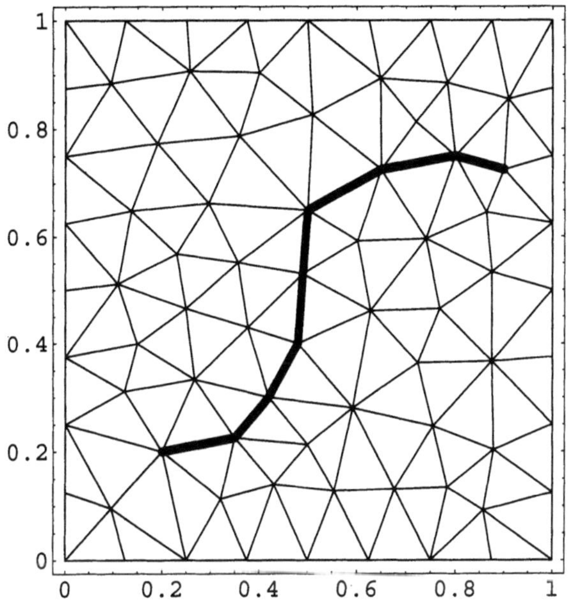


Figure 4: Example 1. Discontinuity set $F$ and triangulation $\mathcal{T}_h$.

with $\mathcal{O} = \{ (x,y) \in (0.15, 0.85) \times (0.3, 0.9) \mid y > \psi(x) \}$. Figures 12 and 14 show the set $F$ and the graph of $f$.

We next simulate Case 3, i.e. we consider that $F$ is unknown. Thus, as a preliminary step, it is necessary to use a discontinuity detection method in order to get a polygonal approximant $F_h$ of $F$ and a compact set $\mathbb{F}$ where $F$ is supposedly contained. We assume that $\mathbb{F}$ and $F_h$ are the sets depicted in Figures 12 and 13.

In order to apply Methods 1 and 2, we then make a triangulation $\mathcal{T}_h$ of $\overline{\Omega'_h}$ (see again Figure 13), we construct the finite element space $V_h$, whose dimension is 618, and we take $A_L = A_{1600} \cap \overline{\Omega} \setminus \mathbb{F}$, which is a set of 1465 points.

Figure 15 shows the $V_h$-discrete smoothing $D^2$-spline $\sigma_{\varepsilon h}$ relative to $\rho$, $\rho f$ and $\varepsilon = 10^{-6}$ yielded by Method 1. Taking $\omega_h = \Omega \setminus \mathbb{F}$ and the same value of $\varepsilon$, Method 2 provides the approximant $\tilde{\sigma}_{\varepsilon h}$ represented in Figure 16. Let us observe that $\sigma_{\varepsilon h}$ and $\tilde{\sigma}_{\varepsilon h}$ have been computed from exactly the same data set, since $A_L \subset \overline{\omega}_h$. Both functions are excellent approximants of $f$.

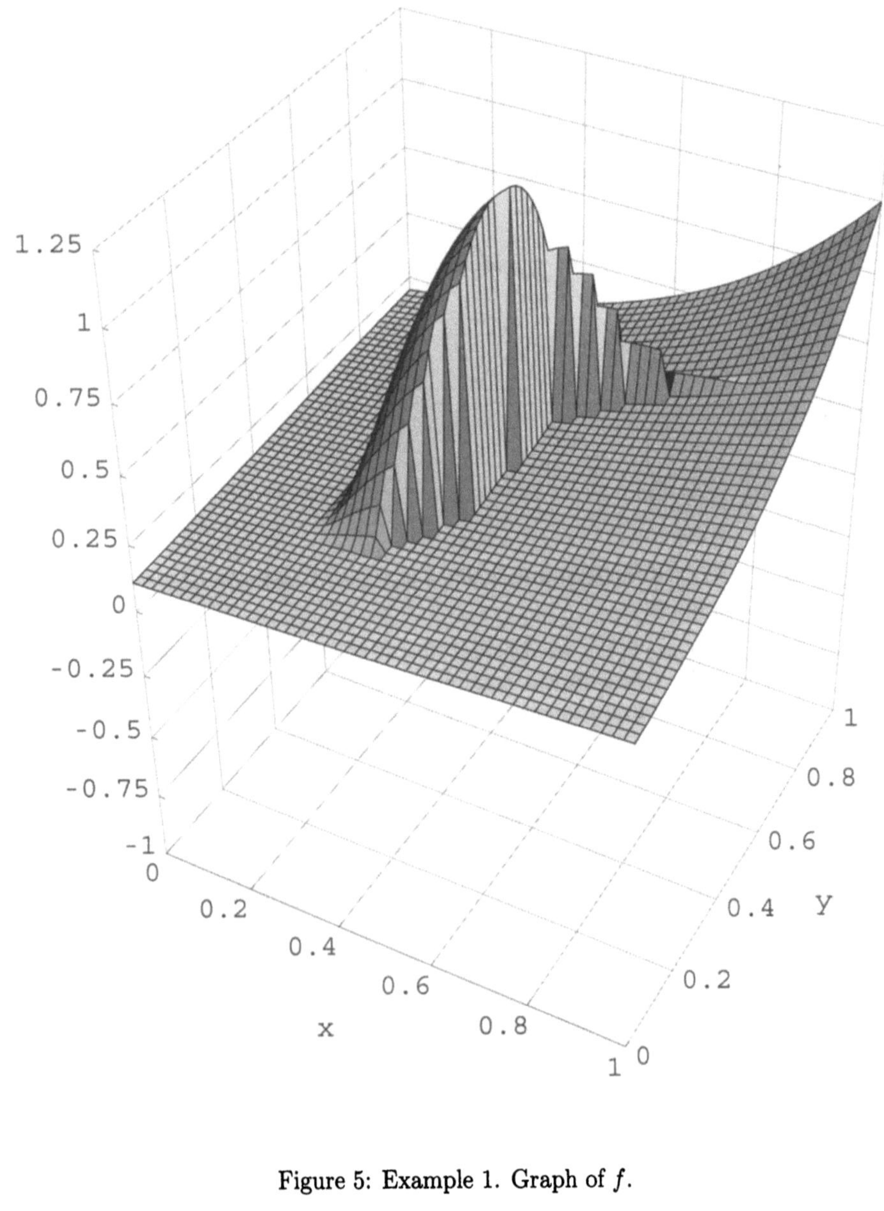


Figure 5: Example 1. Graph of $f$.

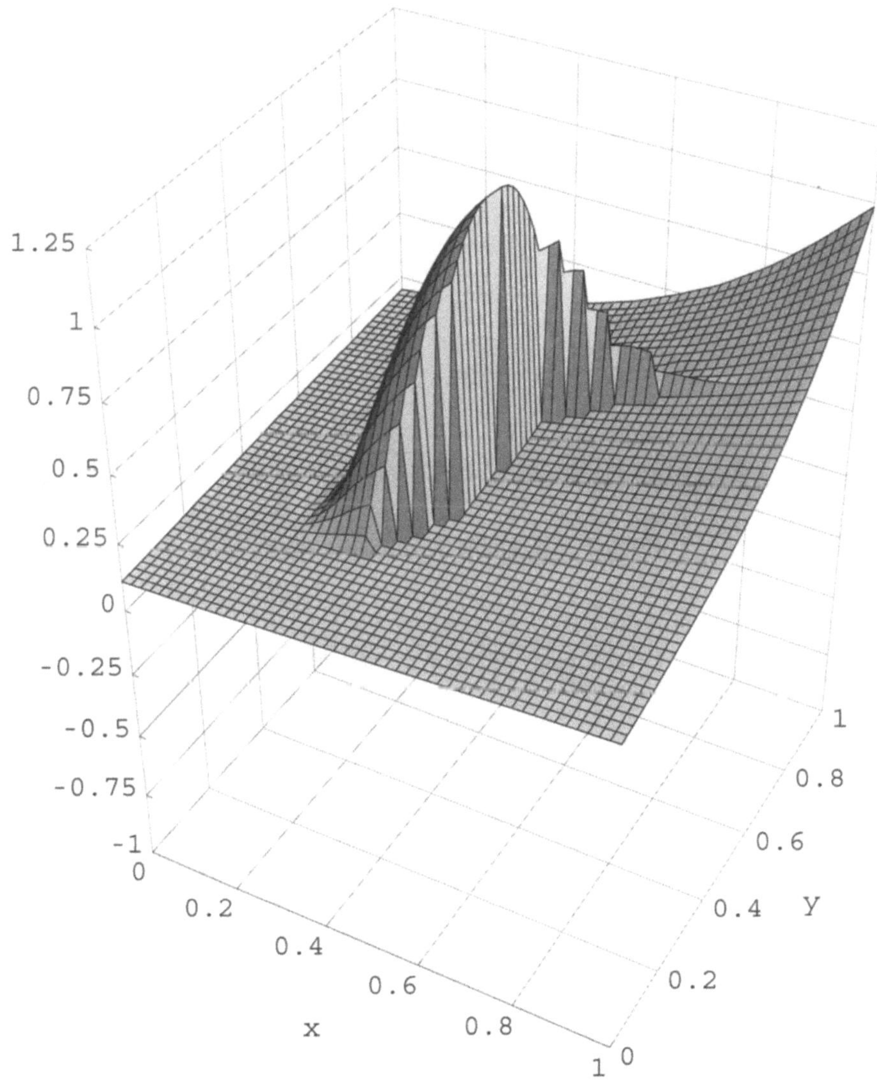


Figure 6: Example 1. Graph of the $V_h$-discrete smoothing $D^2$-spline relative to $\rho$, $\rho f$ and $\varepsilon = 10^{-5}$. Relative error: $r(f) = 0.00696364$.

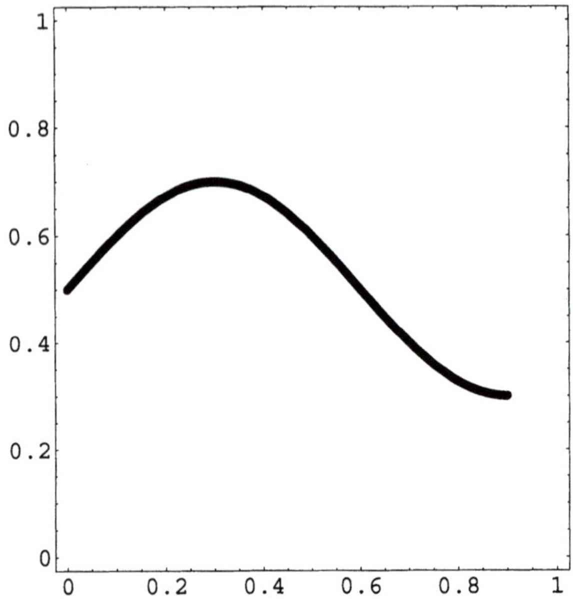


Figure 7: Example 2. Discontinuity set $F$.

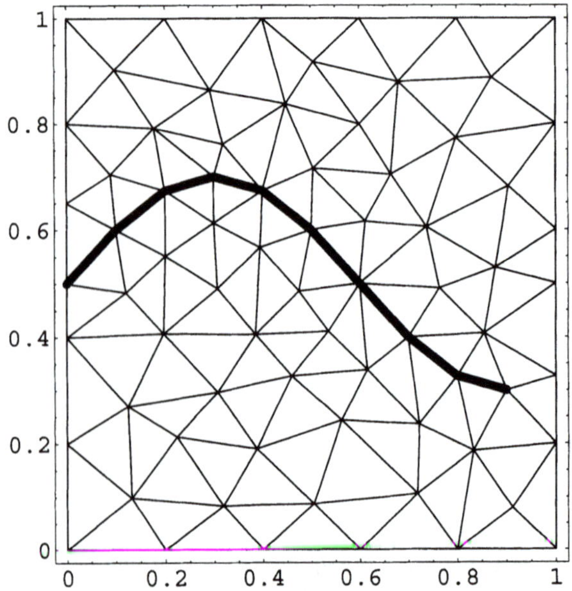


Figure 8: Example 2. Polygonal approximant $F_h$ of $F$ and triangulation $\mathcal{T}_h$.

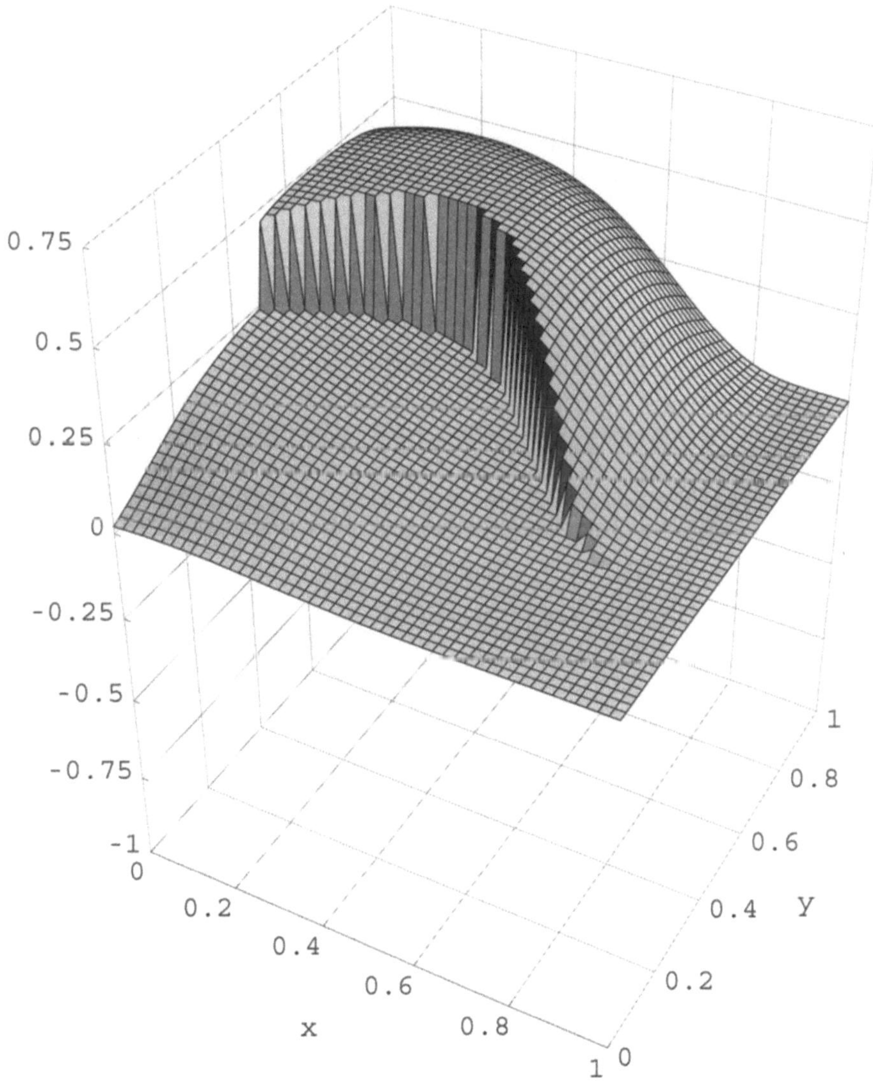


Figure 9: Example 2. Graph of $f$.

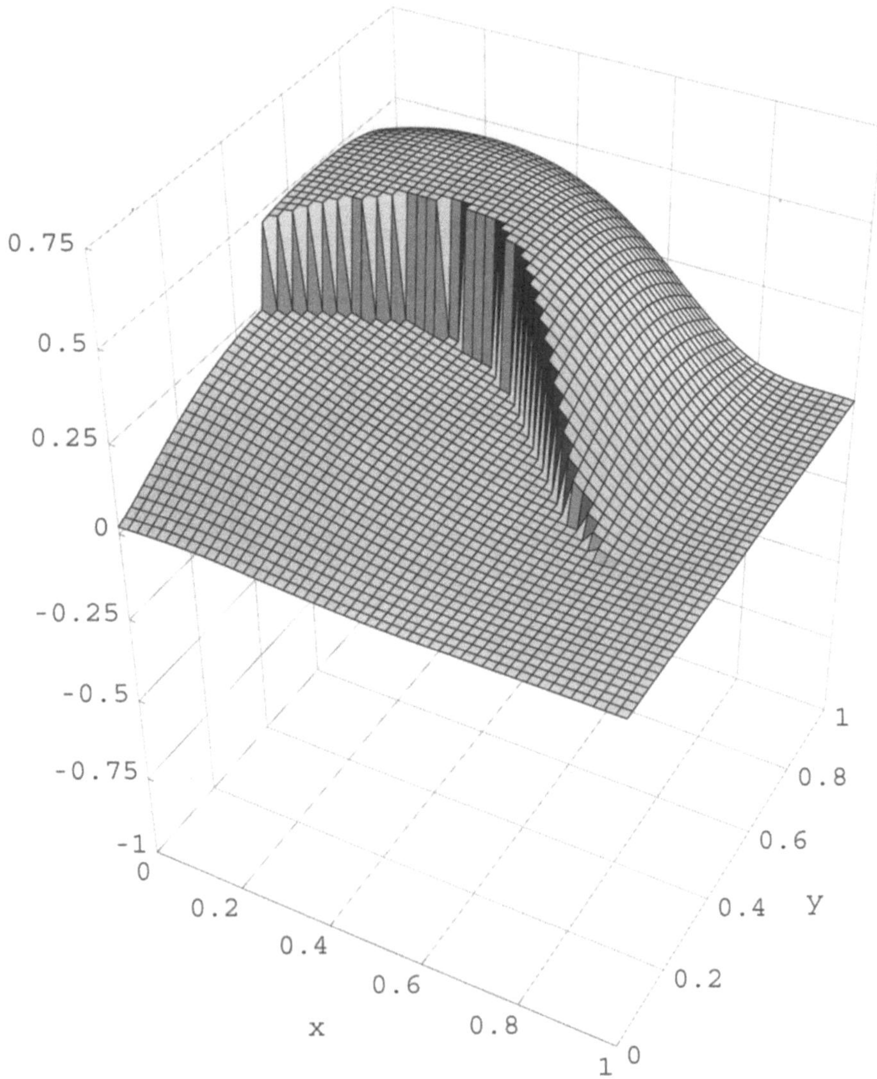


Figure 10: Example 2. Graph of the $V_h$-discrete smoothing $D^2$-spline relative to $\rho$, $\rho f$ and $\varepsilon = 10^{-6}$, taking as $A_L$ the subset of $A_{1600}$ formed by the points that do not lie in between $F$ and $F_h$. Relative error: $r(f) = 0.0699578$.

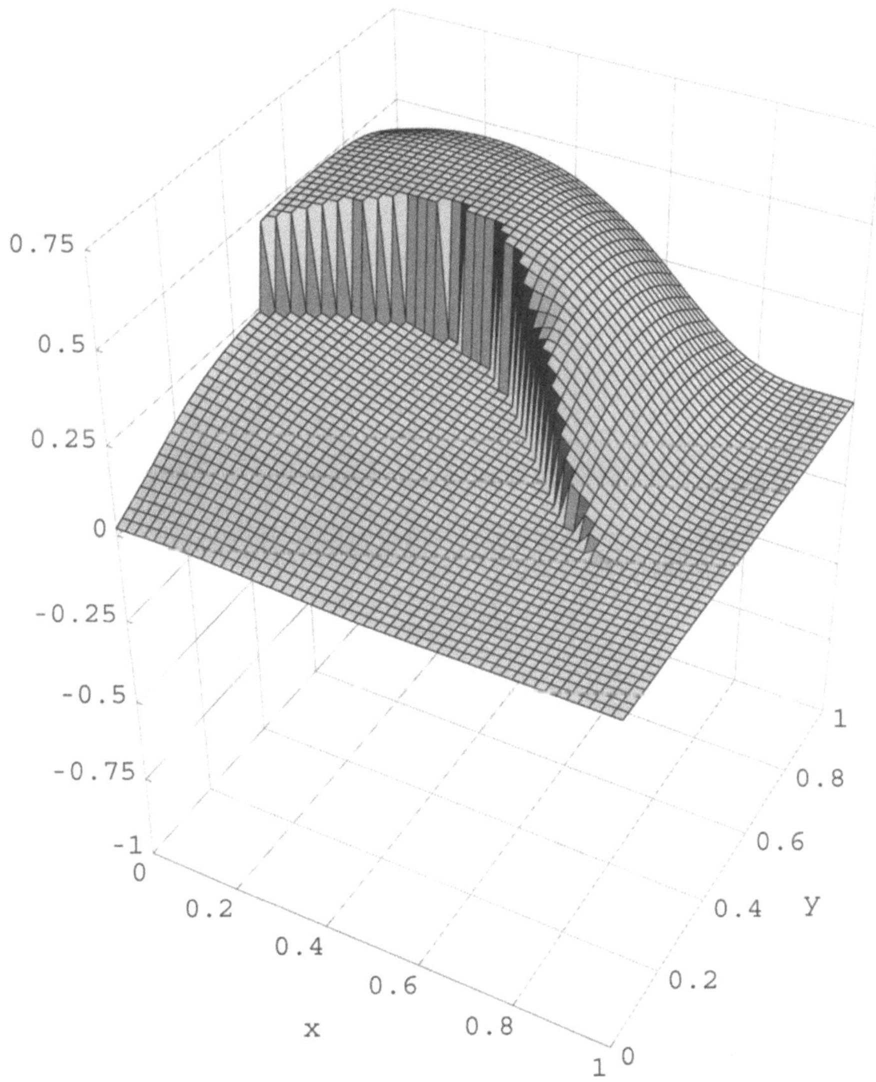


Figure 11: Example 2. Graph of the $V_h$-discrete smoothing $D^2$-spline relative to $\rho$, $\rho f$ and $\varepsilon = 10^{-3}$, with $A_L = A_{1600}$. Relative error: $r(f) = 0.0710063$.

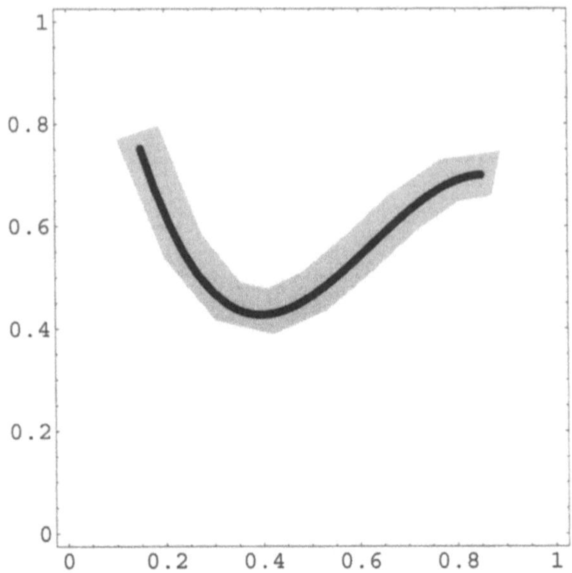


Figure 12: Example 3. Discontinuity set $F$ and, in grey, the compact set $\mathbb{F}$ provided by a discontinuity detection method.

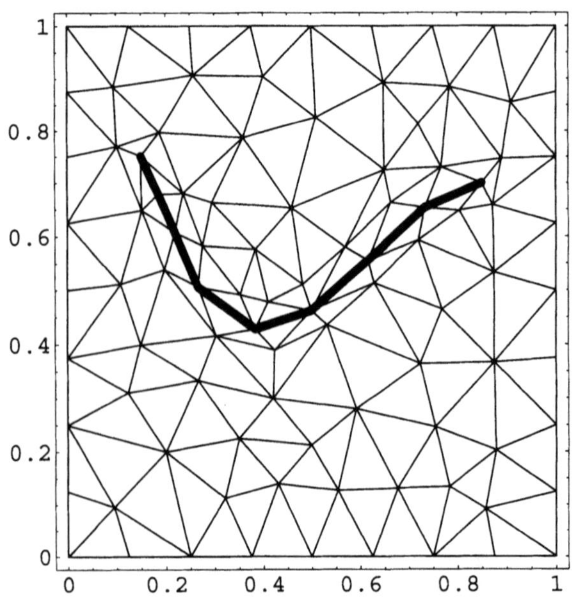


Figure 13: Example 3. Polygonal approximant $F_h$ of $F$ and triangulation $\mathcal{T}_h$.

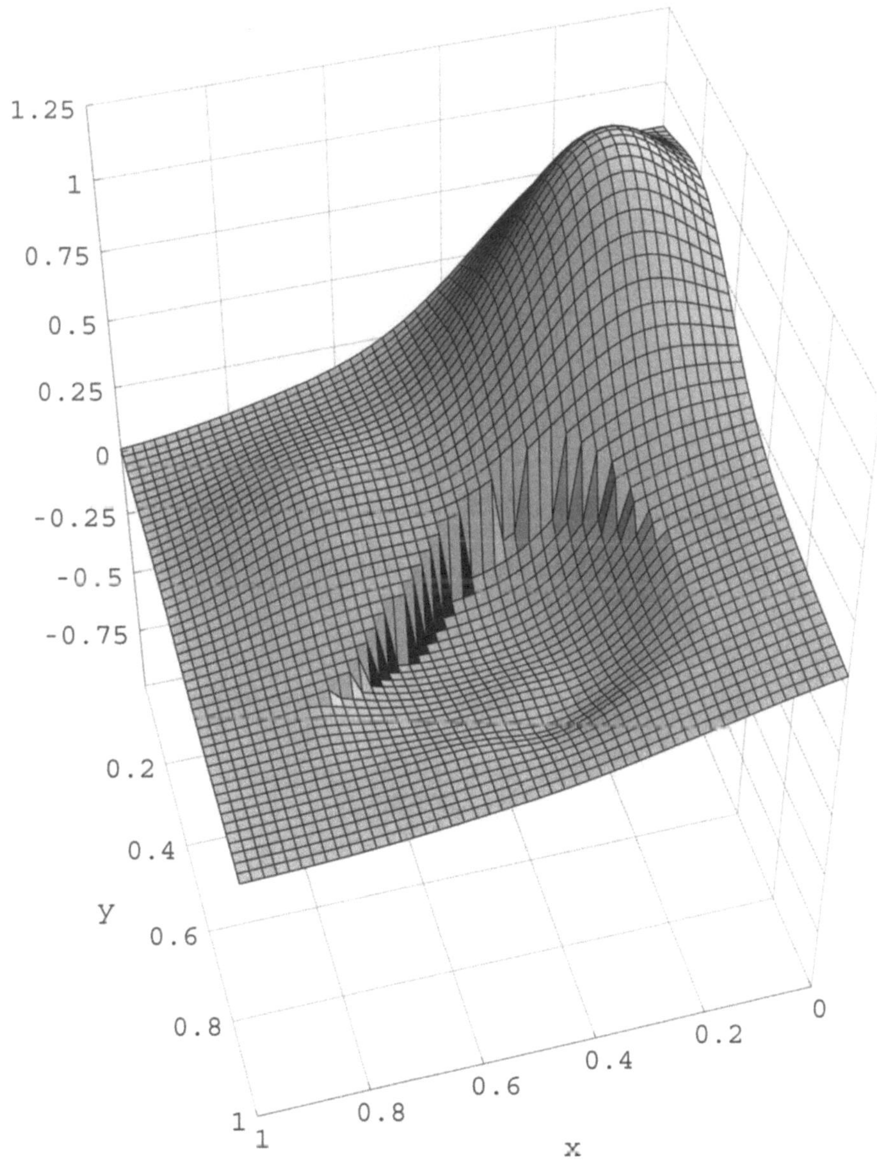


Figure 14: Example 3. Graph of $f$.

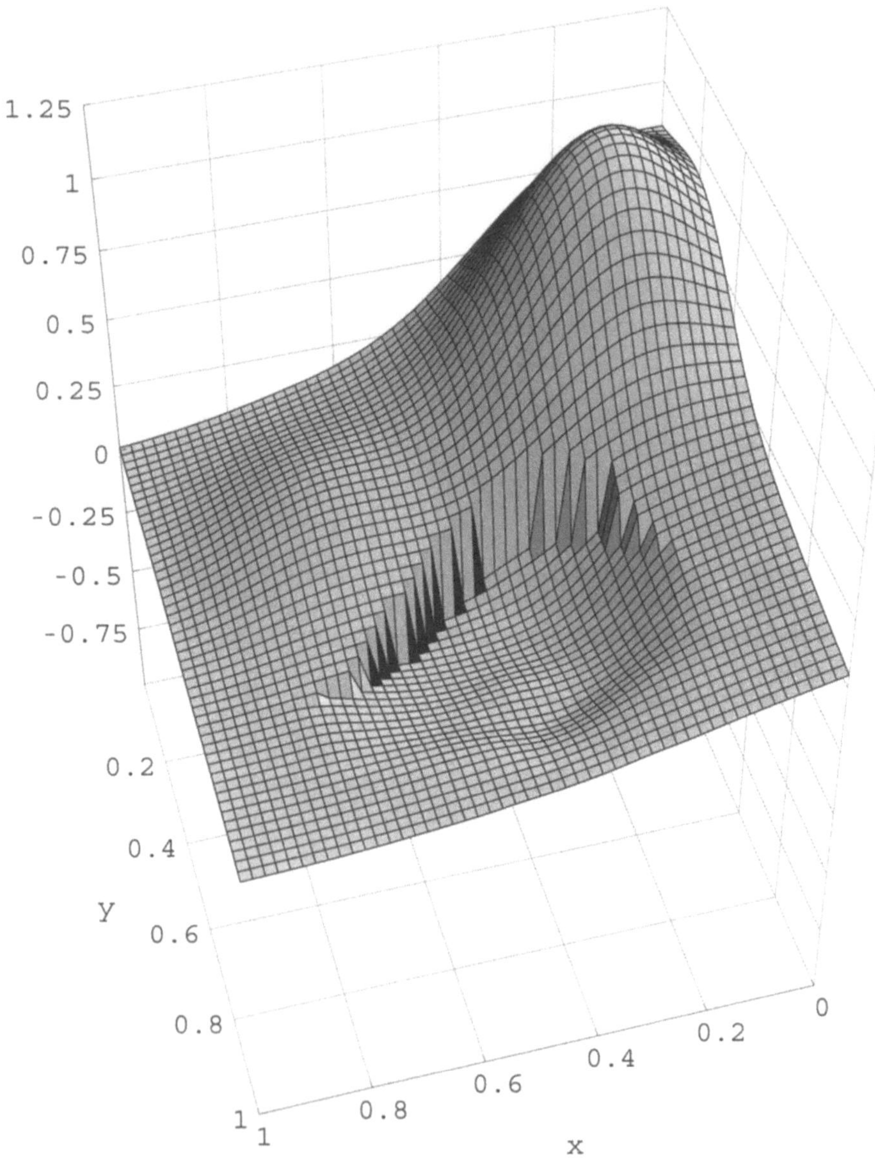


Figure 15: Example 3. Method 1: Graph of the $V_h$-discrete smoothing $D^2$-spline relative to $\rho$, $\rho f$ and $\varepsilon = 10^{-6}$. Relative error: $r(f) = 0.039826$.

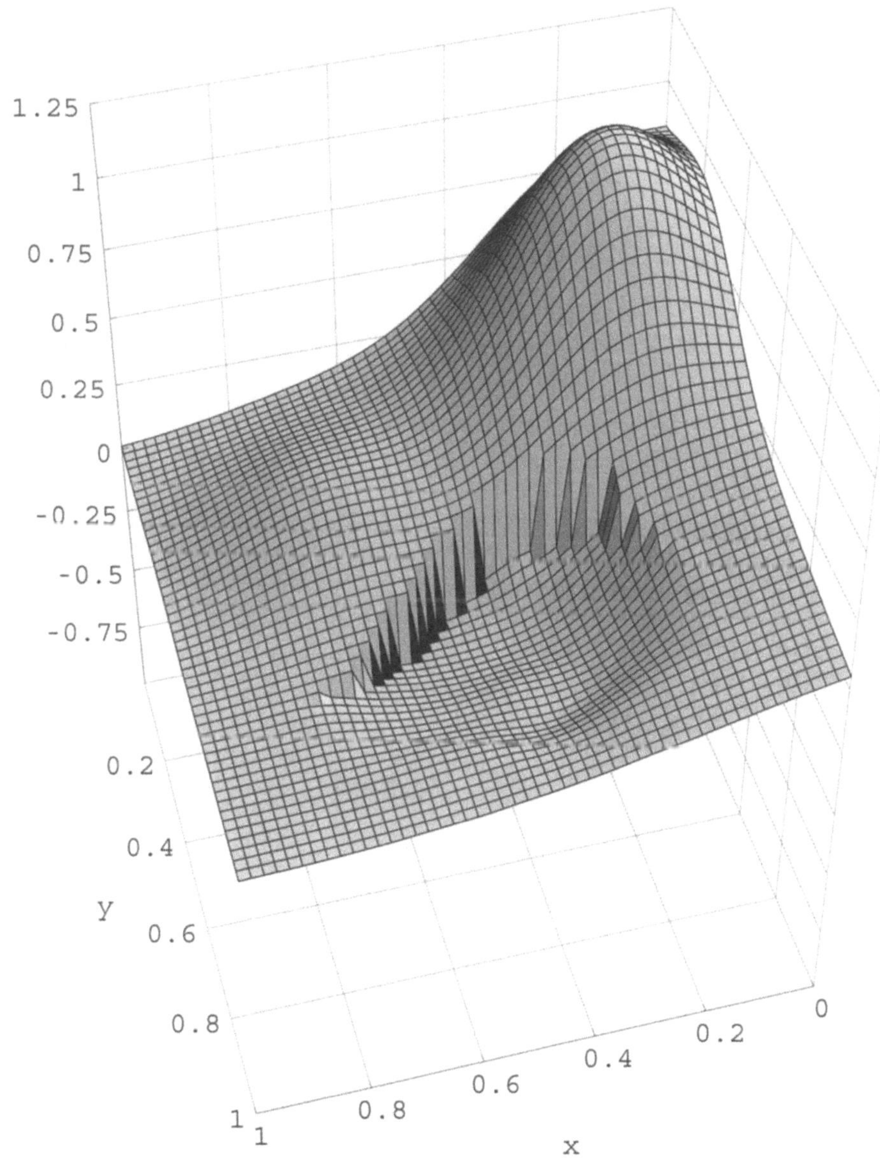


Figure 16: Example 3. Method 2: Graph of the extension to $\Omega'_h$ of the $\widehat{V}_h$-discrete smoothing $D^2$-spline relative to $\rho_h$, $\rho_h f$ and $\varepsilon = 10^{-6}$. Relative error: $r(f) = 0.0399338$.

# CHAPTER X

# FITTING AN EXPLICIT SURFACE ALONG A SET OF CURVES [†]

## 1. FORMULATION OF THE PROBLEM

The problem of construction of surfaces from iso-valued curves (and other problems occurring, for instance, in Geology) leads to the following abstract formulation: given a finite set of curves $F_1, \dots, F_N$ in the closure of a bounded open set $\Omega$ in $\mathbb{R}^2$ (cf. Fig 1) and a function $f$ defined on $F = \bigcup_{j=1}^{N} F_j$, construct a regular function $\phi$ over $\Omega$ that, in some sense, approximates the function $f$ on $F$. More precisely, we suppose that

- $\Omega$ is an open set with a Lipschitz-continuous boundary (cf. Preliminaries),
- for any $j = 1, \dots, N$, there exists an open set $R_j \subset \Omega$ satisfying the conditions:

> (i) $R_j$ is a connected bounded nonempty set with a Lipschitz-continuous boundary,
>
> (ii) $F_j$ is the whole boundary of $R_j$, or a part (open for the trace topology induced by $\mathbb{R}^2$) of this boundary, (1.1)

- for the sake of simplicity, $f$ is the restriction to $F$ of a function, denoted by $f$ as well, that belongs to the usual Sobolev space $H^m(\Omega)$, where $m$ is an integer $\geq 2$, and hence, by Sobolev's Continuous Imbedding Theorem (cf. (4) in Preliminaries), continuous on $\overline{\Omega}$,

and we impose the condition

$$\phi \in H^m(\Omega) \cap C^k(\overline{\Omega}), \text{ with } k = 1 \text{ or } 2.$$

The problem of the approximation of $f$ on $F$ is a fitting problem along the set of curves

$$\{\,(x, y, z) \in \mathbb{R}^3 \mid z = f(x, y),\ (x, y) \in F_j\,\},\ 1 \leq j \leq N.$$

When $m > k+1$, the corresponding *interpolation* problem obviously admits an infinite number of solutions, given that, in that case, $H^m(\Omega) \subset C^k(\overline{\Omega})$. Then, one can get a solution by using J. Duchon's theory of $(m, s)$-splines (cf. [55]).

---

[†]Cf. D. Apprato and R. Arcangéli [8].

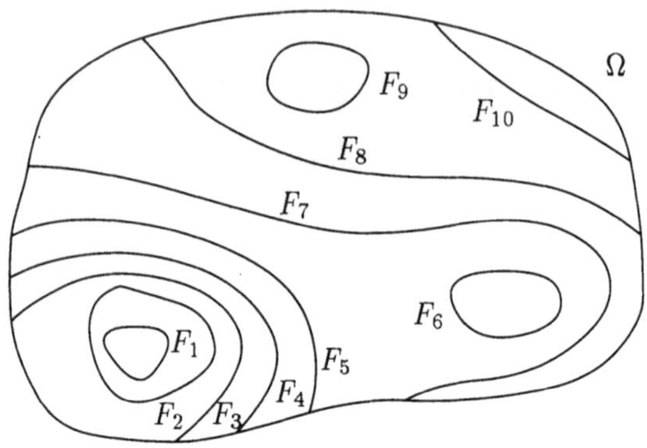


Figure 1: Curves $F_1, \ldots, F_N$, with $N = 10$.

We obtain another solution in the following way. Let

$$\mathcal{K} = \{ v \in H^m(\Omega) \mid v|_F = f|_F \},$$

and consider the problem: find $\sigma$ solution of

$$\begin{cases} \sigma \in \mathcal{K}, \\ \forall v \in \mathcal{K}, \; |\sigma|_{m,\Omega} \leq |v|_{m,\Omega}. \end{cases} \tag{1.2}$$

For any $j = 1, \ldots, N$, let us introduce the Hilbert space $L^2(F_j)$ of all (classes of) square integrable real functions $v$ on $F_j$, endowed with any Hilbertian norm $\|\cdot\|_{0,F_j}$ defining its topology. This makes sense because of hypothesis (1.1) (cf. P. Grisvard [75], J. Nečas [109]). Then, let $L^2(F)$ be the space of all (classes of) real functions $v$ on $F$ such that, for any $j = 1, \ldots, N$, $v|_{F_j} \in L^2(F_j)$, endowed with the norm

$$\|\cdot\|_{0,F} = \left( \sum_{j=1}^{N} \|\cdot\|_{0,F_j}^2 \right)^{1/2}.$$

Then, we have the following result.

***Theorem* 1.1** – *Suppose that the hypotheses formulated before on $\Omega$, $F$, $m$ and $f$ are verified and, moreover, that the following condition is satisfied:*

$$\forall p \in P_{m-1}(\overline{\Omega}), \; (p|_F = 0) \Rightarrow (p = 0), \tag{1.3}$$

*where $P_{m-1}(\overline{\Omega})$ denotes the vector space of restrictions to $\overline{\Omega}$ of the bivariate polynomial functions of degree $\leq m-1$ with respect to the set of variables. Then, problem (1.2) has a unique solution.*

**Proof** –

1) Reasoning by compactness (cf. J. Nečas [109, Theorem 2.7.1]) and taking (1.3) into account, it is shown that the mapping $[\![ \cdot ]\!]$ on $H^m(\Omega)$, defined by

$$[\![ v ]\!] = \left( \|v\|_{0,F}^2 + |v|_{m,\Omega}^2 \right)^{1/2},$$

is a norm on $H^m(\Omega)$ equivalent to the usual norm $\|\cdot\|_{m,\Omega}$.

2) The solution $\sigma$ of problem (1.2) is just the element of minimal norm $[\![\cdot]\!]$ of the convex closed nonempty subset $\mathcal{K}$ in $H^m(\Omega)$. □

So, when $m > k+1$, we can take $\phi = \sigma$. Unfortunately, except in very particular cases, it is impossible to *find $\sigma$ explicitly* (as well as any other interpolant) *in a form which can be used for computing.* Of course, in a finite-dimensional space, it is impossible to satisfy an infinite number of interpolation conditions.

Next, we propose to construct a "discrete smoothing $D^m$-spline" by proceeding as we did in Section VI–3 in the case of point conditions. In order to do this, we introduce the functional $J^\eta_{\varepsilon h}$, defined on the space $H^m(\Omega_h)$ by

$$J^\eta_{\varepsilon h}(v) = l^\eta\big((v-f)^2\big) + \varepsilon |v|^2_{m,\Omega_h}, \tag{1.4}$$

where we have written $l^\eta\big((v-f)^2\big)$ in place of $l^\eta\big((v|_F - f|_F)^2\big)$. In this definition, $l^\eta\big((v-f)^2\big)$ represents an approximant of $\|v-f\|^2_{0,F}$ (which is necessary, because this term cannot be calculated exactly), $\varepsilon$ is any strictly positive parameter, $\Omega_h$ stands for a polygonal open subset that approximates $\Omega$, and $|\cdot|_{m,\Omega_h}$ denotes the usual semi-norm of order $m$ in $H^m(\Omega_h)$. The function that we are looking for will be obtained by minimizing $J^\eta_{\varepsilon h}$ on a suitable finite-dimensional space $V_h$.

## 2. APPROXIMATION OF $\|\cdot\|^2_{0,F}$

Suppose we are given

- a bounded subset $\mathbb{E}$ of $(0,+\infty)$ such that $0 \in \overline{\mathbb{E}}$,
- for any $\eta \in \mathbb{E}$ and for any $j = 1,\ldots,N$, a set $\{\xi_i\}_{1\le i\le L}$ of $L = L(\eta,j)$ distinct points $\xi_i = \xi_i(\eta,j)$ of $\overline{F}_j$ and a set $\{\lambda_i\}_{1\le i\le L}$ of numbers $\lambda_i = \lambda_i(\eta,j) > 0$.

Then we write

$$\forall \eta \in \mathbb{E},\ \forall j = 1,\ldots,N,\ \forall v \in C^0(\overline{F}_j),\ l^\eta_j(v) = \sum_{i=1}^{L} \lambda_i v(\xi_i) \tag{2.1}$$

and

$$\forall \eta \in \mathbb{E},\ \forall v \in C^0(\overline{F}),\ l^\eta(v) = \sum_{j=1}^{N} l^\eta_j(v). \tag{2.2}$$

We assume that

$$\left|\begin{aligned} &\exists C > 0,\ \exists t > 0,\ \forall \eta \in \mathbb{E},\ \forall j = 1,\ldots,N,\\ &\forall v \in H^m(\Omega),\ \left|l^\eta_j(v^2) - \|v\|^2_{0,F_j}\right| \le C\eta^t \|v\|^2_{m,\Omega}, \end{aligned}\right. \tag{2.3}$$

where we have written, as we shall do from now on, $v$ instead of $v|_{F_j}$ or $v|_F$. Likewise, we shall consider $l^\eta$ as a continuous linear form on $C^0(\overline{\Omega})$ or on $C^0(\overline{\Omega'})$, for any open subset $\Omega'$ containing $\Omega$.

**Remark 2.1** – When hypothesis (2.3) is verified, the relation

$$\|v\|_{0,F}^2 \sim l^\eta(v^2),\ \eta \to 0,$$

constitutes an (abstract) numerical integration formula for $\|\cdot\|_{0,F}^2$. Of course, the nodes $\xi_i$ cannot be any points of $\overline{F}$. The convergence of this formula as $\eta \to 0$ implies, for example, that

$$(\eta \to 0) \Rightarrow \Big(\max_{j=1,\dots,N} \sup_{x\in F_j} \delta(x, \Xi_j) \to 0\Big),$$

where $\delta(\,\cdot\,,\,\cdot\,)$ denotes the Euclidean distance in $\mathbb{R}^2$ and $\Xi_j$ is the set of nodes $\xi_i$ in $\overline{F}_j$. It also implies, when the points $\xi_i$ are numbered in a monotone way on every curve $\overline{F}_j$, that

$$(\eta \to 0) \Rightarrow \Big(\max_{j=1,\dots,N} \max_{1\le i\le L-1} \delta(\xi_i, \xi_{i+1}) \to 0\Big).$$

How can we obtain explicit formulae? Let us study an example. Suppose, to simplify, that $N = 1$, that $F(= F_1)$ is represented by a single equation

$$y = a(x),\ x \in I,$$

where $I$ denotes some real interval, that the norm $\|\cdot\|_{0,F}$ is defined by

$$\|v\|_{0,F}^2 = \int_I v(x, a(x))^2 \left(1 + a'(x)^2\right)^{1/2} dx,$$

and that the points $\xi_i$ are numbered in a monotone way on $\overline{F}$. Then, if we take

$$l^\eta(v) = \sum_{i=1}^{L-1} \delta(\xi_i, \xi_{i+1}) v(\xi_i), \tag{2.4}$$

or

$$l^\eta(v) = \sum_{i=1}^{L-1} \delta(\xi_i, \xi_{i+1}) v(\xi_{i+1}), \tag{2.5}$$

or also

$$\begin{aligned} l^\eta(v) &= \frac{1}{2}\delta(\xi_1, \xi_2) v(\xi_1) \\ &+ \frac{1}{2}\sum_{i=2}^{L-1} \left(\delta(\xi_{i-1}, \xi_i) + \delta(\xi_i, \xi_{i+1})\right) v(\xi_i) + \frac{1}{2}\delta(\xi_{L-1}, \xi_L) v(\xi_L), \end{aligned} \tag{2.6}$$

we can prove that hypothesis (2.3) is satisfied in the three cases with $t = 1$ and $m = 2$, provided $a$ is regular enough (of class $C^2$, for instance) and the relation

$$\exists C_1 > 0, \exists C_2 > 0, \forall \eta \in \mathbb{E}, \forall i = 1, \dots, L-1,\ C_1\eta \le \delta(\xi_i, \xi_{i+1}) \le C_2\eta$$

is verified. Thus, the relation $\int_F v(x)\,ds \sim l^\eta(v)$ appears to be, for the curvilinear integral $\int_F v(x)\,ds$, respectively, a "Left-sided Rectangle Formula", a "Right-sided Rectangle Formula" and a "Trapezoidal Rule".

In the general case, the formulae which can be obtained are more complicated, since they depend explicitly on the local systems of coordinates that define the curves $F_j$. However, in real problems, more often than not, the curves $F_j$ are *polygonal.* Then, univariate classical numerical integration formulae are available. It is also the case when the $F_j$'s are *arcs of a circle.* □

# 3. *SPLINE FITTING*

Suppose we are given

- a bounded subset $\mathbb{H}$ of $(0, +\infty)$ such that $0 \in \overline{\mathbb{H}}$,
- a bounded polygonal open set $\widetilde{\Omega}$ in $\mathbb{R}^2$ which contains $\Omega$,
- for any $h \in \mathbb{H}$, a triangulation $\widetilde{\mathcal{T}}_h$ of $\overline{\widetilde{\Omega}}$ by means of elements $K$ with diameters $h_K \leq h$ and a finite element space $\widetilde{V}_h$, constructed on $\widetilde{\mathcal{T}}_h$, such that

$$\widetilde{V}_h \text{ is a finite-dimensional subspace of } H^m(\widetilde{\Omega}) \cap C^k(\overline{\widetilde{\Omega}}), \text{ with } k = 1 \text{ or } 2. \tag{3.1}$$

We also suppose that P. Clément's result (VI–1.7) is verified with $q = m$. Taking (3.1) into account, this result can be stated here in the following restricted form:

> There exists a constant $C > 0$ and, for any $h \in \mathbb{H}$, a linear operator $\widetilde{\Pi}_h : H^m(\widetilde{\Omega}) \to \widetilde{V}_h$ such that
>
> $$\forall l = 0, \ldots, m,\ \forall v \in H^m(\widetilde{\Omega}),\ |v - \widetilde{\Pi}_h v|_{l,\widetilde{\Omega}} \leq C h^{m-l} |v|_{m,\widetilde{\Omega}}. \tag{3.2}$$
>
> Moreover,
> $$\forall v \in H^m(\widetilde{\Omega}),\ \lim_{h \to 0} |v - \widetilde{\Pi}_h v|_{m,\widetilde{\Omega}} = 0.$$

**Remark 3.1** – (Cf. Remark VI–1.1). We recall that the result (3.2) assumes implicitly the following conditions:

- the generic finite element $(K, P_K, \Sigma_K)$ of the family $(\widetilde{V}_h)_{h \in \mathbb{H}}$ verifies the inclusions $P_m(K) \subset P_K \subset H^m(K)$,
- the family $(\widetilde{\mathcal{T}}_h)_{h \in \mathbb{H}}$ is regular,
- for any $h \in \mathbb{H}$, the basis functions of $\widetilde{V}_h$ satisfy the uniformity property (VI–1.9) with $q = m$. □

Concerning the choice of the generic finite element for the construction of the spaces $\widetilde{V}_h$, we refer to Section VIII–3. Let us remember that, in general, the implementation of rectangular finite elements is cheaper than that of triangular ones.

Then we proceed as in Section VI–1. For any $h \in \mathbb{H}$, we consider the open set $\Omega_h$ defined by (VI–1.2), that is, $\Omega_h$ is the interior of the union of all elements $K \in \widetilde{\mathcal{T}}_h$ which intersect $\Omega$ (cf. Figure VI–1). Let us recall that the family $(\Omega_h)_{h\in\mathbb{H}}$ verifies the relations

$$\forall h \in \mathbb{H},\ \Omega \subset \Omega_h \subset \widetilde{\Omega}$$

and

$$\lim_{h\to 0} \operatorname{meas}(\Omega_h \setminus \Omega) = 0$$

(cf. (VI–1.3) and (VI–1.4)). For any $h \in \mathbb{H}$, we also consider the space $V_h$ given by (VI–1.5), i.e. the space of restrictions to $\Omega_h$ of the functions belonging to $\widetilde{V}_h$. In virtue of (3.1), $V_h$ is a finite-dimensional subspace of $H^m(\Omega_h) \cap C^k(\overline{\Omega}_h)$. We suppose that the space $V_h$ verifies the condition

$$\forall v_h \in V_h,\ (v_h|_\Omega = 0) \Rightarrow (v_h = 0). \tag{3.3}$$

Notice that, because of (VI–1.2), hypothesis (3.3) is always satisfied in the case of the usual polynomial finite elements.

Then, for any $\varepsilon > 0$, for any $h \in \mathbb{H}$ and for any $\eta \in \mathbb{E}$, we consider the minimization problem: find $\sigma^\eta_{\varepsilon h}$ such that

$$\begin{cases} \sigma^\eta_{\varepsilon h} \in V_h, \\ \forall v_h \in V_h,\ J^\eta_{\varepsilon h}(\sigma^\eta_{\varepsilon h}) \le J^\eta_{\varepsilon h}(v_h), \end{cases} \tag{3.4}$$

where $J^\eta_{\varepsilon h}$ is the functional introduced in (1.4), with $l^\eta$ and $\Omega_h$ defined, respectively, in (2.2) and (VI–1.2). We also consider the variational problem: find $\sigma^\eta_{\varepsilon h}$ solution of

$$\begin{cases} \sigma^\eta_{\varepsilon h} \in V_h, \\ \forall v_h \in V_h,\ l^\eta(\sigma^\eta_{\varepsilon h}\, v_h) + \varepsilon(\sigma^\eta_{\varepsilon h}, v_h)_{m,\Omega_h} = l^\eta(f\, v_h). \end{cases} \tag{3.5}$$

***Theorem* 3.1** – *Suppose that $\Omega$, $F$, $m$ and $f$ are defined as in Section 1, and that hypotheses* (VI–1.2), (VI–1.5), (1.3), (2.3), (3.1) *and* (3.3) *are verified. Then, there exists $\beta > 0$ such that, for any $h \in \mathbb{H}$ and for any $(\eta, \varepsilon) \in \mathbb{E} \times (0, +\infty)$ verifying the condition*

$$\frac{\eta^t}{\min(1,\varepsilon)} < \beta, \tag{3.6}$$

*problems* (3.4) *and* (3.5) *admit a unique common solution $\sigma^\eta_{\varepsilon h}$.*

**Proof** – Let $(\eta, \varepsilon) \in \mathbb{E} \times (0, +\infty)$. According to the definition of $l^\eta$, for any $h \in \mathbb{H}$, the symmetric bilinear form $a^\eta_\varepsilon$ defined by

$$a^\eta_\varepsilon(u_h, v_h) = l^\eta(u_h\, v_h) + \varepsilon(u_h, v_h)_{m,\Omega_h}$$

is continuous on $V_h \times V_h$. On the other hand, since $V_h \subset H^m(\Omega_h)$, we deduce from (VI–1.3) and (2.3) that

$$\forall h \in \mathbb{H},\ \forall v_h \in V_h,\ a_\varepsilon^\eta(v_h, v_h) \geq \min(1,\varepsilon)\Big(\|v_h\|_{0,F}^2 + |v_h|_{m,\Omega}^2\Big) - NC\eta^t\|v_h\|_{m,\Omega}^2, \quad (3.7)$$

where $C$ is the constant occurring in (2.3). Since the norms $[\![ \cdot ]\!]$ (introduced in the proof of Theorem 1.1) and $\| \cdot \|_{m,\Omega}$ are equivalent on $H^m(\Omega)$, it follows that there exists $C' > 0$ such that

$$\forall h \in \mathbb{H},\ \forall v_h \in V_h,\ a_\varepsilon^\eta(v_h, v_h) \geq \big(C' \min(1,\varepsilon) - NC\eta^t\big)\|v_h\|_{m,\Omega}^2.$$

Let us take $\beta = C'/NC$. Therefore, for any $h \in \mathbb{H}$ and for any $(\eta, \varepsilon) \in \mathbb{E} \times (0, +\infty)$ verifying (3.6), by (3.3), we have

$$\forall v_h \in V_h,\ v_h \neq 0,\ a_\varepsilon^\eta(v_h, v_h) > 0.$$

Since $V_h$ is a finite-dimensional space, we deduce that the bilinear form $a_\varepsilon^\eta$ is $V_h$-elliptic. Then, the Lax-Milgram Lemma (cf. P. G. Ciarlet [45, Theorem 1.1.3]) yields the result. □

The function $\sigma_{\varepsilon h}^\eta$ is called *$V_h$-discrete smoothing $D^m$-spline of $f$ relative to $F$, $\eta$ and $\varepsilon$*. Its definition is similar to that of the discrete smoothing $D^m$-splines introduced in Section VI–3.

**Remark 3.2** – Let $h \in \mathbb{H}$. Let us denote by $M$ and $w_1, \ldots, w_M$, respectively, the dimension and the basis functions of the space $V_h$. Then, $\sigma_{\varepsilon h}^\eta$ can be expressed as

$$\sigma_{\varepsilon h}^\eta = \sum_{j=1}^{M} \alpha_j w_j,$$

with $\alpha_j \in \mathbb{R}$, for $1 \leq j \leq M$. Defining the matrices

$$\mathcal{B} = \big(l^\eta(w_j w_i)\big)_{1 \leq i, j \leq M} \quad \text{and} \quad \mathcal{R} = \big((w_j, w_i)_{m,\Omega_h}\big)_{1 \leq i, j \leq M}$$

and the vector

$$\mathcal{F} = \big(l^\eta(f w_i)\big)_{1 \leq i \leq M},$$

we see that (3.5) is equivalent to the problem: find $\alpha = (\alpha_i)_{1 \leq i \leq M}$ solution of the linear system of order $M$

$$(\mathcal{B} + \varepsilon\mathcal{R})\alpha = \mathcal{F}.$$

In fact, the matrix $\mathcal{B}$ and the vector $\mathcal{F}$ can be written in a more appealing form. Suppose that $N = 1$ and let $\xi_1, \ldots, \xi_L$ and $\lambda_1, \ldots, \lambda_L$ be the nodes and the coefficients of the linear form $l_1^\eta$ (cf. (2.1)). Then, we have

$$\mathcal{B} = \mathcal{A}^T \Lambda \mathcal{A} \quad \text{and} \quad \mathcal{F} = \mathcal{A}^T \Lambda \beta, \quad (3.8)$$

where

$$\mathcal{A} = \big(w_j(\xi_i)\big)_{1 \leq i \leq L,\, 1 \leq j \leq M}, \quad \Lambda = \operatorname{diag}(\lambda_1, \ldots, \lambda_L) \quad \text{and} \quad \beta = \big(f(\xi_i)\big)_{1 \leq i \leq L}. \quad (3.9)$$

If $N > 1$, the relation (3.8) still holds, adding to $\mathcal{A}$, $\Lambda$ and $\beta$ the rows associated with the corresponding nodes and coefficients of $l_2^\eta, \ldots, l_N^\eta$. □

**Remark 3.3** – The value of the smoothing parameter $\varepsilon$ can be chosen by any of the methods cited in Remark VI–3.2. In particular, one can apply the GCV method, detailed in Remark VI–3.3. We point out that, with the notations of Remark 3.2, the influence matrix $Q_\varepsilon$ is now the matrix $\mathcal{A}(\mathcal{A}^T\Lambda\mathcal{A}+\varepsilon\mathcal{R})^{-1}\mathcal{A}^T\Lambda$. The expressions of the approximated GCV function $\overline{\mathcal{V}}$ given in (VI–3.5) and (VI–3.6) are still valid, taking into account that

- $N$ should be replaced by the total number of nodes, say $L$, used to define $l^\eta$,
- $\langle \cdot \rangle$ denotes here the Euclidean norm in $\mathbb{R}^L$,
- $u$ is a vector, as in Remark VI–3.3, whose elements take the values 1 and $-1$ with probability 1/2,
- $\mathcal{A}$ and $\beta$ are defined in (3.9), and
- $\alpha$ and $\tilde{\alpha}$ are the solutions of the linear systems $(\mathcal{A}^T\Lambda\mathcal{A}+\varepsilon\mathcal{R})\alpha = \mathcal{A}^T\Lambda\beta$ and $(\mathcal{A}^T\Lambda\mathcal{A}+\varepsilon\mathcal{R})\tilde{\alpha} = \mathcal{A}^T\Lambda u$, with $\Lambda$ given in (3.9). □

Returning to the initial problem, we propose to take for approximant of $f$ the function $\phi = \sigma^\eta_{\varepsilon h}|_\Omega$ which, by (VI–1.3), (VI–1.5) and (3.1), belongs to $H^m(\Omega)\cap C^k(\overline{\Omega})$. It remains to see how $\phi$ approaches $f$. The following result, given without a proof (which can be established by means of compactness arguments and some results of the finite element theory), shows that $\sigma^\eta_{\varepsilon h}$ converges to $f$ on $F$ under certain conditions. Notice that, by (3.6), $\sigma^\eta_{\varepsilon h}$ is defined for $\varepsilon \in (0,1]$ and $\eta^t/\varepsilon < \beta$.

***Theorem* 3.2** – *Suppose that the conditions of Theorem 3.1 are verified and, moreover, that hypothesis* (3.2) *is satisfied. Then, the solution* $\sigma^\eta_{\varepsilon h}$ *of* (3.4) *and* (3.5) *verifies the relations:*

*(i) for any* $\theta \in (0, m-1)$,

$$\lim_{\substack{\varepsilon\to 0,\, \eta^t/\varepsilon<\beta \\ h^{2(m-1-\theta)}/\varepsilon\to 0}} \|\sigma^\eta_{\varepsilon h} - \sigma\|_{m,\Omega} = 0,$$

*where* $\sigma$ *denotes the solution of problem* (1.2) *and* $\beta$ *is the constant introduced in* (3.6),

*(ii) for any* $\theta \in (0, m-1)$, *there exists a constant* $C > 0$ *such that, as* $\varepsilon \to 0$, $\eta^t/\varepsilon < \beta$ *and* $h^{2(m-1-\theta)}/\varepsilon \to 0$,

$$\|\sigma^\eta_{\varepsilon h} - f\|^2_{0,F} \leq C\big(\varepsilon + \eta^t o(1) + h^{2(m-1-\theta)}\big).$$

In fact, to study the convergence of $\sigma^\eta_{\varepsilon h}$ to $f$ *over the open set* $\Omega$, we have to examine the classical problem of the *convergence of the approximation* as the number of curves $F_j$ tends to infinity, in such a way that the family of the corresponding sets $F$ fills up $\overline{\Omega}$ (in a sense to be specified).

# 4. CONVERGENCE OF THE APPROXIMATION

Let us modify the situation of the previous sections. Suppose we are given

- a subset $\mathbb{D}$ of $(0,+\infty)$ such that $0 \in \overline{\mathbb{D}}$;
- for any $d \in \mathbb{D}$, an ordered system of $N = N(d)$ connected nonempty subsets $F_j^d$ of $\overline{\Omega}$, with $1 \le j \le N(d)$, verifying hypothesis (1.1); we write

$$\forall d \in \mathbb{D},\ F^d = \bigcup_{j=1}^{N} F_j^d.$$

We suppose that

$$\exists F_0 \subset \overline{\Omega},\ \forall d \in \mathbb{D},\ \exists j_0 \in \{1,\dots,N(d)\},\ F_0 = F_{j_0}^d \tag{4.1}$$

and that

$$\forall p \in P_{m-1}(\overline{\Omega}),\ (p|_{F_0}) \Rightarrow (p = 0). \tag{4.2}$$

We keep the notations introduced in Section 2, but with $L = L(d,\eta,j)$, $\xi_i = \xi_i(d,\eta,j)$ and $\lambda_i = \lambda_i(d,\eta,j)$. It is agreed that, hereafter, $F_j^d$, $F^d$, $l_j^{d\eta}$, $l^{d\eta}$ and $J_{\varepsilon h}^{d\eta}$ will respectively replace $F_j$, $F$, $l_j^\eta$, $l^\eta$ and $J_{\varepsilon h}^\eta$.

In order to study the convergence of the fitting, we consider the following additional assumptions (where the same letter $C$ denotes various strictly positive constants):

$$\exists C > 0,\ \forall d \in \mathbb{D},\ \forall j = 1,\dots,N(d),\ \operatorname{meas} F_j^d \le C, \tag{4.3}$$

$$\left| \begin{array}{l} \exists C > 0,\ \exists t > 0,\ \forall (d,\eta) \in \mathbb{D} \times \mathbb{E},\ \forall j = 1,\dots,N(d), \\ \forall v \in H^m(\Omega),\ |l_j^{d\eta}(v^2) - \|v\|_{0,F_j^d}^2| \le C\eta^t \|v\|_{m,\Omega}^2, \end{array} \right. \tag{4.4}$$

$$\lim_{d\to 0} \sup_{x\in\Omega} \delta(x, F^d) = 0, \tag{4.5}$$

$$N = O(d^{-1}),\ d \to 0, \tag{4.6}$$

$$\left| \begin{array}{l} \forall \omega \ne \emptyset,\ \omega \text{ open set } \subset \Omega,\ \exists C_\omega > 0,\ \forall \nu \in \mathbb{N}^*,\ \exists \mu > 0, \\ \forall d \in \mathbb{D},\ d \le \mu,\ \exists \{j_1,\dots,j_\nu\} \in \{1,\dots,N(d)\}, \\ \forall k = 1,\dots,\nu,\ \operatorname{meas}(F_{j_k}^d \cap \omega) \ge C_\omega. \end{array} \right. \tag{4.7}$$

In these hypotheses, for any $d \in \mathbb{D}$, any $j = 1,\dots,N(d)$ and any $F^* \subset F_j^d$, $\operatorname{meas} F^*$ stands for the *measure* of $F^*$ in the measure space $F_j^d$ (which is well defined due to hypothesis (1.1)), i.e., in colloquial words, $\operatorname{meas} F^*$ is the "length" of $F^*$.

**Remark 4.1** – Hypothesis (4.1) is a simplifying assumption which is probably not essential.

Hypothesis (4.3) implies that the sets $F_j^d$ are not too "irregular".

Hypothesis (4.4), which is just condition (2.3) *with C independent of d*, is not easy to be verified in the general case. Fortunately, owing to (4.3), the situation is simpler when the curves $F_j^d$ are *polygonal.*

Hypothesis (4.5) means that the Hausdorff distance $\sup_{x\in\Omega}\delta(x, F^d)$ from $\overline{F^d}$ to $\overline{\Omega}$ tends to 0 as $d \to 0$. This is a classical condition for the convergence of spline functions.

Hypothesis (4.6) expresses a property of asymptotic regularity of the distribution of the curves $F_j^d$ in $\overline{\Omega}$. It is the analogue of condition (III–3.11) introduced in the case of the smoothing splines relative to point data.

Finally, hypothesis (4.7) is an assumption of "minoration" of the lengths of the curves $F_j^d$. Hypothesis (4.7) has no equivalent in the case of point conditions. By means of counterexamples, one can prove that neither (4.6) nor (4.7) are consequences of (4.5).

The different conditions (1.1), (1.3), (4.1)–(4.7) concerning the curves $F_j^d$ seem to be verified in the modelling of real problems, in particular, in Geology. □

For any $(d, \eta, \varepsilon, h) \in \mathbb{D} \times \mathbb{E} \times (0, +\infty) \times \mathbb{H}$, we again consider problems (3.4) and (3.5). Assuming that $\Omega$, $F^d$, $m$ and $f$ are defined as in Section 1, and that hypotheses (VI–1.2), (VI–1.5), (3.1), (3.3), (4.1), (4.2) and (4.4) are verified, a reasoning analogous to that in the proof of Theorem 3.1 shows that these problems admit, under hypothesis (3.6), a common unique solution, now denoted by $\sigma_{\varepsilon h}^{d\eta}$. We remark that, in (3.6), one can take the constant $\beta$ *independently of d.* To see this, it suffices to observe that, in the proof of Theorem 3.1, the relation (3.7) can be now replaced, for any $d \in \mathbb{D}$ and any $(\eta, \varepsilon) \in \mathbb{E} \times (0, +\infty)$, by

$$\forall h \in \mathbb{H},\ \forall v_h \in V_h,\ a_\varepsilon^\eta(v_h, v_h) \geq \min(1,\varepsilon)\Big(\|v_h\|_{0,F_0}^2 + |v_h|_{m,\Omega}^2\Big) - C\eta^t\|v_h\|_{m,\Omega}^2,$$

where $C$ is the constant (independent of $d$) introduced in (4.4). But (4.1) and (4.2) imply that the mapping $\left(\|\cdot\|_{0,F_0}^2 + |\cdot|_{m,\Omega}^2\right)^{1/2}$ is a norm on $H^m(\Omega)$ equivalent to the usual norm $\|\cdot\|_{m,\Omega}$. Hence, there exists a constant $C_0'$ such that

$$\forall h \in \mathbb{H},\ \forall v_h \in V_h,\ a_\varepsilon^\eta(v_h, v_h) \geq \Big(C_0'\min(1,\varepsilon) - C\eta^t\Big)\|v_h\|_{m,\Omega}^2.$$

Therefore, one can take $\beta = C_0'/C$, which is independent of $d$.

We have the following convergence result.

**Theorem 4.1** – *Suppose that $\Omega$, $m$ and $f$ are defined as in Section 1 and that hypotheses* (VI–1.2), (VI–1.5), (3.1), (3.2), (3.3), (4.1)–(4.4), (4.6) *and* (4.7) *are satisfied. Then, for any $\gamma > 0$ and for any $\varepsilon_0 \in (0, m-1)$, the solution $\sigma_{\varepsilon h}^{d\eta}$ of* (3.4) *and* (3.5) *verifies*

$$\lim_{\substack{d\to 0,\ \eta^t/d\leq\gamma,\ 0<\varepsilon\leq\varepsilon_0 \\ \eta^t/\min(1,\varepsilon)<\beta,\ h^{2(m-1-\theta)}/(d\varepsilon)\to 0}} \|\sigma_{\varepsilon h}^{d\eta} - f\|_{m,\Omega} = 0.$$

**Proof** – The same letter $C$ denotes various strictly positive constants.

From now on, condition (3.6) is supposed to be verified, although it is not explicitly mentioned (remember that, under hypotheses (4.1) and (4.2), $\beta$ can be chosen independently of $d$).

1) Let $\tilde{f} \in H^m(\widetilde{\Omega})$ be any extension of $f$ (the existence of $\tilde{f}$ is justified by (5) in Preliminaries). Taking $v_h = (\widetilde{\Pi}_h \tilde{f})|_{\Omega_h}$, where $\widetilde{\Pi}_h$ is the operator introduced in (3.2), we have

$$\forall (d, \eta, \varepsilon, h) \in \mathbb{D} \times \mathbb{E} \times (0, +\infty) \times \mathbb{H},\ J_{\varepsilon h}^{d\eta}(\sigma_{\varepsilon h}^{d\eta}) \leq l^{d\eta}((\widetilde{\Pi}_h \tilde{f} - \tilde{f})^2) + \varepsilon |\widetilde{\Pi}_h \tilde{f}|_{m,\Omega_h}^2.$$

Taking into account that $\mathbb{E}$ is bounded, it follows from relations (4.3) and (4.4) that

$$\exists C,\ \forall (d, \eta) \in \mathbb{D} \times \mathbb{E},\ \forall j = 1, \ldots, N(d),\ l_j^{d\eta}(1) \leq C.$$

Hence,

$$\exists C,\ \forall (d, \eta) \in \mathbb{D} \times \mathbb{E},\ l^{d\eta}((\widetilde{\Pi}_h \tilde{f} - \tilde{f})^2) \leq C N(d) \|\widetilde{\Pi}_h \tilde{f} - \tilde{f}\|_{C^0(\overline{\widetilde{\Omega}})}^2.$$

From Sobolev's Continuous Imbedding Theorem (cf. (4) in Preliminaries), we have

$$\forall \theta > 0,\ H^{1+\theta}(\widetilde{\Omega}) \hookrightarrow C^0(\overline{\widetilde{\Omega}})$$

(let us recall that, when $\theta$ is noninteger, $H^{1+\theta}(\widetilde{\Omega})$ is the Sobolev space of noninteger order $1+\theta$, whose norm is denoted by $\| \cdot \|_{1+\theta,\widetilde{\Omega}}$). Using an interpolation method between the spaces $H^1(\widetilde{\Omega})$ and $H^m(\widetilde{\Omega})$ (cf., for instance, J. Peetre [116]), we deduce from (3.2) that

$$\forall \theta \in (0, m-1),\ \exists C,\ \forall h \in \mathbb{H},\ \|\widetilde{\Pi}_h \tilde{f} - \tilde{f}\|_{1+\theta,\widetilde{\Omega}} \leq C h^{m-1-\theta} \|\tilde{f}\|_{m,\widetilde{\Omega}}.$$

Summing up the previous results and using (4.6), we obtain

$$\begin{gathered}\forall \theta \in (0, m-1),\ \exists C,\ \exists d_0 > 0,\ \forall (d, \eta, \varepsilon, h) \in \mathbb{D} \times \mathbb{E} \times (0, +\infty) \times \mathbb{H},\ d \leq d_0, \\ J_{\varepsilon h}^{d\eta}(\sigma_{\varepsilon h}^{d\eta}) \leq C h^{2(m-1-\theta)}/d + \varepsilon |\widetilde{\Pi}_h \tilde{f}|_{m,\Omega_h}^2.\end{gathered} \tag{4.8}$$

Now, let us point out that, from (VI–1.3) and (3.2),

$$\exists C,\ \forall h \in \mathbb{H},\ |\widetilde{\Pi}_h \tilde{f}|_{m,\Omega_h} \leq C. \tag{4.9}$$

On the other hand, using (4.1), (4.4) and the definition of $J_{\varepsilon h}^{d\eta}$, we obtain

$$\|\sigma_{\varepsilon h}^{d\eta} - f\|_{0,F_0}^2 \leq J_{\varepsilon h}^{d\eta}(\sigma_{\varepsilon h}^{d\eta}) + C\eta^t \|\sigma_{\varepsilon h}^{d\eta} - f\|_{m,\Omega}^2.$$

Taking (VI–1.3) into account, we have

$$|\sigma_{\varepsilon h}^{d\eta} - f|_{m,\Omega}^2 \leq (2/\varepsilon) J_{\varepsilon h}^{d\eta}(\sigma_{\varepsilon h}^{d\eta}) + 2|f|_{m,\Omega}^2.$$

Then, it follows from (4.8) and (4.9) that, for any $\theta \in (0, m-1)$,

$$\begin{aligned}\forall (d,\eta,\varepsilon,h) \in \mathbb{D}\times\mathbb{E}\times(0,+\infty)\times\mathbb{H},\ d\le d_0,\ &\|\sigma_{\varepsilon h}^{d\eta} - f\|_{0,F_0}^2 + |\sigma_{\varepsilon h}^{d\eta} - f|_{m,\Omega}^2 \\ &\le C(\varepsilon+2)\big(h^{2(m-1-\theta)}/(d\varepsilon)+1\big) + 2|f|_{m,\Omega}^2 + C\eta^t\|\sigma_{\varepsilon h}^{d\eta} - f\|_{m,\Omega}^2.\end{aligned} \tag{4.10}$$

Since the mapping $(\|\cdot\|_{0,F_0}^2 + |\cdot|_{m,\Omega}^2)^{1/2}$ is a norm on $H^m(\Omega)$ equivalent to the natural norm $\|\cdot\|_{m,\Omega}$, we derive from (4.10) that the family $(\sigma_{\varepsilon h}^{d\eta})_{(d,\eta,\varepsilon,h)\in\mathbb{D}\times\mathbb{E}\times(0,+\infty)\times\mathbb{H}}$ is bounded in $H^m(\Omega)$ for $d$ sufficiently small, provided $\varepsilon$ and $h^{2(m-1-\theta)}/(d\varepsilon)$ remain bounded.

Let us suppose from now on that

$$\left|\begin{array}{l}\gamma>0,\ \varepsilon_0>0,\ \theta\in(0,m-1),\ d\to 0,\\ \eta^t/d\le\gamma,\ 0<\varepsilon\le\varepsilon_0,\ h^{2(m-1-\theta)}/(d\varepsilon)\to 0.\end{array}\right. \tag{4.11}$$

In this situation, by Corollary 1 in Preliminaries, there exists a sequence $(\sigma_{\varepsilon_n h_n}^{d_n\eta_n})_{n\in\mathbb{N}}$, with $\lim_{n\to+\infty} d_n = 0$, $(\eta_n^t/d_n)_{n\in\mathbb{N}} \subset (0,\gamma]$, $(\varepsilon_n)_{n\in\mathbb{N}} \subset (0,\varepsilon_0]$, $(\eta_n^t/\min(1,\varepsilon_n))_{n\in\mathbb{N}} \subset (0,\beta)$ and $\lim_{n\to+\infty} h_n^{2(m-1-\theta)}/(d_n\varepsilon_n) = 0$, extracted from the family $(\sigma_{\varepsilon h}^{d\eta})$, and there also exists an element $f^* \in H^m(\Omega)$ such that, as $n\to+\infty$,

$$\sigma_{\varepsilon_n h_n}^{d_n\eta_n} \to f^*, \quad \text{weakly in } H^m(\Omega).$$

2) Let us prove that $f^* = f$ reasoning by contradiction. Suppose that $f^* \ne f$. Then, there exists a nonempty open set $\omega$ contained in $\Omega$ and a real number $\alpha^* > 0$ such that

$$\forall x\in\omega,\ |f^*(x) - f(x)| > \alpha^*.$$

But $(\sigma_{\varepsilon_n h_n}^{d_n\eta_n})_{n\in\mathbb{N}}$ tends weakly to $f^*$ in $H^m(\Omega)$, and therefore converges strongly in $C^0(\overline{\Omega})$. Consequently,

$$\exists\alpha>0,\ \exists n'\in\mathbb{N},\ \forall n\in\mathbb{N},\ n\ge n',\ \forall x\in\omega,\ |\sigma_{\varepsilon_n h_n}^{d_n\eta_n}(x) - f(x)| > \alpha.$$

Furthermore, we have from (4.7)

$$\begin{aligned}&\exists C_\omega>0,\ \forall\nu\in\mathbb{N}^*,\ \exists n''\in\mathbb{N},\ \forall n\in\mathbb{N},\ n\ge n'',\\ &\qquad \exists\{j_1,\ldots,j_\nu\}\subset\{1,\ldots,N(d_n)\},\ \forall k=1,\ldots,\nu,\ \operatorname{meas}(F_{j_k}^{d_n}\cap\omega)\ge C_\omega.\end{aligned}$$

Let

$$\forall\nu\in\mathbb{N}^*,\ \forall n\in\mathbb{N},\ n\ge n'',\ \Psi(\nu,n) = \sum_{k=1}^{\nu} \|\sigma_{\varepsilon_n h_n}^{d_n\eta_n} - f\|_{0,F_{j_k}^{d_n}}^2.$$

We infer that

$$\forall n\in\mathbb{N},\ n\ge\max(n',n''),\ \Psi(\nu,n) > \nu\alpha^2 C_\omega.$$

On the other hand, under conditions (4.11), it follows from (4.4), (4.6), (4.8) and (4.9) that

$$\exists C,\ \forall\nu\in\mathbb{N}^*,\ \forall n\in\mathbb{N},\ n\ge n'',\ \Psi(\nu,n)\le C(\varepsilon_0+\gamma).$$

Hence, we get the relation

$$\forall n \in \mathbb{N},\ n \geq \max(n', n''),\ \nu\alpha^2 C_\omega < C(\varepsilon_0 + \gamma),$$

which yields a contradiction as $\nu \to +\infty$.

3) Let us now show that

$$\lim_{n\to+\infty} \|\sigma_{\varepsilon_n h_n}^{d_n \eta_n} - f\|_{m,\Omega} = 0.$$

Since the injection from $H^m(\Omega)$ into $H^{m-1}(\Omega)$ is compact, we derive from points 1) and 2) that, as $n \to +\infty$,

$$\sigma_{\varepsilon_n h_n}^{d_n \eta_n} \to f, \ \text{ strongly in } H^{m-1}(\Omega).$$

On the other hand, under conditions (4.11), we have from condition (4.8)

$$|\sigma_{\varepsilon_n h_n}^{d_n \eta_n}|_{m,\Omega}^2 \leq |\widetilde{\Pi}_{h_n} \tilde{f}|_{m,\Omega_{h_n}}^2 + o(1),\ n \to +\infty.$$

Now, for any $h \in \mathbb{H}$,

$$\left| |\widetilde{\Pi}_h \tilde{f}|_{m,\Omega_h} - |\tilde{f}|_{m,\Omega_h} \right| \leq |\tilde{f} - \widetilde{\Pi}_h \tilde{f}|_{m,\Omega_h}$$

and the right-hand member tends to 0, as $h \to 0$, because of (3.2) and (VI–1.3). Using (VI–1.4), we get

$$\lim_{h\to 0} |\widetilde{\Pi}_h \tilde{f}|_{m,\Omega_h} = |f|_{m,\Omega}.$$

Hence,

$$|\sigma_{\varepsilon_n h_n}^{d_n \eta_n}|_{m,\Omega}^2 \leq |f|_{m,\Omega}^2 + o(1),\ n \to +\infty,$$

from which we deduce that

$$|\sigma_{\varepsilon_n h_n}^{d_n \eta_n} - f|_{m,\Omega}^2 \leq 2|f|_{m,\Omega}^2 - 2\left(\sigma_{\varepsilon_n h_n}^{d_n \eta_n}, f\right)_{m,\Omega} + o(1),\ n \to +\infty,$$

Then,

$$\lim_{n\to+\infty} |\sigma_{\varepsilon_n h_n}^{d_n \eta_n} - f|_{m,\Omega} = 0,$$

and the result follows.

4) Finally, let us prove that

$$\lim \|\sigma_{\varepsilon h}^{d\eta} - f\|_{m,\Omega} = 0,$$

where the limit is to be taken in the sense of the conditions (4.11). Let us reason by contradiction. Suppose that the previous relation does not hold under condition (4.11). This comes down to saying that there exists $\alpha > 0$, $\gamma' > 0$, $\varepsilon_0' > 0$, $\theta' \in (0, m-1)$ and $\left((d_n', \eta_n', \varepsilon_n', h_n')\right)_{n\in\mathbb{N}}$, a sequence in $\mathbb{D} \times \mathbb{E} \times (0, \varepsilon_0'] \times \mathbb{H}$, such that $d_n' \to 0$, ${\eta_n'}^t / d_n' \leq \gamma'$, ${\eta_n'}^t / \min(1, \varepsilon_n') < \beta$ and ${h_n'}^{2(m-1-\theta')} / d_n' \varepsilon_n' \to 0$, verifying

$$\forall n \in \mathbb{N},\ \|\sigma_{\varepsilon_n' h_n'}^{d_n' \eta_n'} - f\|_{m,\Omega} > \alpha.$$

But such a sequence is bounded in $H^m(\Omega)$ and the previous reasoning leads to a contradiction. □

**Remark 4.2** – Theorem 4.1 can be stated in the following way: the solution $\sigma_{\varepsilon h}^{d\eta}$ of (3.4) and (3.5) converges to $f$ in $H^m(\Omega)$ through the filter basis $\mathcal{B} = \{ B_{r,s} \mid r > 0,\ s > 0 \}$, with $B_{r,s} = \{ (d,\eta,\varepsilon,h) \in \mathbb{D} \times \mathbb{E} \times (0,\varepsilon_0] \times \mathbb{H} \mid d \leq r,\ \eta^t/d \leq \gamma,\ \eta^t/\min(1,\varepsilon) < \beta,\ h^{2(m-1-\theta)}/d\varepsilon \leq s \}$. □

**Remark 4.3** – When $\varepsilon \to 0$ (which is, in certain respects, a restrictive condition), we can obtain a convergence result which does not require hypothesis (4.7), but uses hypothesis (4.5). We reason as follows.

For any $d \in \mathbb{D}$, we consider the "interpolating spline" $\sigma^d$ solution of the problem (analogous to (1.2)): find $\sigma^d$ solution of

$$\begin{cases} \sigma^d \in \mathcal{K}^d, \\ \forall v \in \mathcal{K}^d,\ |\sigma^d|_{m,\Omega} \leq |v|_{m,\Omega}, \end{cases}$$

where

$$\mathcal{K}^d = \{ v \in H^m(\Omega) \mid v|_{F^d} = f|_{F^d} \}.$$

It is clear that the family $(\sigma^d)_{d\in\mathbb{D}}$ is bounded in $H^m(\Omega)$. Moreover, it follows from (4.5) that, for any $x \in \Omega$,

$$\exists (x^d)_{d\in\mathbb{D}},\ \left(\forall d \in \mathbb{D},\ x^d \in \overline{F^d}\right) \text{ and } \left(x = \lim_{d\to 0} x^d\right).$$

Then, reasoning as in the proof of Theorem II–5.1, we show that

$$\lim_{d\to 0} \|\sigma^d - f\|_{m,\Omega} = 0.$$

This result, together with that of Theorem 3.2, proves that under hypotheses (VI–1.2), (VI–1.5), (3.1), (3.2), (3.3) and (4.1)–(4.6), the solution $\sigma_{\varepsilon h}^{d\eta}$ of (3.4) and (3.5) verifies

$$\lim_{\substack{\varepsilon\to 0,\ \eta^t/\varepsilon<\beta \\ h^{2(m-1-\theta)}/\varepsilon\to 0}} \|\sigma_{\varepsilon h}^{d\eta} - f\|_{m,\Omega} = 0. \quad \square$$

## 5. NUMERICAL RESULTS

To illustrate the smoothing method developed in this chapter, we have considered several sets $F$ of data curves, depicted in Figures 2–5. They allow us to exemplify the fitting of a surface from either a geometrically simple set of curves, or a set of parallel cross-sections, or a set of iso-valued curves. Figures 2–5 also show, marked with dots, the nodes $\xi_1, \ldots, \xi_L$ that we have chosen on every curve $F_j \subset F$ in order to define the linear form $l_j^\eta$, introduced in (2.1). We have applied the theory of Sections 1, 2 and 3, setting $m = 2$ and taking as $l_j^\eta$ the expression (2.6), corresponding to the Trapezoidal Rule. Expressions (2.4) and (2.5) also verify (2.3) and are simpler, but the Trapezoidal Rule is probably numerically better than the Rectangle Formulae. We indicate in Table 1 the total number of nodes $\xi_i$ used to compute $l^\eta$, as well as the number $N$ of simple curves (open arcs or closed curves) contained in $F$.

As in Section VIII–6, the test function $f$ is either Franke's function (cf. (VIII–6.1) and Figure VIII–8) or Nielson's function (cf. (VIII–6.2) and Figure VIII–13), both defined

| $F$ | Figure 2 | Figure 3 | Figure 4 | Figure 5 |
|---|---|---|---|---|
| $N$ | 10 | 11 | 14 | 28 |
| No. nodes | 282 | 220 | 278 | 528 |

Table 1: Number $N$ of the curves that compose $F$ and number of nodes $\xi_i$ fixed on $F$ to define $l^\eta$.

on $\Omega = (0,1)\times(0,1)$. Likewise, we have set $\widetilde{\Omega} = \Omega$, we have taken a triangulation $\widetilde{\mathcal{T}}_h$ of $\overline{\widetilde{\Omega}}$ made up of $7\times 7$ equal squares and we have constructed on $\widetilde{\mathcal{T}}_h$ a finite element space $\overline{\widetilde{\Omega}}$ from the Bogner-Fox-Schmit rectangle of class $C^1$. Once again we have $\Omega_h = \Omega$, $\mathcal{T}_h = \widetilde{\mathcal{T}}_h$ and $V_h = \widetilde{V}_h$. The dimension $M$ of $V_h$ is 256.

For every test function $f$, we have computed a $V_h$-discrete smoothing $D^2$-spline $\sigma^\eta_{\varepsilon h}$ for every one of the sets $F$ represented in Figures 2, 3 and 4, if $f$ is Franke's function, or in Figures 2, 3 and 5, if $f$ is Nielson's function. All these splines are shown in Figures 6–11. We have written at the foot of each figure the specific value of $\varepsilon$ corresponding to every spline, as well as the associated relative error, given by (VIII–6.3) with $\sigma^\eta_{\varepsilon h}$ instead of $\sigma_{\varepsilon h}$.

In every case, the value of $\varepsilon$ has been fixed with the help of the GCV method (cf. Remark 3.3). First, we have set $\varepsilon$ as an estimate of the minimum of the approximated GCV function $\overline{\mathcal{V}}$, which is around $10^{-7}$ in most of these examples. Then, if needed, we have progressively increased the parameter $\varepsilon$, without entering the interval on which $\overline{\mathcal{V}}$ has steep slopes, until we have found a visually more acceptable approximation. For the sake of completeness, we include the plot, in a logarithmic scale, of every function $\overline{\mathcal{V}}$ (cf. Figure 12).

These examples show that the quality of the fitting is influenced not only by the quantity of data at our disposal, but also, even more strongly, by the distribution of the data. This fact can be observed in the approximants of Franke's function, obtained from data sets with almost the same number of curves and a similar number of nodes. The three approximants of Nielson's function exhibit much more apparent differences, mainly due to the diverse locations of the data curves.

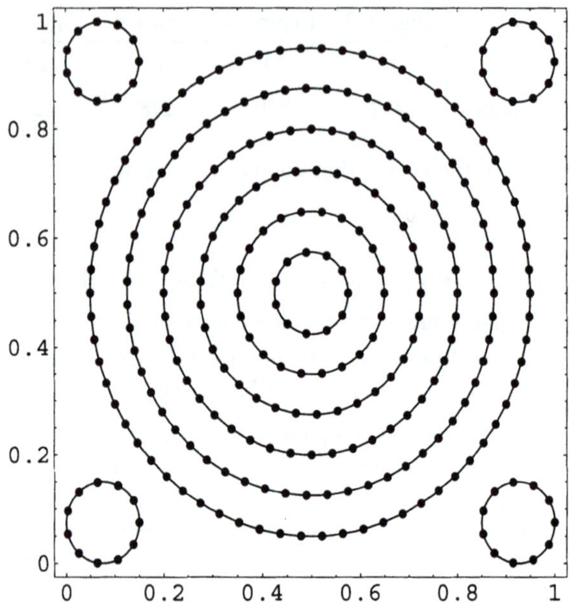


Figure 2: Set $F$ of data curves and nodes $\xi_i$. First case: circles of centre $(0.5, 0.5)$ and radius $(1+i)r$, for $r = 0.075$ and $i = 0, \ldots, 5$, and circles of radius $r$ and centre $(r, r)$, $(r, 1-r)$, $(1-r, 1-r)$ and $(1-r, r)$.

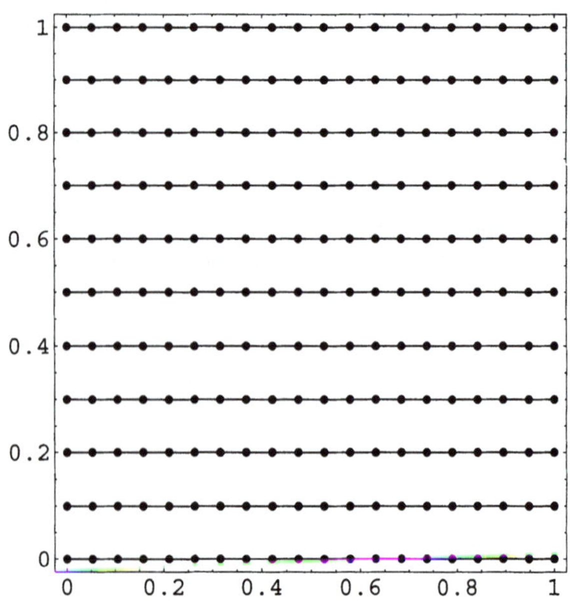


Figure 3: Set $F$ of data curves and nodes $\xi_i$. Second case: segments of the form $[0,1] \times \{i/10\}$, for $i = 0, \ldots, 10$.

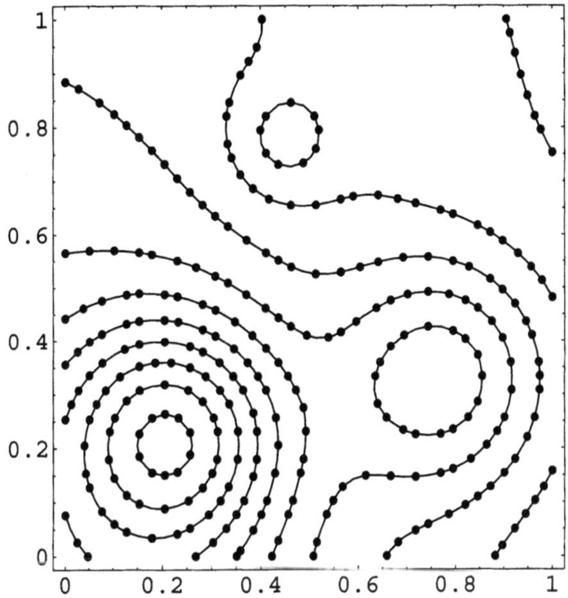


Figure 4: Set $F$ of data curves and nodes $\xi_i$. Third case: level curves of Franke's function corresponding to the values $0.05 + 0.125\,i$, for $i = 0, \ldots, 9$.

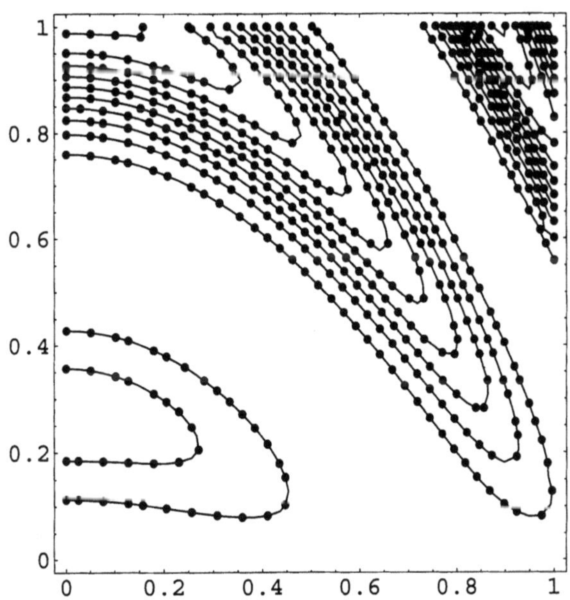


Figure 5: Set $F$ of data curves and nodes $\xi_i$. Fourth case: level curves of Nielson's function corresponding to the values $0.04 + 0.05\,i$, for $i = 0, \ldots, 9$.

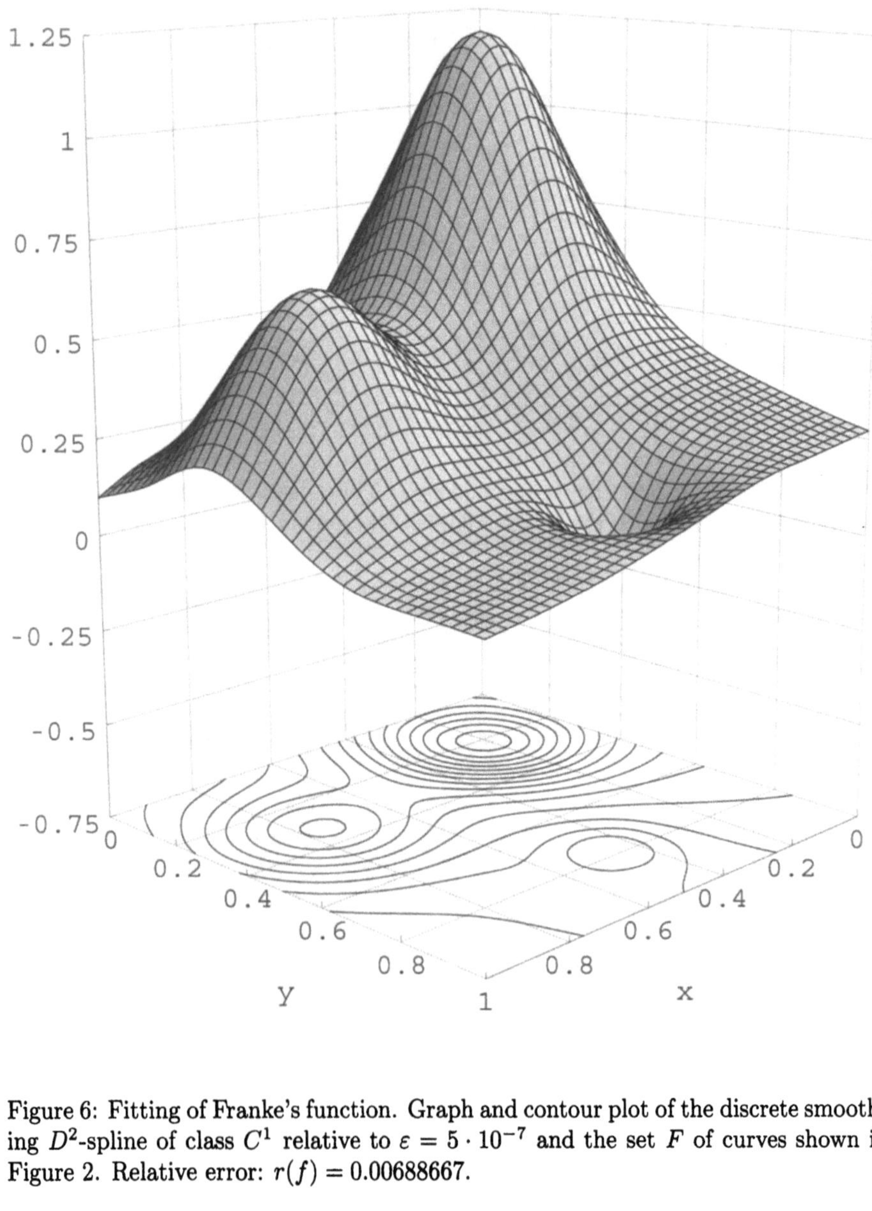


Figure 6: Fitting of Franke's function. Graph and contour plot of the discrete smoothing $D^2$-spline of class $C^1$ relative to $\varepsilon = 5 \cdot 10^{-7}$ and the set $F$ of curves shown in Figure 2. Relative error: $r(f) = 0.00688667$.

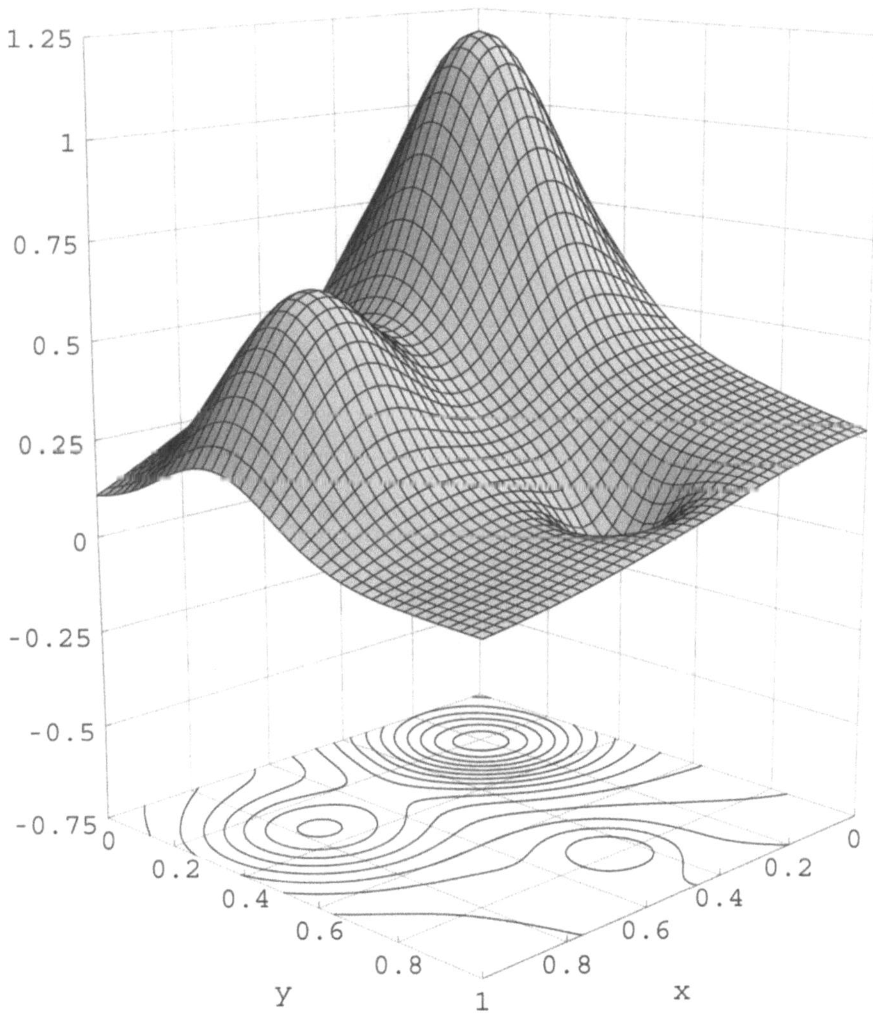


Figure 7: Fitting of Franke's function. Graph and contour plot of the discrete smoothing $D^2$-spline of class $C^1$ relative to $\varepsilon = 10^{-7}$ and the set $F$ of curves shown in Figure 3. Relative error: $r(f) = 0.00274627$.

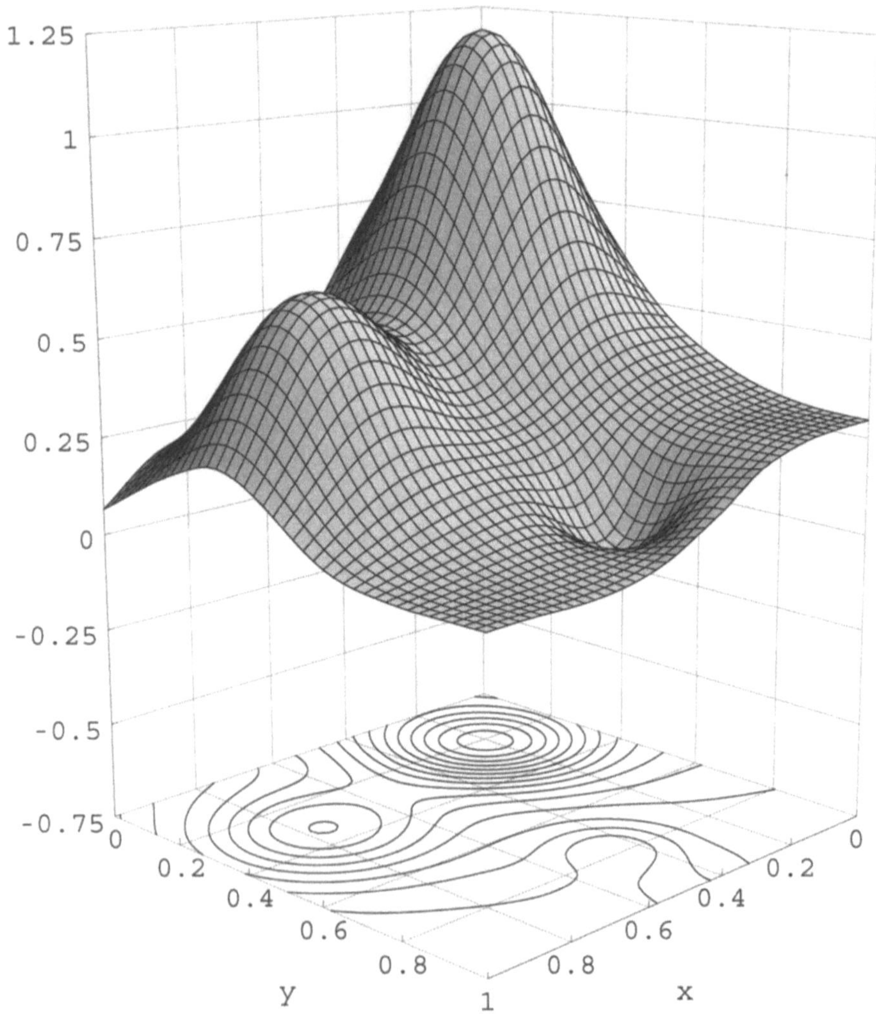


Figure 8: Fitting of Franke's function. Graph and contour plot of the discrete smoothing $D^2$-spline of class $C^1$ relative to $\varepsilon = 10^{-7}$ and the set $F$ of curves shown in Figure 4. Relative error: $r(f) = 0.0256927$.

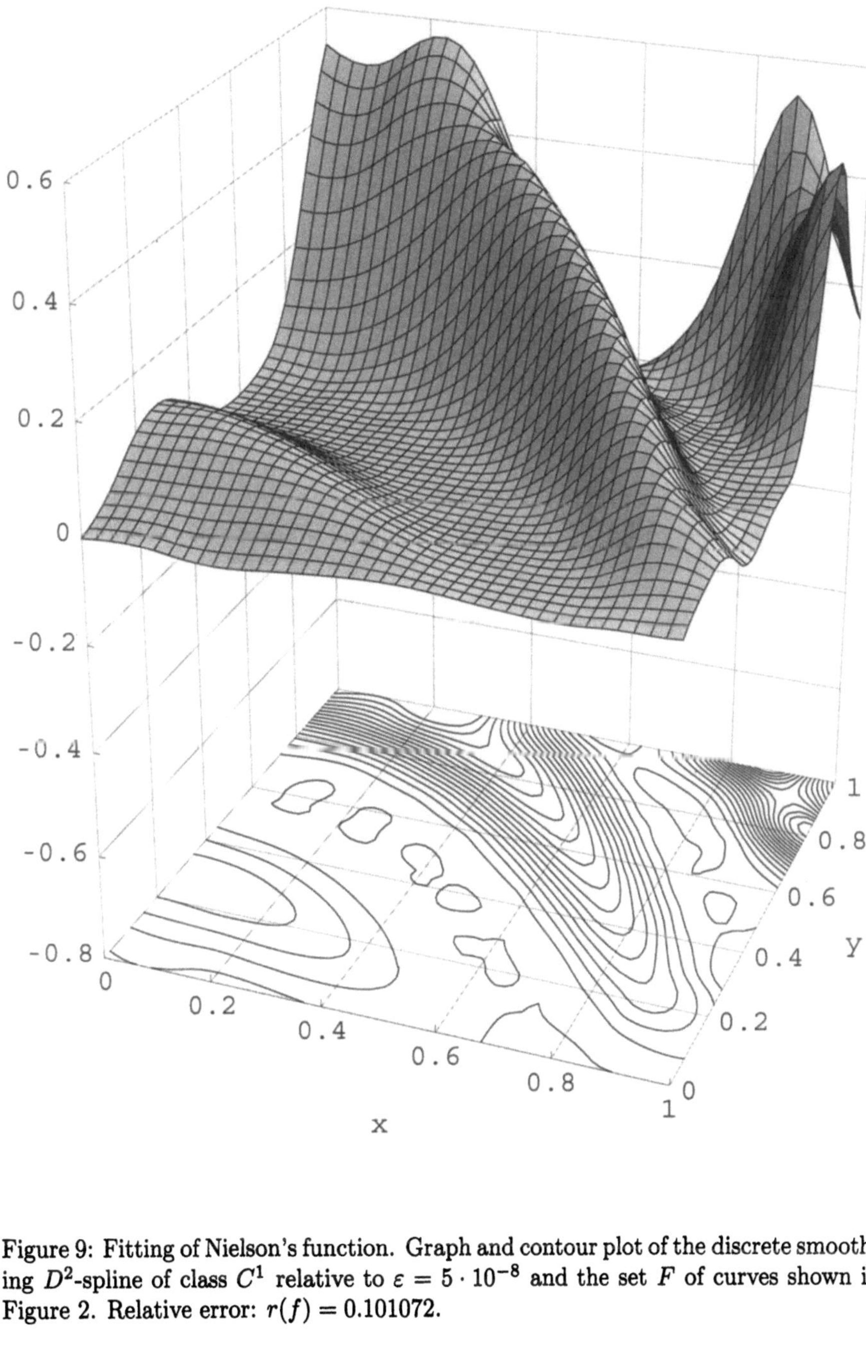


Figure 9: Fitting of Nielson's function. Graph and contour plot of the discrete smoothing $D^2$-spline of class $C^1$ relative to $\varepsilon = 5 \cdot 10^{-8}$ and the set $F$ of curves shown in Figure 2. Relative error: $r(f) = 0.101072$.

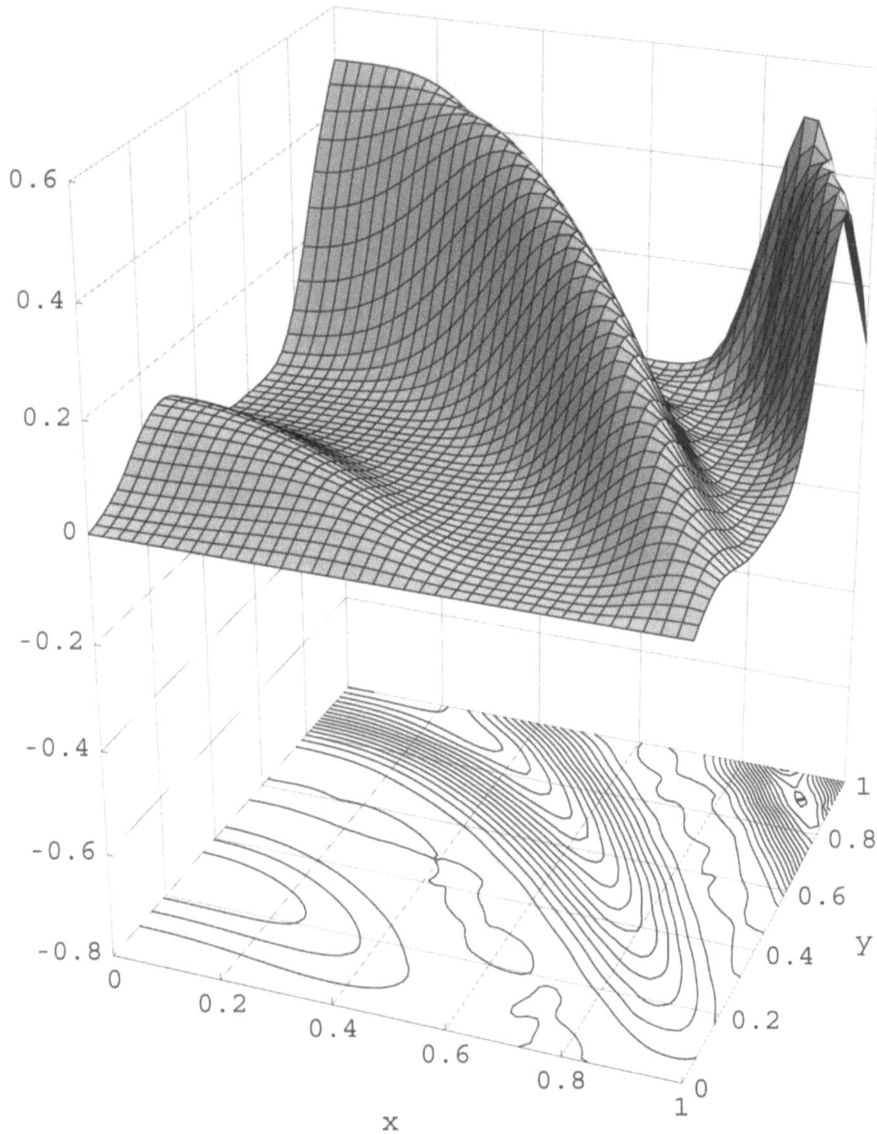


Figure 10: Fitting of Nielson's function. Graph and contour plot of the discrete smoothing $D^2$-spline of class $C^1$ relative to $\varepsilon = 5 \cdot 10^{-7}$ and the set $F$ of curves shown in Figure 3. Relative error: $r(f) = 0.0243512$.

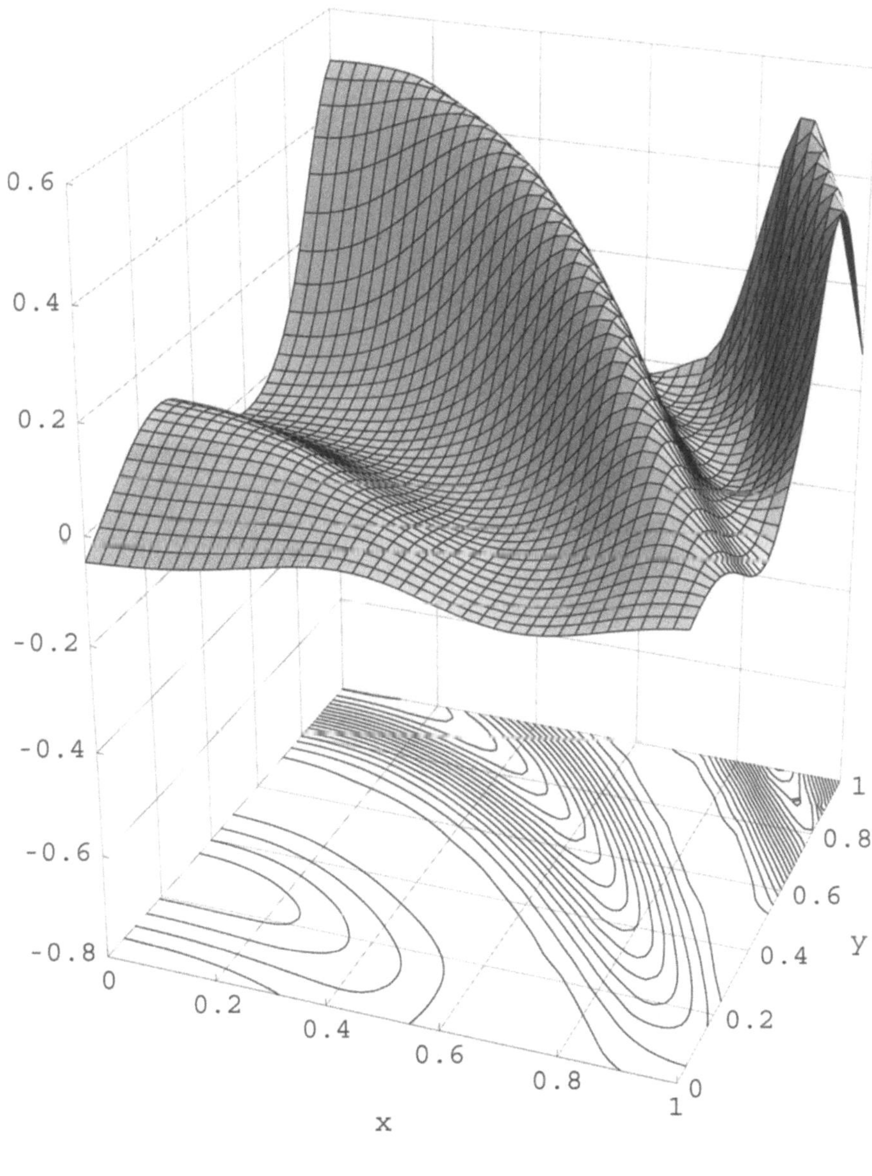


Figure 11: Fitting of Nielson's function. Graph and contour plot of the discrete smoothing $D^2$-spline of class $C^1$ relative to $\varepsilon = 10^{-6}$ and the set $F$ of curves shown in Figure 5. Relative error: $r(f) = 0.0793664$.

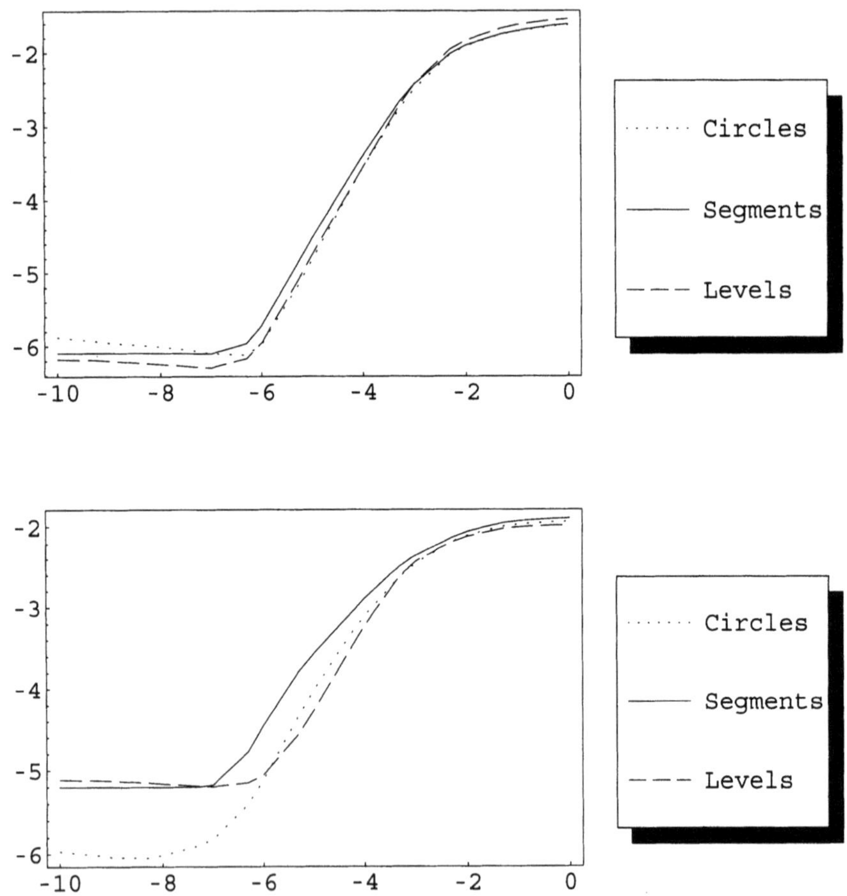


Figure 12: Plot of $\log_{10} \overline{\mathcal{V}}(\varepsilon)$ versus $\log_{10} \varepsilon$. Top: Franke's function, bottom: Nielson's function. The labels **Circles**, **Segments** and **Levels** refer, respectively, to the data sets represented in Figures 2, 3 and 4, for Franke's function, or in Figures 2, 3 and 5, for Nielson's function.

# CHAPTER XI

# FITTING AN EXPLICIT SURFACE OVER AN OPEN SET †

## 1. FORMULATION OF THE PROBLEM

The problem that we propose to study in this chapter can be formulated as follows: given a bounded open set $\Omega$ in $\mathbb{R}^n$ and a regular function $f$ defined on a nonempty open subset $\omega$ of $\Omega$, construct over $\Omega$ a regular function $\phi$ that approximates, in a sense to be defined, the function $f$ on $\omega$. More precisely, we suppose that

- $\Omega$ is an open set in $\mathbb{R}^n$ with a Lipschitz-continuous boundary (cf. Preliminaries),
- for the sake of simplicity, $f$ is the restriction to $\omega$ of a function, denoted by $f$ as well, that belongs to the usual Sobolev space $H^m(\Omega)$, where $m$ is an integer $> n/2$,

and we impose the condition

$$\phi \in H^m(\Omega) \cap C^k(\overline{\Omega}), \text{ with } k = 1 \text{ or } 2.$$

The main example of this situation, when $n = 2$, is that of fitting a surface of explicit type on an open subset of its domain of definition. In the applications, the set $\omega$ is, for instance, a finite union of nonempty pairwise disjoint subdomains of $\Omega$ (cf. Figure 1).

When $m > k + n/2$, the inclusion $H^m(\Omega) \subset C^k(\overline{\Omega})$ holds. It follows that, in this case, the *interpolation* problem corresponding to the general problem considered above obviously admits an infinite number of solutions. Unfortunately, it is quite impossible to *find* $\sigma$ *explicitly* (as well as any other interpolant) *in a form which can be used for computing.*

Next, we propose to construct, as the approximating function $\phi$, a "discrete smoothing $D^m$-spline", adapted from the theory in Section VI–3. We shall proceed in a way quite similar to that of Chapter X.

## 2. SPLINE FITTING

Suppose we are given

- a bounded subset $\mathbb{E}$ of $(0, +\infty)$ such that $0 \in \overline{\mathbb{E}}$,

†The origin of this chapter is an unpublished joint work by D. Apprato and R. Arcangéli. We also refer to C. Gout [72] and D. Apprato and C. Gout [11, 12]. However, we do not agree with the point of view that the latter authors adopt in the study of the convergence.

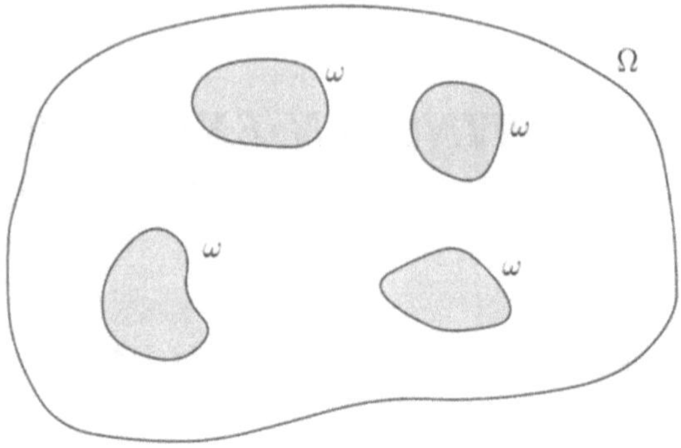


Figure 1: Example of open subset $\omega$, composed of four subdomains.

- for any $\eta \in \mathbb{E}$, a set $\{\xi_i\}_{1 \le i \le L}$ of $L = L(\eta)$ distinct points $\xi_i = \xi_i(\eta)$ of $\overline{\omega}$ and a set $\{\lambda_i\}_{1 \le i \le L}$ of numbers $\lambda_i = \lambda_i(\eta) > 0$.

Then we write

$$\forall \eta \in \mathbb{E},\ \forall v \in C^0(\overline{\Omega}),\ l^\eta(v) = \sum_{i=1}^{L} \lambda_i v(\xi_i). \tag{2.1}$$

From now on, we shall consider $l^\eta$ as a continuous linear form on $C^0(\overline{\Omega})$ or on $C^0(\overline{\Omega'})$, for any open subset $\Omega'$ containing $\Omega$.

We assume that

$$\left| \begin{array}{l} \exists C > 0,\ \exists t > 0,\ \forall \eta \in \mathbb{E}, \\ \forall v \in H^m(\Omega),\ |l^\eta(v^2) - \|v\|_{0,\omega}^2| \le C\eta^t \|v\|_{m,\Omega}^2. \end{array} \right. \tag{2.2}$$

**Remark 2.1** – When hypothesis (2.2) is verified, the relation

$$\|v\|_{0,\omega}^2 \sim l^\eta(v^2),\ \eta \to 0$$

constitutes an (abstract) numerical integration formula for $\|\cdot\|_{0,\omega}^2$. Of course, the nodes $\xi_i$ cannot be any points of $\overline{\omega}$. The convergence of this formula as $\eta \to 0$ implies that

$$(\eta \to 0) \Rightarrow \left(\sup_{x \in \omega} \delta(x, \Xi) \to 0\right),$$

where $\delta(\cdot,\cdot)$ denotes the Euclidean distance in $\mathbb{R}^2$ and $\Xi$ is the set of the nodes $\xi_i$.

There exist explicit formulae verifying (2.2). Suppose, for instance, that $\omega$ is *polyhedral.* For any $\eta \in \mathbb{E}$, let $\mathbb{T}_\eta$ be a triangulation of $\overline{\omega}$ by means of $n$-simplices $K$ with diameters $\le \eta$. There exist numerical $P_{m-1}(K)$-exact quadrature formulae of the type

$$\int_K v(x)\,dx \ \sim\ (\operatorname{meas} K) \sum_{i=1}^{N} c_i v(x_{iK}),$$

where the coefficients $c_i$ and the points $x_{iK} \in K$ classically denote the weights and the nodes of the formula. It is not necessary that the weights be positive, but this is a convenient feature to ensure the numerical stability of the formula. To clarify, it is agreed that $x_{iK} = F_K(\hat{x}_i)$, for $i = 1, \dots, N$, where $F_K$ is an invertible affine mapping such that $K = F_K(\widehat{K})$, $\widehat{K}$ denoting a fixed $n$-simplex in $\mathbb{R}^n$ and $\hat{x}_i \in \widehat{K}$, for $i = 1, \dots, N$.

Let us suppose that $m \geq n$. Let $\mathcal{O}$ be any open set in $\mathbb{R}^n$ and let $v \in H^m(\mathcal{O})$. Using Leibniz's Formula, one verifies that $v^2 \in W^{m,1}(\mathcal{O})$ (cf., for instance, R. A. Adams [1] for the definition of this classical Sobolev space) and that

$$|v^2|_{m,1,\mathcal{O}} \leq C \|v\|^2_{m,\mathcal{O}},$$

where $| \cdot |_{m,1,\mathcal{O}}$ stands for the usual semi-norm of length $m$ in $W^{m,1}(\mathcal{O})$ and $C$ is a constant independent of $\mathcal{O}$. If $\mathcal{O}$ has a Lipschitz-continuous boundary (cf. Preliminaries), then $v \in C^0(\overline{\mathcal{O}})$ (cf. Theorem (3) in Preliminaries) and, therefore, $v^2 \in W^{m,1}(\mathcal{O}) \cap C^0(\overline{\mathcal{O}})$.

Let us denote by $\overset{\circ}{K}$ and $\operatorname{diam} K$ the interior and the diameter of any $n$-simplex $K$. From the above considerations, we deduce that there exists some constant $C$ such that, for any $K \in \mathbb{T}_\eta$ and for any $v \in W^{m,1}(\overset{\circ}{K}) \cap C^0(K)$,

$$\left| \int_K v(x)\,dx - (\operatorname{meas} K) \sum_{i=1}^{N} c_i v(x_{iK}) \right| \leq C |v|_{m,1,K} (\operatorname{diam} K)^m,$$

(cf. P. G. Ciarlet [45, Problem 4.1.4], where a proof of this result, by passing to the reference element $\widehat{K}$ and using the Bramble-Hilbert Lemma –cf. [45, Theorem 4.1.3]–, is proposed).

Now, let us write

$$\forall v \in W^{m,1}(\omega) \cap C^0(\overline{\omega}),\ l^\eta(v) = \sum_{K \in \mathbb{T}_\eta} (\operatorname{meas} K) \sum_{i=1}^{N} c_i v(x_{iK}).$$

Then, one can verify that, when $m \geq n$, the formula (2.2) with $l^\eta$ defined as above, is satisfied for $t = m$. Notice that, in this example, the constant $C$ of (2.2) is *independent* of $\omega$. □

**Remark 2.2** – Condition (2.2) implies that there exists $\eta_0 > 0$ such that, for any $\eta \in \mathbb{E} \cap (0, \eta_0]$, the set $\{\xi_i(\eta)\}_{1 \leq i \leq L(\eta)}$ contains a $P_{m-1}$-unisolvent subset. To prove this, we reason by contradiction. If the result is false, there exists a sequence $(\eta_n)_{n \in \mathbb{N}} \subset \mathbb{E}$ such that $\lim_{n \to +\infty} \eta_n = 0$ and, for any $n \in \mathbb{N}$, the set $\{\xi_i(\eta_n)\}_{1 \leq i \leq L(\eta_n)}$ does not contain a $P_{m-1}$-unisolvent subset. Hence, there exists a sequence $(p_n)_{n \in \mathbb{N}} \subset P_{m-1}(\Omega)$ such that, for any $n \in \mathbb{N}$, $p_n \neq 0$ and, for $i = 1, \dots, L(\eta_n)$, $p_n(\xi_i(\eta_n)) = 0$. It follows from (2.2) that

$$\forall n \in \mathbb{N},\ \|p_n\|^2_{0,\omega} \leq C \eta_n^t \|p_n\|^2_{m,\Omega},$$

and so, since $(p_n)_{n \in \mathbb{N}} \subset P_{m-1}(\Omega)$,

$$\forall n \in \mathbb{N},\ \|p_n\|^2_{0,\omega} + |p_n|^2_{m,\Omega} \leq C \eta_n^t \|p_n\|^2_{m,\Omega}.$$

But the mappings $\left(\|\cdot\|_{0,\omega}^2+|\cdot|_{m,\Omega}^2\right)^{1/2}$ and $\|\cdot\|_{m,\Omega}$ are equivalent norms on $H^m(\Omega)$. Therefore, there exists $C'>0$ such that

$$\forall n \in \mathbb{N},\ C' \leq C\eta_n^t,$$

which leads to a contradiction as $n \to +\infty$. □

Next, suppose we are given, as in Section X–3,

- a bounded subset $\mathbb{H}$ of $(0,+\infty)$ such that $0 \in \overline{\mathbb{H}}$,
- a bounded polygonal open set $\widetilde{\Omega}$ in $\mathbb{R}^n$ which contains $\Omega$,
- for any $h \in \mathbb{H}$, a triangulation $\widetilde{\mathcal{T}}_h$ of $\overline{\widetilde{\Omega}}$ by means of elements $K$ with diameters $h_K \leq h$ and a finite element space $\widetilde{V}_h$, constructed on $\widetilde{\mathcal{T}}_h$, verifying hypotheses (X–3.1) and (X–3.2).

Then, we again proceed as in Section VI–1. For any $h \in \mathbb{H}$, we consider the open set $\Omega_h$ defined by (VI–1.2) (cf. Figure VI–1). We remember that the family $(\Omega_h)_{h\in\mathbb{H}}$ satisfies relations (VI–1.3) and (VI–1.4). Likewise, for any $h \in \mathbb{H}$, we also consider the space $V_h$ defined by (VI–1.5). In addition, we suppose that the family $(V_h)_{h\in\mathbb{H}}$ verifies condition (X–3.3).

Lastly, for any $(\eta,\varepsilon,h) \in \mathbb{E}\times(0,+\infty)\times\mathbb{H}$ and for any $v \in H^m(\Omega_h)$, we write

$$J_{\varepsilon h}^{\eta}(v) = l^{\eta}\left((v-f)^2\right) + \varepsilon|v|_{m,\Omega_h}^2, \tag{2.3}$$

where $l^\eta$ is the linear form defined in (2.1), and $f$ is written instead of $\tilde{f}|_{\Omega_h}$, $\tilde{f} \in H^m(\widetilde{\Omega})$ denoting any extension of $f$ (whose existence follows from (5) in Preliminaries).

Then, for any $\varepsilon > 0$, for any $h \in \mathbb{H}$ and for any $\eta \in \mathbb{E}$, we consider the minimization problem: find $\sigma_{\varepsilon h}^{\eta}$ solution of

$$\begin{cases} \sigma_{\varepsilon h}^{\eta} \in V_h, \\ \forall v_h \in V_h,\ J_{\varepsilon h}^{\eta}(\sigma_{\varepsilon h}^{\eta}) \leq J_{\varepsilon h}^{\eta}(v_h), \end{cases} \tag{2.4}$$

and also the variational problem: find $\sigma_{\varepsilon h}^{\eta}$ solution of

$$\begin{cases} \sigma_{\varepsilon h}^{\eta} \in V_h, \\ \forall v_h \in V_h,\ l^{\eta}(\sigma_{\varepsilon h}^{\eta}\, v_h) + \varepsilon(\sigma_{\varepsilon h}^{\eta}, v_h)_{m,\Omega_h} = l^{\eta}(f\, v_h). \end{cases} \tag{2.5}$$

***Theorem* 2.1** – *Suppose that $\Omega$, $\omega$, $m$ and $f$ are defined as in Section 1 and that hypotheses* (VI–1.2), (VI–1.5), (X–3.1), (X–3.3) *and* (2.2) *are verified. Then, there exists $\beta > 0$ such that, for any $h \in \mathbb{H}$ and for any $(\eta,\varepsilon) \in \mathbb{E}\times(0,+\infty)$ verifying the condition*

$$\frac{\eta^t}{\min(1,\varepsilon)} < \beta, \tag{2.6}$$

*problems* (2.4) *and* (2.5) *admit a unique common solution $\sigma_{\varepsilon h}^{\eta}$.*

**Proof** – Let $(\eta, \varepsilon) \in \mathbb{E} \times (0, +\infty)$. According to the definition of $l^\eta$, the symmetric bilinear form $a_\varepsilon^\eta$ defined by

$$a_\varepsilon^\eta(u_h, v_h) = l^\eta(u_h\, v_h) + \varepsilon(u_h, v_h)_{m,\Omega_h}$$

is continuous on $V_h \times V_h$. On the other hand, since $V_h \subset H^m(\Omega_h)$, we deduce from (VI–1.3) and (2.2) that

$$\forall h \in \mathbb{H},\ \forall v_h \in V_h,\ a_\varepsilon^\eta(v_h, v_h) \geq \min(1, \varepsilon)\big(\|v_h\|_{0,\omega}^2 + |v_h|_{m,\Omega}^2\big) - C\eta^t \|v_h\|_{m,\Omega}^2,$$

where $C$ is the constant occurring in (2.2). Since the norms $\big(\|\cdot\|_{0,\omega}^2 + |\cdot|_{m,\Omega}^2\big)^{1/2}$ and $\|\cdot\|_{m,\Omega}$ are equivalent on $H^m(\Omega)$ (cf., for instance, J. Nečas [109, Theorem 2.7.1]), it follows that there exists $C' > 0$ such that

$$\forall h \in \mathbb{H},\ \forall v_h \in V_h,\ a_\varepsilon^\eta(v_h, v_h) \geq \big(C' \min(1, \varepsilon) - C\eta^t\big)\|v_h\|_{m,\Omega}^2.$$

Let $\beta = C'/C$. Therefore, for any $h \in \mathbb{H}$ and for any $(\eta, \varepsilon) \in \mathbb{E} \times (0, +\infty)$ verifying (2.6), by (X–3.3), we have

$$\forall v_h \in V_h,\ v_h \neq 0,\ a_\varepsilon^\eta(v_h, v_h) > 0.$$

Since $V_h$ is a finite-dimensional space, we deduce that the bilinear form $a_\varepsilon^\eta$ is $V_h$-elliptic. Then, the Lax-Milgram Lemma (cf. P. G. Ciarlet [45, Theorem 1.1.3]) yields the result. □

The function $\sigma_{\varepsilon h}^\eta$ is called *$V_h$-discrete smoothing $D^m$-spline of $f$ relative to $\omega$, $\eta$ and $\varepsilon$*. Its definition is similar to that of discrete smoothing $D^m$-splines introduced in Section VI–3 and its computation is done exactly as already explained in Remark X–3.2 for the smoothing $D^m$ spline of $f$ relative to the set of curves $F$, $\eta$ and $\varepsilon$. See also Remark X–3.3 for the application of the GCV method in order to determine the value of $\varepsilon$.

**Remark 2.3** – The result of Theorem 2.1 can be obtained under weaker conditions. Let us assume that only (VI–1.2), (VI–1.5), (X–3.1) and (2.2) hold. It follows from Remark 2.2 that there exists $\eta_0 > 0$ such that, for any $\eta \in \mathbb{E} \cap (0, \eta_0]$, the set $\{\xi_i(\eta)\}_{1 \leq i \leq L(\eta)}$ contains a $P_{m-1}$-unisolvent subset. Hence, one easily shows that, for any $(h, \eta) \in \mathbb{H} \times \mathbb{E}$, with $\eta \leq \eta_0$, the mapping $[\![\cdot]\!]_{\eta,h}$ defined on $V_h$ by

$$[\![v]\!]_{\eta,h} = \big(l^\eta(v^2) + |v|_{m,\Omega_h}^2\big)^{1/2},$$

is a Hilbertian norm on the (finite-dimensional) space $V_h$. Then, it is deduced from the Lax-Milgram Lemma (cf. P. G. Ciarlet [45, Theorem 1.1.3]) that, for any $(h, \eta, \varepsilon) \in \mathbb{H} \times \mathbb{E} \times (0, +\infty)$, with $\eta \leq \eta_0$, problems (2.4) and (2.5) admit a common unique solution $\sigma_{\varepsilon h}^\eta$. □

Then, we propose to take as a solution of the initial problem the function $\phi = \sigma_{\varepsilon h}^\eta|_\Omega$. Before studying the problem of the convergence of the approximant $\phi$ to $f$ on $\Omega$ when the open subset $\omega$ tends to $\Omega$ (in a sense to be stated precisely), let us give without a proof a convergence result of $\sigma_{\varepsilon h}^\eta$ to $f$ *on $\omega$ supposed fixed.* Notice that, by (2.6), $\sigma_{\varepsilon h}^\eta$ is defined for $\varepsilon \in (0, 1]$ and $\eta^t/\varepsilon < \beta$.

***Theorem 2.2*** – *Suppose that the conditions of Theorem 2.1 are verified and, moreover, that hypothesis* (X–3.2) *is satisfied. Then, the solution* $\sigma^\eta_{\varepsilon h}$ *of* (2.4) *and* (2.5) *verifies the relations*

*(i)* $\lim\limits_{\substack{\varepsilon\to 0,\, \eta^t/\varepsilon<\beta \\ h^{2m}/\varepsilon\to 0}} \|\sigma^\eta_{\varepsilon h} - \sigma\|_{m,\Omega} = 0$, *where* $\sigma$ *denotes the unique element of minimal semi-norm* $|\cdot|_{m,\Omega}$ *of the set* $\{v \in H^m(\Omega) \mid v = f \text{ on } \omega\}$ *and* $\beta$ *is the constant introduced in* (2.6),

*(ii) there exists a constant* $C > 0$ *such that, as* $\varepsilon \to 0$, $\eta^t/\varepsilon < \beta$ *and* $h^{2m}/\varepsilon \to 0$,

$$\|\sigma^\eta_{\varepsilon h} - f\|^2_{0,\omega} \leq C(\varepsilon + \eta^t o(1) + h^{2m}).$$

## 3. CONVERGENCE OF THE APPROXIMATION

Let us modify the previous situation. Suppose we are given

- a subset $\mathbb{D}$ of $(0, +\infty)$ such that $0 \in \overline{\mathbb{D}}$,
- a family $(\omega^d)_{d\in\mathbb{D}}$ of nonempty open subsets of $\Omega$ such that

$$\lim_{d\to 0} \operatorname{meas}(\Omega \setminus \omega^d) = 0. \tag{3.1}$$

We have the following result.

***Theorem 3.1*** – *Suppose that* (3.1) *holds. Then, there exists* $d_0 > 0$ *such that the mapping* $[\![\cdot]\!]_d$, *defined for any* $d \in \mathbb{D}$ *by*

$$\forall v \in H^m(\Omega),\ [\![v]\!]_d = \left(\|v\|^2_{0,\omega^d} + |v|^2_{m,\Omega}\right)^{1/2},$$

*is a norm on* $H^m(\Omega)$ *uniformly equivalent over* $\mathbb{D} \cap (0, d_0]$ *to the norm* $\|\cdot\|_{m,\Omega}$.

**Proof** – For any $d \in \mathbb{D}$, it is obvious that

$$\forall v \in H^m(\Omega),\ [\![v]\!]_d \leq \|v\|_{m,\Omega}.$$

On the other hand, for any $d \in \mathbb{D}$ and any $v \in H^m(\Omega)$, one has

$$\begin{aligned} \|v\|^2_{0,\omega^d} &= |v|^2_{0,\Omega} - \int_{\Omega\setminus\omega^d} v^2(x)\,dx \\ &\geq |v|^2_{0,\Omega} - \operatorname{meas}(\Omega \setminus \omega^d) \sup_{x\in\Omega} |v^2(x)|. \end{aligned}$$

By Sobolev's Continuous Imbedding Theorem (cf. (4) in Preliminaries), there exists a constant $C > 0$ such that

$$\forall v \in H^m(\Omega),\ \sup_{x\in\Omega}|v(x)| \leq C\|v\|_{m,\Omega}.$$

Since the norms $\left(|\cdot|_{0,\Omega}^2 + |\cdot|_{m,\Omega}^2\right)^{1/2}$ and $\|\cdot\|_{m,\Omega}$ are equivalent on $H^m(\Omega)$ (cf., for instance, J. Nečas [109, Theorem 2.7.1]), there exists a constant $C' > 0$ such that

$$\forall v \in H^m(\Omega),\ \left(|v|_{0,\Omega}^2 + |v|_{m,\Omega}^2\right)^{1/2} \geq C'\|v\|_{m,\Omega}.$$

From this, we deduce that there exist constants $C > 0$ and $C' > 0$ such that

$$\forall d \in \mathbb{D},\ \forall v \in H^m(\Omega),\ [\![v]\!]_d^2 \geq \left(C'^2 - C^2 \operatorname{meas}(\Omega \setminus \omega^d)\right)\|v\|_{m,\Omega}^2.$$

Let $C''$ be any constant belonging to $(0, C')$. Then, taking (3.1) into account, we can choose $d_0$ such that

$$(d \leq d_0) \Rightarrow (C'^2 - C^2 \operatorname{meas}(\Omega \setminus \omega^d) \geq C''^2).$$

This completes the proof. □

We keep the notations introduced in Section 2, but with $L = L(d,\eta)$, $\xi_i = \xi_i(d,\eta)$ and $\lambda_i = \lambda_i(d,\eta)$. It is agreed that, hereafter, $\omega^d$, $l^{d\eta}$ and $J_{\varepsilon h}^{d\eta}$ will respectively replace $\omega$, $l^\eta$ and $J_{\varepsilon h}^\eta$.

Now, we formulate the hypothesis

$$\left|\begin{array}{l}\exists C > 0,\ \exists t > 0,\ \forall (d,\eta) \in \mathbb{D} \times \mathbb{E}, \\ \forall v \in H^m(\Omega),\ |l^{d\eta}(v^2) - \|v\|_{0,\omega^d}^2| \leq C\eta^t \|v\|_{m,\Omega}^2,\end{array}\right. \tag{3.2}$$

which is just condition (2.2) with $C$ *independent of* $d$. Notice that (3.2) is realistic (cf. Remark 2.1).

We consider again problems (2.4) and (2.5). Assume that $\Omega$, $m$ and $f$ are defined as in Section 1, and that hypotheses (2.2), (X–3.1), (VI–1.2), (VI–1.5) and (X–3.3) are verified. Then, under condition (2.6), both problems admit a common unique solution, now denoted by $\sigma_{\varepsilon h}^{d\eta}$. Notice that the constant $\beta = C'/C$ in (2.6) is *independent of* $d$, owing to Theorem 3.1 and hypothesis (2.2).

**Theorem 3.2** – *Suppose that $\Omega$, $m$ and $f$ are defined as in Section 1 and that hypotheses* (VI–1.2), (VI–1.5), (X–3.1), (X–3.2), (X–3.3), (3.1) *and* (3.2) *are satisfied. Then, the solution $\sigma_{\varepsilon h}^{d\eta}$ of* (2.4) *and* (2.5) *verifies*

$$\lim_{\substack{d\to 0,\ \varepsilon \to 0 \\ \eta^t/\varepsilon < \beta,\ h^{2m}/\varepsilon \to 0}} \|\sigma_{\varepsilon h}^{d\eta} - f\|_{m,\Omega} = 0,$$

*where $\beta$ is the constant introduced in* (2.6)

**Proof** – Let us recall that, by (2.6), $\sigma_{\varepsilon h}^{d\eta}$ is defined for $\varepsilon \in (0,1]$ and $\eta^t/\varepsilon < \beta$.

1) Let $\tilde{f} \in H^m(\tilde{\Omega})$ be any extension of $f$. Let us consider the expression (2.3) defining $J_{\varepsilon h}^{d\eta}$ and take $v = (\tilde{\Pi}_h \tilde{f})|_{\Omega_h}$, where $\tilde{\Pi}_h$ is the operator introduced in (X–3.2). Then, using (VI–1.3), we have

$$\forall (d,\eta,\varepsilon,h) \in \mathbb{D} \times \mathbb{E} \times (0,+\infty) \times \mathbb{H},\ J_{\varepsilon h}^{d\eta}(\tilde{\Pi}_h \tilde{f}) = l^{d\eta}((\tilde{\Pi}_h \tilde{f} - \tilde{f})^2) + \varepsilon |\tilde{\Pi}_h \tilde{f}|_{m,\Omega_h}^2.$$

It follows from (3.2) that

$$J_{\varepsilon h}^{d\eta}(\widetilde{\Pi}_h \tilde{f}) \leq \|\widetilde{\Pi}_h \tilde{f} - \tilde{f}\|_{0,\omega^d}^2 + C\eta^t \|\widetilde{\Pi}_h \tilde{f} - \tilde{f}\|_{m,\Omega}^2 + \varepsilon |\widetilde{\Pi}_h \tilde{f}|_{m,\Omega_h}^2.$$

Given that, for any $h \in \mathbb{H}$,

$$\left| |\widetilde{\Pi}_h \tilde{f}|_{m,\Omega_h} - |\tilde{f}|_{m,\Omega_h} \right| \leq |\widetilde{\Pi}_h \tilde{f} - \tilde{f}|_{m,\Omega_h},$$

we infer from (VI–1.3), (X–3.2) and (VI–1.4) that

$$\lim_{h \to 0} |\widetilde{\Pi}_h \tilde{f}|_{m,\Omega_h} = |f|_{m,\Omega}. \tag{3.3}$$

Then, we derive from (2.4), (VI–1.3) and (X–3.2) that, for any $(d, \eta, \varepsilon, h) \in \mathbb{D} \times \mathbb{E} \times (0, +\infty) \times \mathbb{H}$ such that $\varepsilon \in (0, 1]$ and $\eta^t / \varepsilon < \beta$,

$$J_{\varepsilon h}^{d\eta}(\sigma_{\varepsilon h}^{d\eta}) \leq h^{2m} O(1) + \eta^t o(1) + \varepsilon(|f|_{m,\Omega}^2 + o(1)), \ h \to 0. \tag{3.4}$$

Likewise, from (3.2), we get

$$J_{\varepsilon h}^{d\eta}(\sigma_{\varepsilon h}^{d\eta}) \geq \|\sigma_{\varepsilon h}^{d\eta} - f\|_{0,\omega^d}^2 - C\eta^t \|\sigma_{\varepsilon h}^{d\eta} - f\|_{m,\Omega}^2.$$

Then, we deduce from Theorem 3.1 that there exists a constant $C' > 0$, independent of $d$, such that

$$(1 - C'\eta^t) \|\sigma_{\varepsilon h}^{d\eta} - f\|_{0,\omega^d}^2 \leq J_{\varepsilon h}^{d\eta}(\sigma_{\varepsilon h}^{d\eta}) + C'\eta^t |\sigma_{\varepsilon h}^{d\eta} - f|_{m,\Omega}^2.$$

Let $\eta_0$ be any constant verifying the condition $C'\eta_0^t < 1$. Taking (3.4) into account, we see that there exist constants $C'' > 0$ and $C''' > 0$ such that, for any $(d, \eta, \varepsilon, h) \in \mathbb{D} \times \mathbb{E} \times (0, +\infty) \times \mathbb{H}$ such that $\varepsilon \in (0, 1]$, $\eta^t / \varepsilon < \beta$ and $\eta \leq \eta_0$, the following relation holds:

$$\begin{aligned} \|\sigma_{\varepsilon h}^{d\eta} - f\|_{0,\omega^d}^2 \leq & C''\eta^t |\sigma_{\varepsilon h}^{d\eta} - f|_{m,\Omega}^2 + h^{2m} O(1) \\ & + \eta^t o(1) + \varepsilon(C''' |f|_{m,\Omega}^2 + o(1)), \ h \to 0. \end{aligned} \tag{3.5}$$

Suppose now that the conditions of the Theorem are satisfied, i.e. that $d \to 0$, $\varepsilon \to 0$, $\eta^t / \varepsilon < \beta$, $h^{2m} / \varepsilon \to 0$. Moreover, assume that $\eta \leq \eta_0$. It follows from (3.4), (3.5) and the definition (2.3) of $J_{\varepsilon h}^{d\eta}$ that the terms $|\sigma_{\varepsilon h}^{d\eta}|_{m,\Omega}$ and $\|\sigma_{\varepsilon h}^{d\eta} - f\|_{0,\omega^d}$ are bounded. Then, Theorem 3.1 shows that the family $(\sigma_{\varepsilon h}^{d\eta})$ is bounded in $H^m(\Omega)$. Therefore, by Corollary 1 in Preliminaries, there exists a sequence $(\sigma_{\varepsilon_n h_n}^{d_n \eta_n})_{n \in \mathbb{N}}$, with $\lim_{n \to +\infty} d_n = 0$, $\lim_{n \to +\infty} \varepsilon_n = 0$, $(\eta_n^t / \varepsilon_n)_{n \in \mathbb{N}} \subset (0, \beta)$, $\eta_n \leq \eta_0$, and $\lim_{n \to +\infty} h_n^{2m} / \varepsilon_n = 0$, extracted from the family $(\sigma_{\varepsilon h}^{d\eta})$, and an element $f^* \in H^m(\Omega)$ such that, as $n \to +\infty$,

$$\sigma_{\varepsilon_n h_n}^{d_n \eta_n} \to f^*, \quad \text{weakly in } H^m(\Omega).$$

2) Let us show that $f^* = f$. Given that $H^m(\Omega) \overset{c}{\subset} L^2(\Omega)$ (resp. $H^m(\Omega) \overset{c}{\subset} C^0(\overline{\Omega})$), according to the Rellich-Kondrašov Compact Imbedding Theorems (cf. (2) and (3)

in Preliminaries), we have

$$\|f^* - f\|_{0,\Omega}^2 = \lim_{n\to+\infty} \|\sigma_{\varepsilon_n h_n}^{d_n\eta_n} - f\|_{0,\Omega}^2$$
$$\leq \lim_{n\to+\infty} \left( \|\sigma_{\varepsilon_n h_n}^{d_n\eta_n} - f\|_{0,\omega^{d_n}}^2 + \|\sigma_{\varepsilon_n h_n}^{d_n\eta_n} - f\|_{0,\Omega\setminus\omega^{d_n}}^2 \right).$$

The two terms in the right-hand member tend to 0 as $n \to +\infty$, the first, by (3.5), and the second, because it is bounded by the quantity

$$\|\sigma_{\varepsilon_n h_n}^{d_n\eta_n} - f\|_{C^0(\overline{\Omega})}^2 \operatorname{meas}(\Omega \setminus \omega^{d_n}),$$

which, by (3.1), goes to 0 as $n \to +\infty$. Thus $f^* = f$.

3) Now, let us show that

$$\lim_{n\to+\infty} \|\sigma_{\varepsilon_n h_n}^{d_n\eta_n} - f\|_{m,\Omega} = 0.$$

Since the injection of $H^m(\Omega)$ into $H^{m-1}(\Omega)$ is compact (cf. (2) in Preliminaries), it follows from points 1) and 2) that

$$\sigma_{\varepsilon_n h_n}^{d_n\eta_n} \to f, \text{ strongly in } H^{m-1}(\Omega).$$

On the other hand, we deduce from (3.4) that

$$|\sigma_{\varepsilon_n h_n}^{d_n\eta_n}|_{m,\Omega}^2 \leq |f|_{m,\Omega}^2 + o(1), \; n \to +\infty.$$

Then, we have

$$|\sigma_{\varepsilon_n h_n}^{d_n\eta_n} - f|_{m,\Omega}^2 \leq 2|f|_{m,\Omega}^2 - 2\left(\sigma_{\varepsilon_n h_n}^{d_n\eta_n}, f\right)_{m,\Omega} + o(1), \; n \to +\infty.$$

Hence,

$$\lim_{n\to+\infty} |\sigma_{\varepsilon_n h_n}^{d_n\eta_n} - f|_{m,\Omega} = 0,$$

and the result follows.

4) Finally, let us prove that, as $d \to 0$, $\varepsilon \to 0$, $\eta^t/\varepsilon < \beta$, $h^{2m}/\varepsilon \to 0$,

$$\lim \|\sigma_{\varepsilon h}^{d\eta} - f\|_{m,\Omega} = 0.$$

Let us assume that the relation does not hold under the above conditions. This comes down to saying that there exists $\alpha > 0$ and $((d'_n, \eta'_n, \varepsilon'_n, h'_n))_{n\in\mathbb{N}}$, a sequence in $\mathbb{D} \times \mathbb{E} \times (0,+\infty) \times \mathbb{H}$, such that $d'_n \to 0$, $\eta'_n \to 0$, ${\eta'_n}^t/\varepsilon'_n < \beta$ and ${h'_n}^{2m}/\varepsilon'_n \to 0$, verifying

$$\forall n \in \mathbb{N}, \; \|\sigma_{\varepsilon'_n h'_n}^{d'_n\eta'_n} - f\|_{m,\Omega} > \alpha.$$

But such a sequence is bounded in $H^m(\Omega)$ and the previous reasoning leads to a contradiction. □

**Remark 3.1** – (Cf. Remark X–4.2). Theorem 3.2 can be stated in the following way: the solution $\sigma_{\varepsilon h}^{d\eta}$ of (2.4) and (2.5) converges to $f$ in $H^m(\Omega)$ through the filter basis $\mathcal{B} = \{ B_{q,r,s} \mid q > 0, \; r > 0, \; s > 0 \}$, with $B_{q,r,s} = \{ (d,\eta,\varepsilon,h) \in \mathbb{D} \times \mathbb{E} \times (0,+\infty) \times \mathbb{H} \mid d \leq q, \; \varepsilon \leq r, \; \eta^t/\varepsilon < \beta, \; h^{2m}/\varepsilon \leq s \}$. □

| $\omega$ | Figure 2 | Figure 3 | Figure 4 | Figure 5 |
|---|---|---|---|---|
| $\operatorname{card} \mathbb{T}_\eta$ | 320 | 288 | 106 | 222 |
| $\operatorname{meas} \omega$ | 0.215095 | 0.307692 | 0.144556 | 0.296289 |

Table 1: Number of triangles in the triangulation $\mathbb{T}_\eta$ of $\overline{\omega}$ and area of $\omega$.

# 4. NUMERICAL RESULTS

We conserve the notations $f$, $\Omega = \widetilde{\Omega} = \Omega_h$, $\widetilde{\mathcal{T}_h} = \mathcal{T}_h$ and $\widetilde{V_h} = V_h$ introduced in Section X–5.

We shall show the behaviour of the smoothing method presented in this chapter by fitting Franke's and Nielson's functions (cf., respectively, (VIII–6.1) and (VIII–6.2), and also Figures VIII–8 and VIII–13), from three different polygonal data sets $\omega$, two of which are shared by both test functions. The geometry of these sets, which can be seen in Figures 2–5, resembles that of the sets $F$ used in Section X–5.

We have applied the theory of Section 2, setting $m = 2$ and defining the linear form $l^\eta$ introduced in (2.1) as follows:

$$l^\eta(v) = \sum_{K \in \mathbb{T}_\eta} \frac{\operatorname{meas} K}{60} \left( 3 \sum_{i=1}^{3} v(x_{iK}) + 8 \sum_{i=4}^{6} v(x_{iK}) + 27 v(x_{7K}) \right),$$

where $\mathbb{T}_\eta$ is a triangulation of $\overline{\omega}$ and, for any triangle $K \in \mathbb{T}_\eta$, $x_{1K}, \ldots, x_{7K}$ are, in increasing order, the three vertices, the three midpoints of the sides and the barycenter of $K$. Table 1 indicates the number of triangles in $\mathbb{T}_\eta$ and also the area of $\omega$ (let us observe that $\operatorname{meas} \Omega = 1$).

It is clear that the above expression of $l^\eta$ has been obtained, as detailed in Remark 2.1, from a $P_3(K)$-exact quadrature formula. According to this remark, since $m = 2$, it would be sufficient to choose a $P_1(K)$-exact quadrature formula, such as

$$\int_K v(x)dx \sim (\operatorname{meas} K) v(a_K),$$

where $a_K$ denotes the barycenter of $K$. However, the implementation of this formula has given slightly poorer results.

We have computed six $V_h$-discrete smoothing $D^2$-splines $\sigma^\eta_{\varepsilon h}$, one per test function and data set, represented in Figures 6–11. In every case, we have chosen the smoothing parameter $\varepsilon$ following the strategy described in Section X–5. Finally, we present the plots of the approximated GCV functions $\overline{\mathcal{V}}$ in Figure 12.

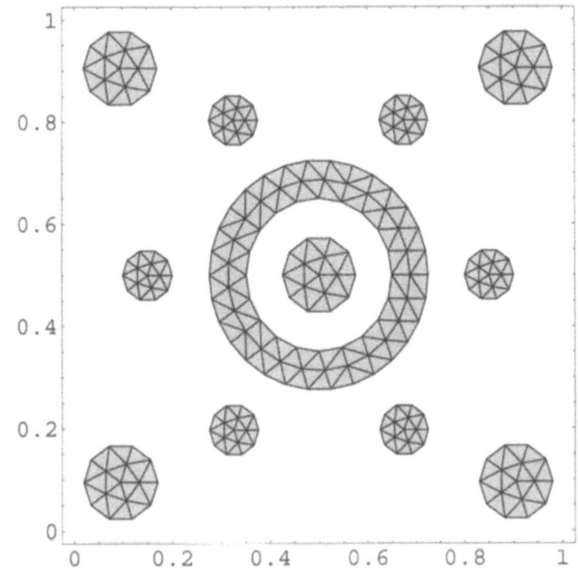


Figure 2: Set $\omega$ and triangulation $\mathbb{T}_\eta$. First case.

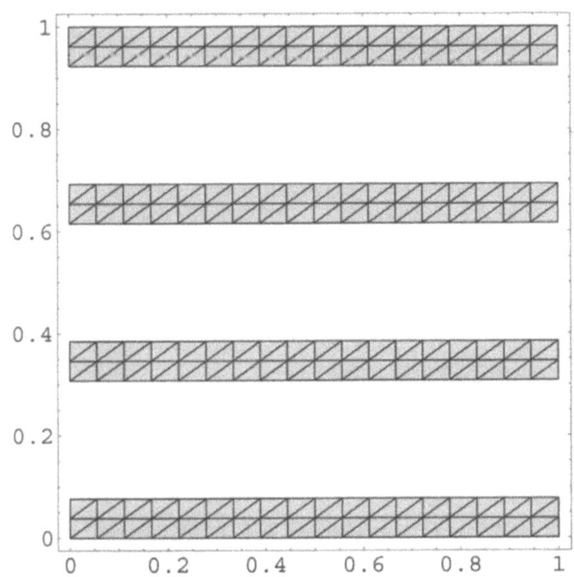


Figure 3: Set $\omega$ and triangulation $\mathbb{T}_\eta$. Second case.

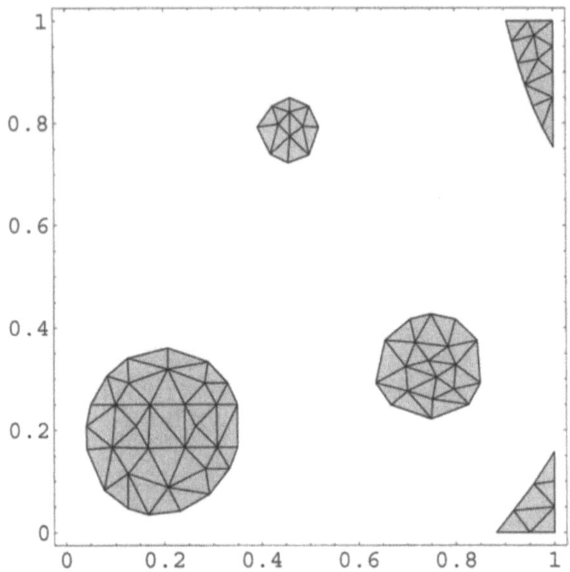


Figure 4: Set $\omega$ and triangulation $\mathbb{T}_\eta$. Third case.

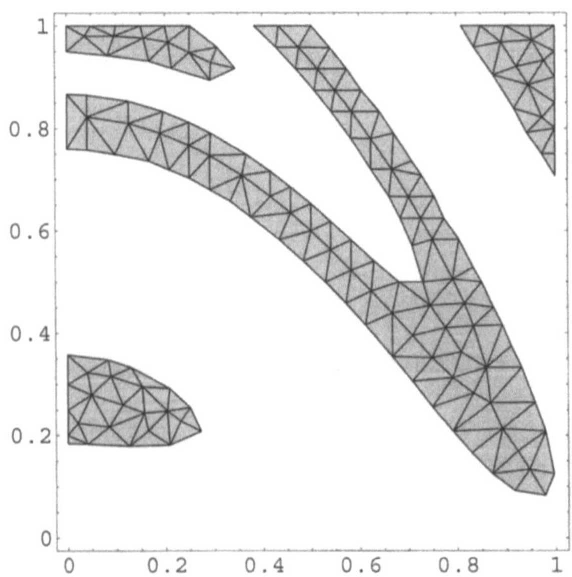


Figure 5: Set $\omega$ and triangulation $\mathbb{T}_\eta$. Fourth case.

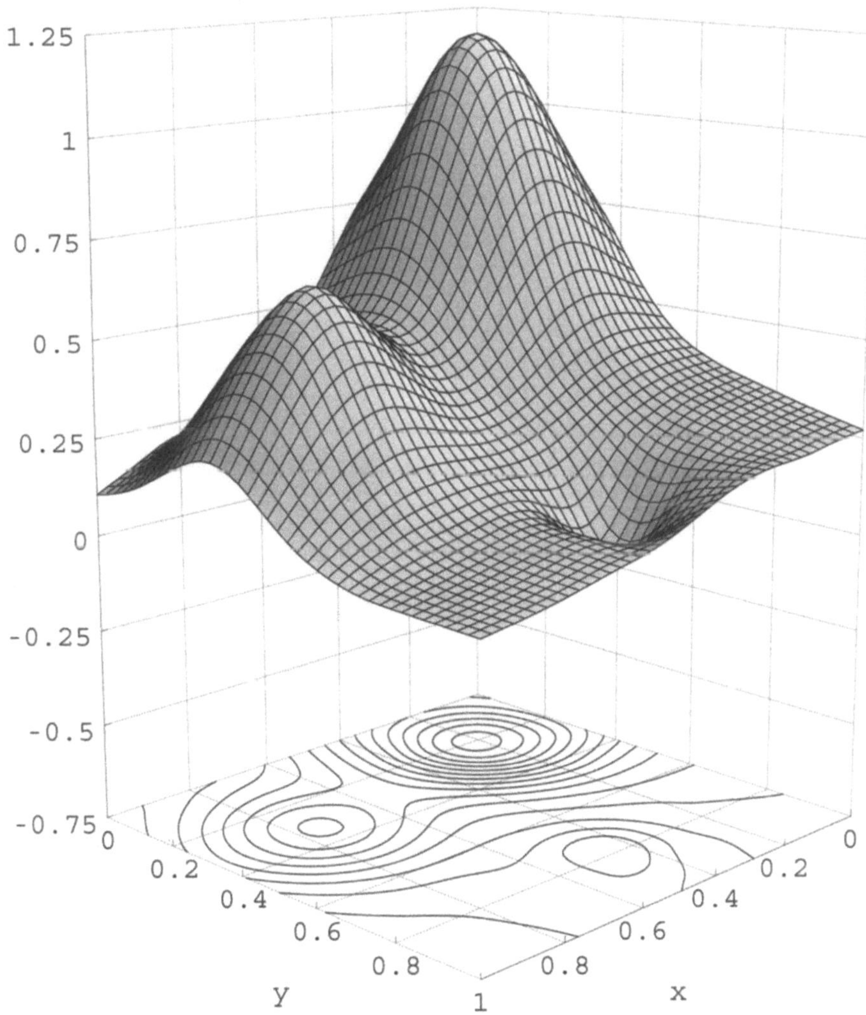


Figure 6: Fitting of Franke's function. Graph and contour plot of the discrete smoothing $D^2$-spline of class $C^1$ relative to $\varepsilon = 10^{-10}$ and the set $\omega$ shown in Figure 2. Relative error: $r(f) = 0.0169289$.

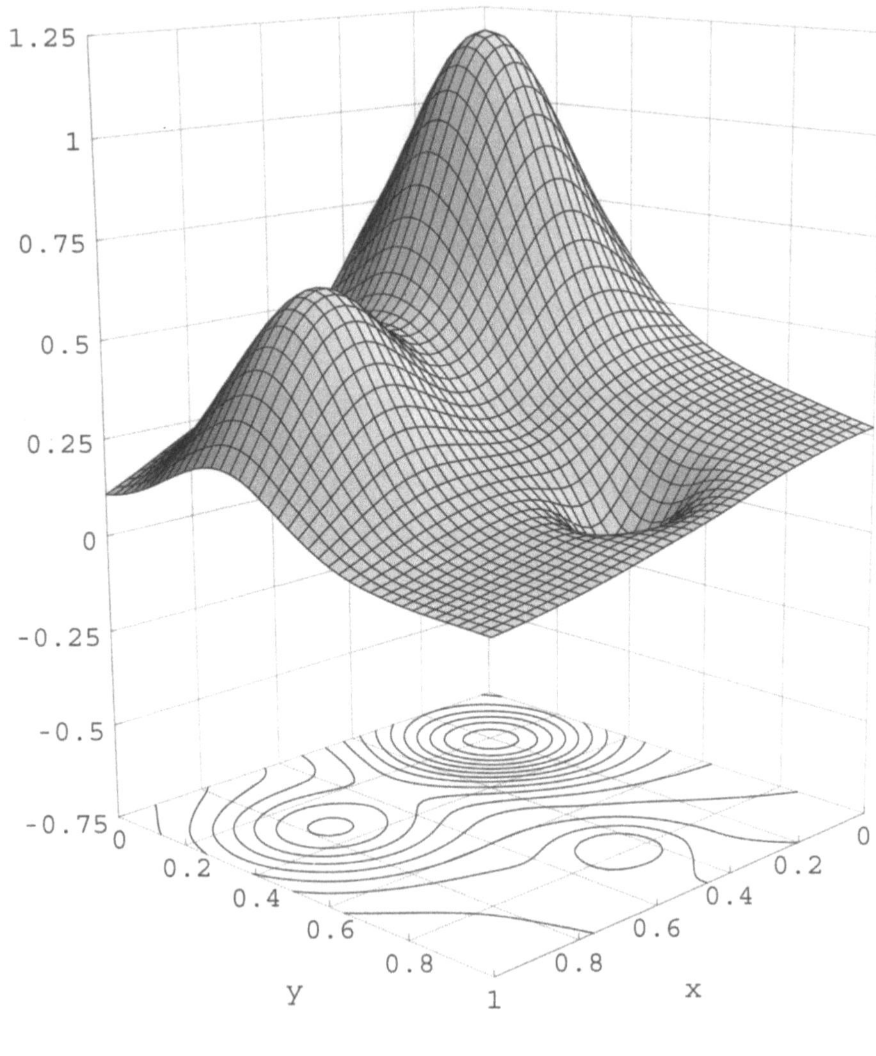


Figure 7: Fitting of Franke's function. Graph and contour plot of the discrete smoothing $D^2$-spline of class $C^1$ relative to $\varepsilon = 5 \cdot 10^{-11}$ and the set $\omega$ shown in Figure 3. Relative error: $r(f) = 0.00444455$.

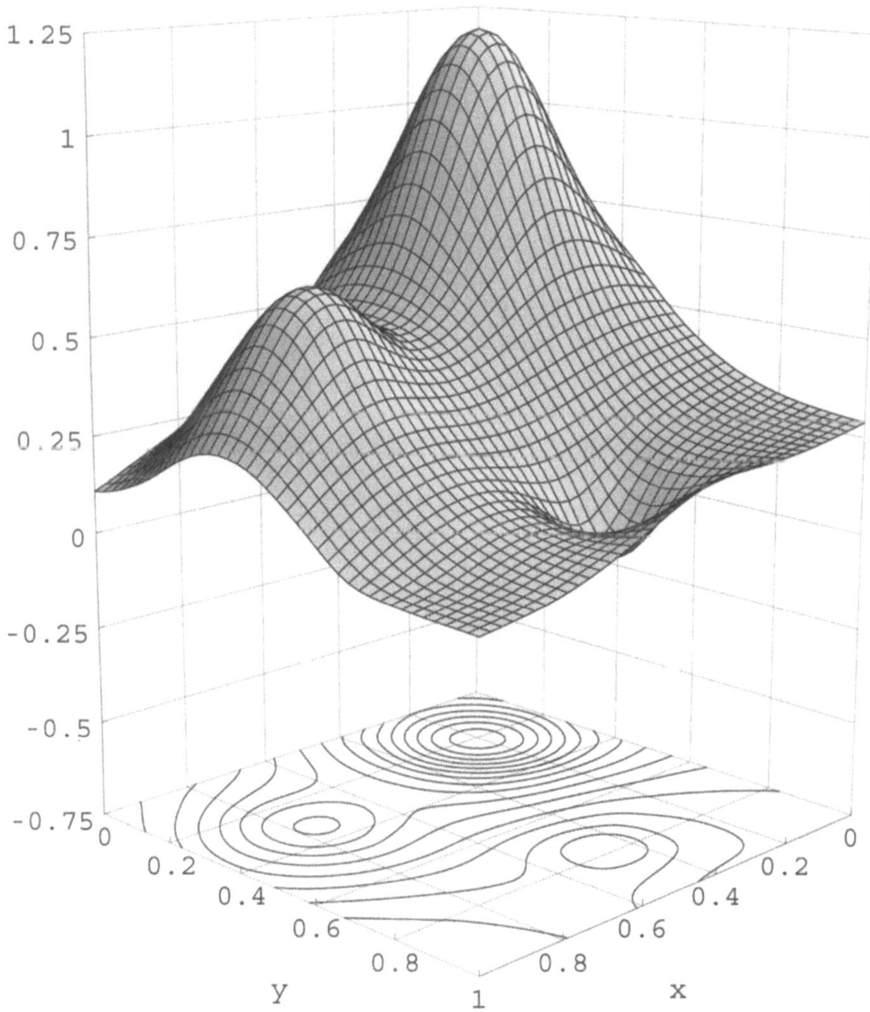


Figure 8: Fitting of Franke's function. Graph and contour plot of the discrete smoothing $D^2$-spline of class $C^1$ relative to $\varepsilon = 5 \cdot 10^{-10}$ and the set $\omega$ shown in Figure 4. Relative error: $r(f) = 0.0527779$.

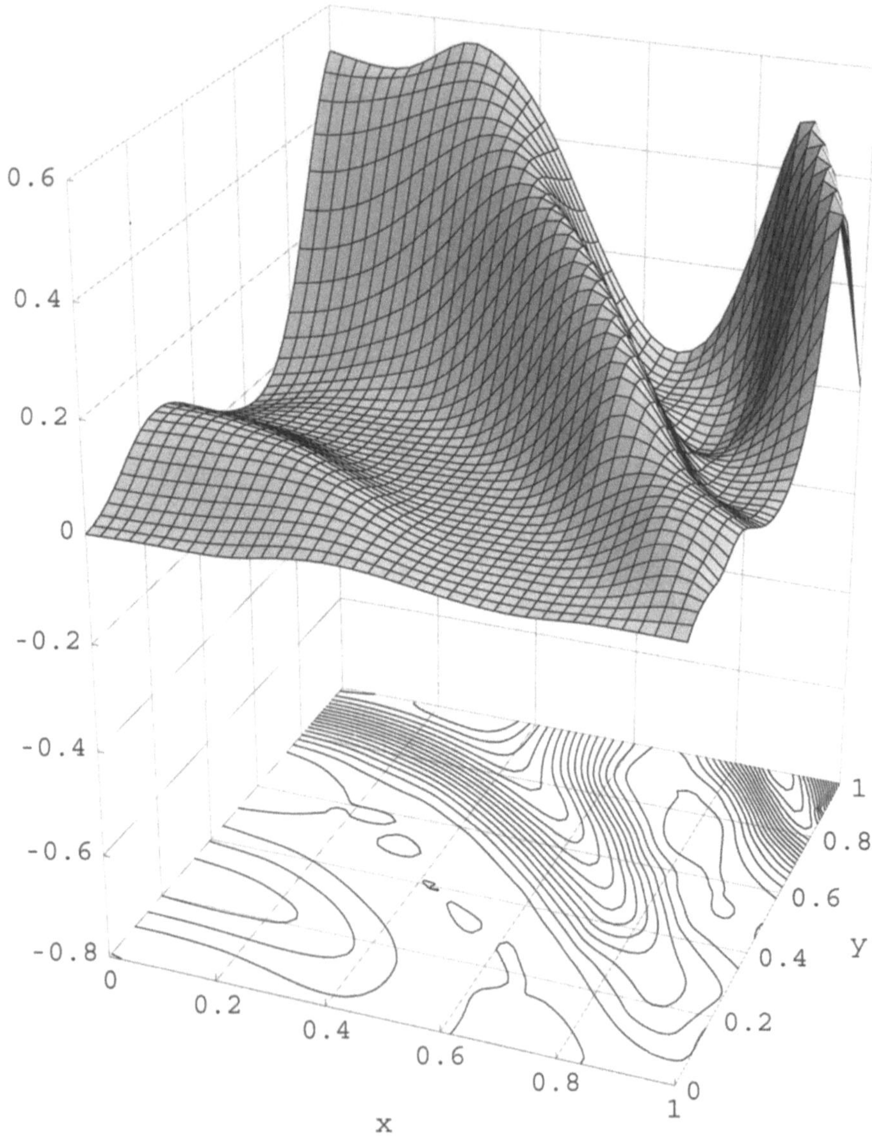


Figure 9: Fitting of Nielson's function. Graph and contour plot of the discrete smoothing $D^2$-spline of class $C^1$ relative to $\varepsilon = 5 \cdot 10^{-10}$ and the set $\omega$ shown in Figure 2. Relative error: $r(f) = 0.168407$.

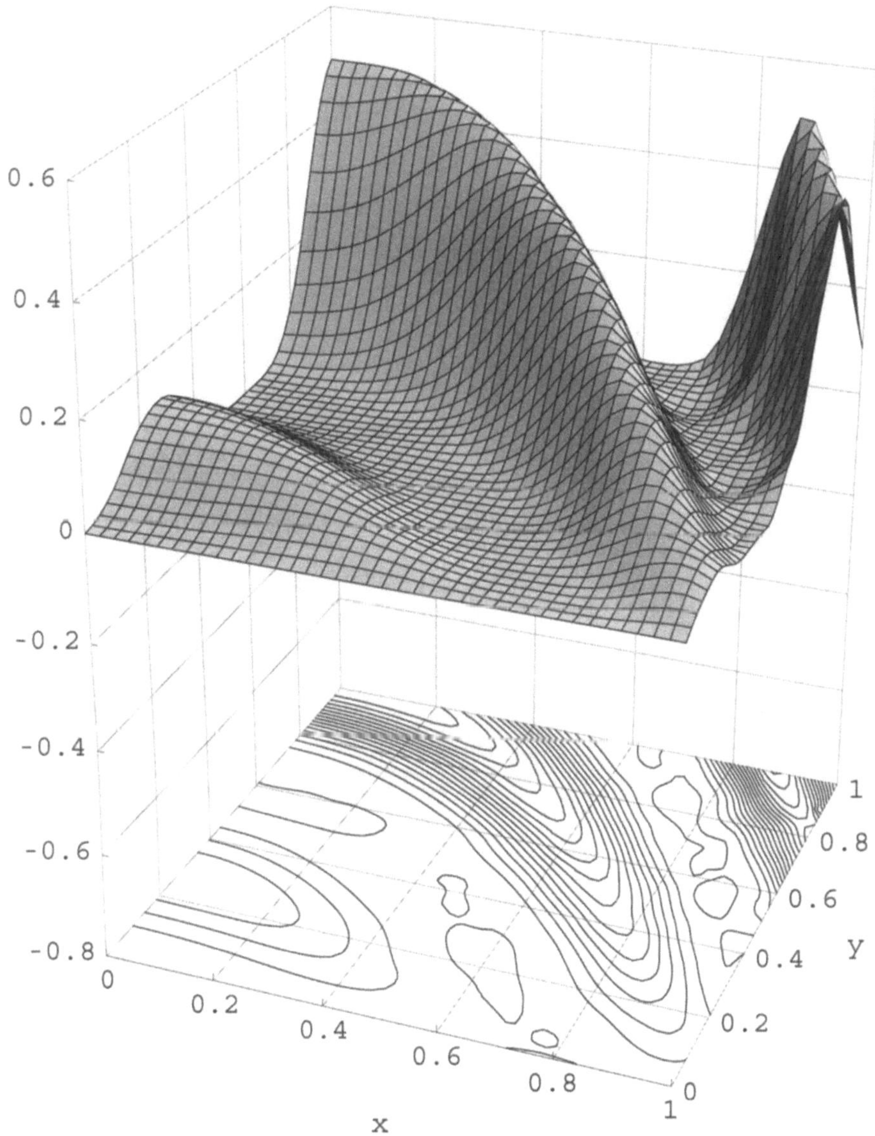


Figure 10: Fitting of Nielson's function. Graph and contour plot of the discrete smoothing $D^2$-spline of class $C^1$ relative to $\varepsilon = 10^{-10}$ and the set $\omega$ shown in Figure 3. Relative error: $r(f) = 0.0291339$.

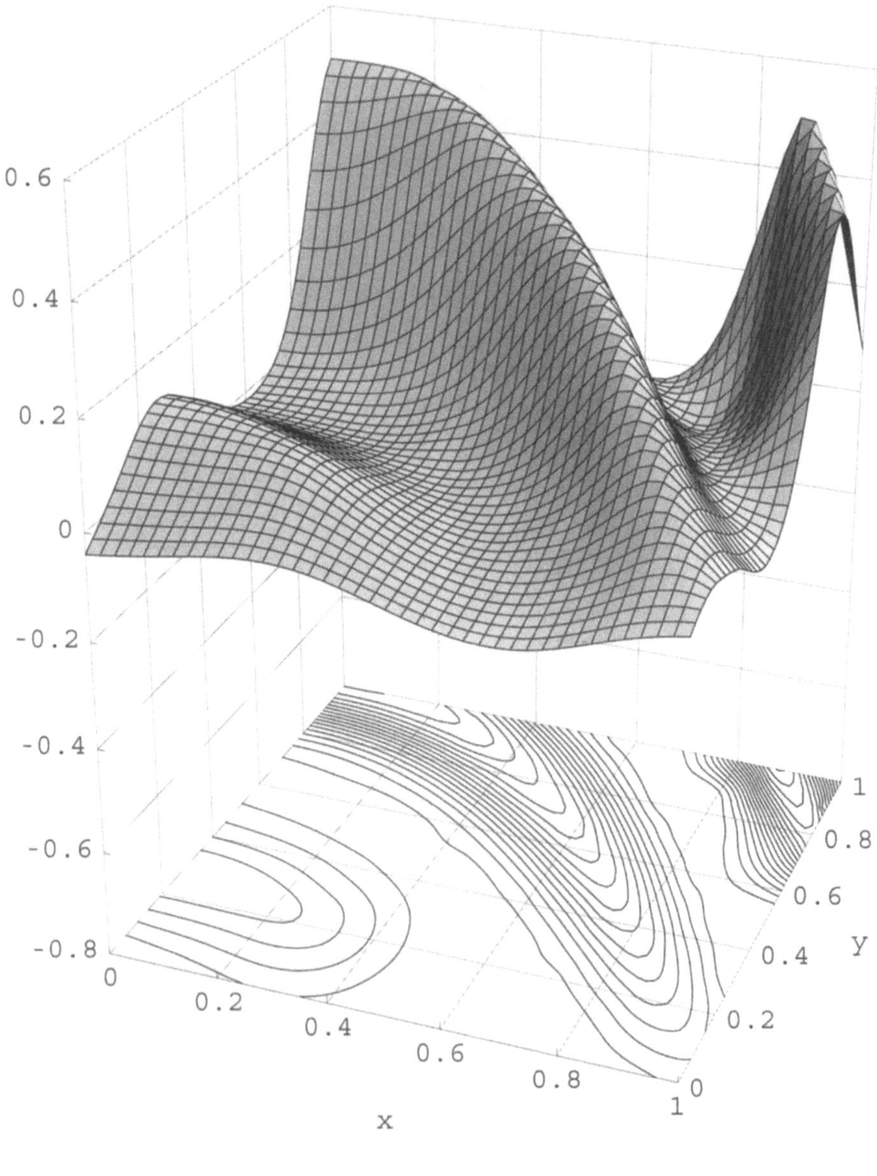


Figure 11: Fitting of Nielson's function. Graph and contour plot of the discrete smoothing $D^2$-spline of class $C^1$ relative to $\varepsilon = 5 \cdot 10^{-9}$ and the set $\omega$ shown in Figure 5. Relative error: $r(f) = 0.105684$.

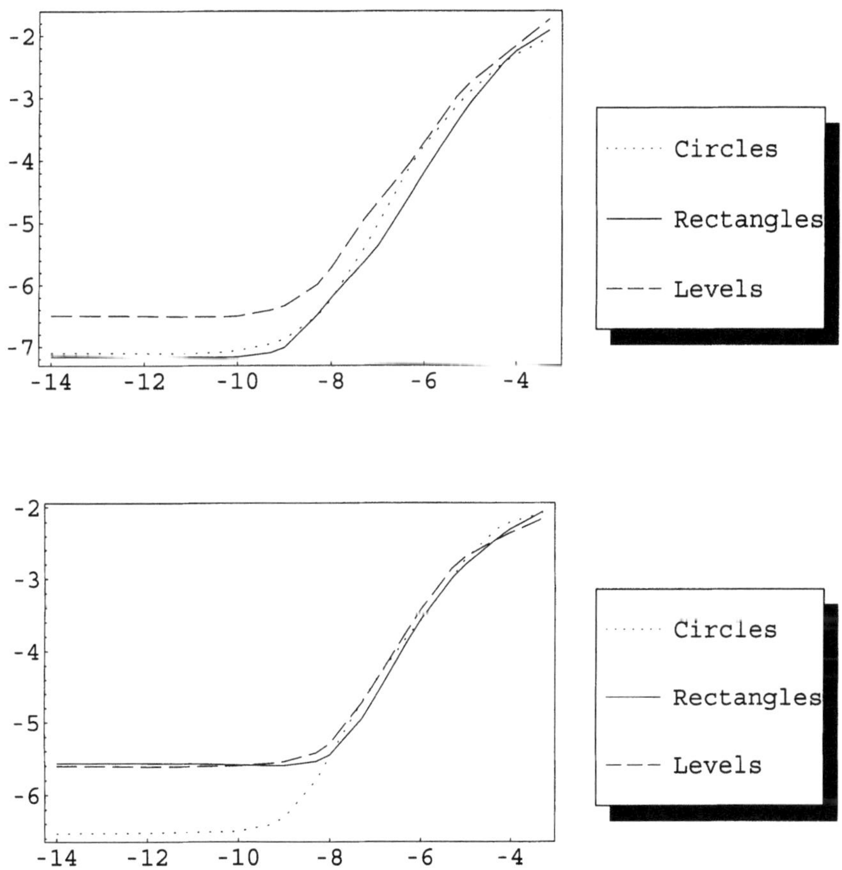


Figure 12: Plot of $\log_{10} \overline{\mathcal{V}}(\varepsilon)$ versus $\log_{10} \varepsilon$. Top: Franke's function, bottom: Nielson's function. The labels **Circles**, **Rectangles** and **Levels** refer, respectively, to the data sets represented in Figures 2, 3 and 4, for Franke's function, or in Figures 2, 3 and 5, for Nielson's function.

# *CHAPTER XII*

# *APPROXIMATION OF PARAMETRIC SURFACES*

## 1. *INTRODUCTION*

The previous chapters in Part C dealt with the approximation of a real function over an open set in $\mathbb{R}^2$ or, more generally, in $\mathbb{R}^n$. The case $n = 2$ corresponds, from a geometric point of view, to the approximation of explicit surfaces. However, this type of surfaces does not suit the modelling of a lot of real problems. Let us quote, for example, the reconstruction of certain geological structures (gas or oil fields, overlapping folds, etc.) or the design of industrial components, coachworks, fuselages, etc. In these situations, we have to represent in a parametric form the surfaces occurring in the problem.

We shall give here a brief introduction to the approximation problems for parametric surfaces. In Section 2, the problems that we are going to face will be formulated precisely. Moreover, an intrinsic difficulty of these problems, namely the lack of a natural parameter system for the data points, will be considered. In Section 3, we shall adapt the discrete smoothing $D^m$-spline theory developed in Section VI–3. Section 4 will be devoted to the convergence study and, finally, in Section 5, some numerical examples will be given.

Hereafter, we shall need some additional notations. We denote by $p_1$, $p_2$ and $p_3$ the canonical projections from $\mathbb{R}^3$ onto $\mathbb{R}$, given by $p_1(x, y, z) = x$, $p_2(x, y, z) = y$ and $p_3(x, y, z) = z$. For any $l \in \mathbb{N}^*$, $\mathbb{R}^{l,3}$ stands for the space of matrices of real numbers having $l$ rows and 3 columns.

Let $\omega$ be a nonempty open set in $\mathbb{R}^2$. For any $l \in \mathbb{N}$, we denote by $H^l(\omega, \mathbb{R}^3)$ the Sobolev space of (classes of) all $\mathbb{R}^3$-valued functions $v$ defined over $\omega$ such that, for $i = 1, 2, 3$, $p_i \circ v \in H^l(\omega)$. Equipped with the norm

$$\|v\|_{l,\omega,\mathbb{R}^3} = \left( \sum_{i=1}^{3} \|p_i \circ v\|_{l,\omega}^2 \right)^{1/2},$$

where $\| \cdot \|_{l,\omega}$ denotes the norm in $H^l(\omega)$ (cf. Preliminaries), the space $H^l(\omega, \mathbb{R}^3)$ is a Hilbert space. Similarly, for $j = 0, \ldots, l$, one defines the scalar semi-products $( \cdot , \cdot )_{j,\omega,\mathbb{R}^3}$ and the associated semi-norms $| \cdot |_{j,\omega,\mathbb{R}^3}$. We shall write $L^2(\omega, \mathbb{R}^3)$ instead of $H^0(\omega, \mathbb{R}^3)$.

When $\omega$ is bounded, for any $\mu \in \mathbb{N}$, we denote by $C^\mu(\overline{\omega}, \mathbb{R}^3)$ the Banach space of all $\mathbb{R}^3$-valued functions $v$ such that, for $i = 1, 2, 3$, $p_i \circ v \in C^\mu(\overline{\omega})$, equipped with the norm

$$\|v\|_{C^\mu(\overline{\omega},\mathbb{R}^3)} = \max_{1 \le i \le 3} \|p_i \circ v\|_{C^\mu(\overline{\omega})}.$$

It follows from (2) and (3) in Preliminaries that, when $\omega$ has a Lipschitz-continuous boundary, the following compact injection theorems hold:

$$\forall l, l' \in \mathbb{N},\ l > l',\ H^l(\omega, \mathbb{R}^3) \overset{c}{\subset} H^{l'}(\omega, \mathbb{R}^3), \tag{1.1}$$

$$\forall l, \mu \in \mathbb{N},\ l > \mu + 1,\ H^l(\omega, \mathbb{R}^3) \overset{c}{\subset} C^{\mu}(\overline{\omega}, \mathbb{R}^3). \tag{1.2}$$

## 2. FORMULATION OF THE PROBLEM

Consider a surface $\mathcal{S}$ verifying the following assumption:

> there exists a bounded open set $U \subset \mathbb{R}^2$ with a Lipschitz-continuous boundary (cf. Preliminaries), a function $f : \overline{U} \to \mathbb{R}^3$, injective on $U$, and an integer $m' \geq 2$ such that $\mathcal{S} = f(\overline{U})$ and $f \in H^{m'}(U, \mathbb{R}^3)$. (2.1)

Therefore, the function $f$ is a *parametrization* of the surface $\mathcal{S}$ which, by (1.2), belongs to $C^0(\overline{U}, \mathbb{R}^3)$. The preceding hypothesis is satisfied by most of the parametric surfaces that appear in practice, including, of course, *disk-like surfaces* (i.e. surfaces homeomorphic to the unit closed ball $\overline{B}((0,0),1)$). We shall not consider surfaces with arbitrary topology here.

Let $\mathbb{D}$ be a bounded subset of $(0, +\infty)$ such that $0 \in \overline{\mathbb{D}}$. Then, we pose the following problem: *given a family* $(\mathcal{P}^d)_{d\in\mathbb{D}}$ *of finite sets of points of* $\mathcal{S}$, *construct a family* $(\mathfrak{S}^d)_{d\in\mathbb{D}}$ *of approximating surfaces of* $\mathcal{S}$, *which are "sufficiently regular"* (parametrized, for example, by a vector function of class $C^1$ or $C^2$). So, for any $d \in \mathbb{D}$, $\mathfrak{S}^d$ has to approach $\mathcal{S}$ on $\mathcal{P}^d$ and, as $d \to 0$, the surface $\mathcal{S}$ must be the limit of $\mathfrak{S}^d$ in a way to be stated precisely.

Let us first point out that the family $(\mathfrak{S}^d)_{d\in\mathbb{D}}$ cannot be obtained by approximation of the parametrization $f$ of $\mathcal{S}$. Although the sets $\mathcal{P}^d$ of values of $f$ are known, the domain $U$ and the points of $\overline{U}$ whose images under $f$ are the points of $\mathcal{P}^d$ are almost always unknown. That is the reason why any approximation process of $\mathcal{S}$ must introduce, at first, a *parametrization method* of the data points. Such a method consists in choosing, in a suitable way, an open set $\Omega$ in $\mathbb{R}^2$ and, for any $d \in \mathbb{D}$, a subset $A^d$ of $\overline{\Omega}$ having the same cardinal number as $\mathcal{P}^d$ and a bijection $\gamma^d : A^d \to \mathcal{P}^d$ (i.e. a "rule" which links, in a one-to-one way, any point of $\mathcal{P}^d$ to a point of $A^d$). Of course, the choice of $\Omega$, $(A^d)_{d\in\mathbb{D}}$ and $(\gamma^d)_{d\in\mathbb{D}}$ cannot be made arbitrarily: it must be compatible with the existence of a family $(g^d)_{d\in\mathbb{D}}$ of parametrizations of $\mathcal{S}$ defined over $\overline{\Omega}$ such that, for any $d \in \mathbb{D}$, $g^d|_{A^d} = \gamma^d$.

It is clear that the definition of these parametrization algorithms requires a prior modelling of the geometry of the surface $\mathcal{S}$, which may come from some real knowledge of the surface (for instance, when one digitalizes a physical object), or which can be deduced from supplementary information or from a particular spatial structure of the data points (for example, when these are the vertices of a curvilinear grid lying on $\mathcal{S}$).

**Remark 2.1** – Among the parametrization methods commonly used, in the first place one finds the methods of projection onto a reference surface (cf. W. Ma and J. P. Kruth [104]), which apply when the data points have no particular structure.

In these methods, a parametric surface $\mathcal{S}_R$ "close" to $\mathcal{S}$, called *base or reference surface*, is introduced. For any data point $P$, one associates with $P$ the parameters corresponding to the point of $\mathcal{S}_R$ obtained by projection (for instance, orthogonal) of $P$ onto $\mathcal{S}_R$. When $\mathcal{S}$ is sufficiently flat, a standard choice of $\mathcal{S}_R$ is the plane calculated by a least squares approximation of the data points.

There also exist parametrization methods, mostly restricted to disk-like surfaces, which are based on the "development" in $\mathbb{R}^2$ of 3D polygonal meshes (cf. G. Greiner and K. Hormann [73, 81], M. S. Floater [65] and references therein). These methods usually map a surface triangulation of $\mathcal{S}$, whose vertices are the data points, onto a topologically equivalent triangulation of a planar domain. The vertices of the 2D triangulation are then the parameter values of the data points. Some of these methods have been extended to clouds of unorganized data points (cf. M. S. Floater and M. Reimers [66]). □

Suppose that a suitable parametrization method of the data points is available. Then we assume that

$$\left|\begin{array}{l}\text{there exists a bounded open set } \Omega \subset \mathbb{R}^2 \text{ with a Lipschitz-continuous} \\ \text{boundary (cf. Preliminaries), a family } (A^d)_{d\in\mathbb{D}} \text{ of finite subsets of} \\ \overline{\Omega} \text{ and a family } (g^d)_{d\in\mathbb{D}} \subset H^{m'}(\Omega,\mathbb{R}^3) \text{ such that, for any } d \in \mathbb{D}, \\ g^d(\overline{\Omega}) = \mathcal{S} \text{ and } g^d(A^d) = \mathcal{P}^d,\end{array}\right. \tag{2.2}$$

where $m'$ denotes the integer introduced in (2.1). In practice, the open set $\Omega$ and the family $(A^d)_{d\in\mathbb{D}}$ are explicitly determined. Therefore, the initial problem of approximation of $\mathcal{S}$ is transformed into the following problem: for any $d \in \mathbb{D}$, find an approximant of the function $g^d$ from the set of parameters $A^d$ and the set of points $\{g^d(a)\}_{a\in A^d} = \mathcal{P}^d$. To solve this problem, we shall propose a fitting method by means of discrete smoothing $D^m$-splines. Furthermore, in order to study the convergence of these splines, we suppose, as usual, that

$$\forall d \in \mathbb{D}, \ \sup_{x\in\Omega} \delta(x, A^d) = d, \tag{2.3}$$

and that the family $(g^d)_{d\in\mathbb{D}}$ converges in a weak sense to a parametrization $\mathfrak{g}$ of $\mathcal{S}$, i.e.

$$\left|\begin{array}{l}\text{there exists a function } \mathfrak{g} \in H^{m'}(\Omega,\mathbb{R}^3) \text{ such that } \mathfrak{g}(\overline{\Omega}) = \mathcal{S} \text{ and, for} \\ \text{any } v \in H^{m'}(\Omega,\mathbb{R}^3),\ \lim_{d\to 0}((g^d, v))_{m',\Omega,\mathbb{R}^3} = ((\mathfrak{g}, v))_{m',\Omega,\mathbb{R}^3},\end{array}\right. \tag{2.4}$$

where $((\cdot\,,\,\cdot))_{m',\Omega,\mathbb{R}^3}$ denotes the scalar product associated with the norm $\|\cdot\|_{m',\Omega,\mathbb{R}^3}$.

**Remark 2.2** – It follows from (2.4) that there exists $d_0$ such that $(g^d)_{d\in\mathbb{D},\, d\leq d_0}$ is a bounded family in $H^{m'}(\Omega,\mathbb{R}^3)$. From now on, we shall suppose that $d_0 = \sup \mathbb{D}$. □

**Remark 2.3** – Suppose that $\mathcal{S}$ admits an explicit representation in the form $z = \phi(x,y)$, where $\phi$ is a function that belongs to $H^{m'}(\Omega)$, $\Omega$ being, in turn, an open set as defined in (2.2). For any $d \in \mathbb{D}$, let $A^d$ be the projection of $\mathcal{P}^d$ onto the plane $xy$, and let $g^d$ be the function defined on $\overline{\Omega}$ by $g^d(x,y) = \big(x, y, \phi(x,y)\big)$. Then, hypotheses (2.2) and (2.4) are easily verified. With a suitable distribution of the points of $\mathcal{P}^d$ on $\mathcal{S}$, hypothesis (2.3) is satisfied too.

In the general case, these hypotheses can be justified by considering a problem of existence of families of parameter changes for the surface $\mathcal{S}$. In [6] and [139] (cf. also D. Apprato and R. Arcangéli [7] and J. J. Torrens [140]), it is supposed that, for any $d \in \mathbb{D}$, the surface $\mathcal{S}$ is a union of skew quadrilaterals whose vertices make up the set $\mathcal{P}^d$. More precisely, taking (2.1) into account, it is assumed that $\mathcal{P}^d$ is the image under $f$ of the set $B^d$ of the vertices of a triangulation of $\overline{U}$ by means of rectangles. This implies that $U$ has a rectangular geometry, i.e. that the boundary of $U$ is a finite union of segments parallel to the coordinate axes. In this situation, it is possible to take an open set $\Omega$ with a rectangular geometry and parametrize uniformly the data points by attaching them, for any $d \in \mathbb{D}$, to the set $A^d$ of the vertices of a triangulation of $\overline{\Omega}$ by means of equal rectangles. Then, one can explicitly construct bijections $\varphi^d : \overline{\Omega} \to \overline{U}$ of class $C^{m'}$ such that $\varphi^d(A^d) = B^d$ and, under suitable assumptions (implying, in particular, (2.3)), show that, as $d \to 0$, $(\varphi^d)_{d\in\mathbb{D}}$ converges uniformly on $\overline{\Omega}$ to a homeomorphism $\varphi$ from $\overline{\Omega}$ onto $\overline{U}$. This yields (2.2) and (2.4) taking $\mathfrak{g} = f \circ \varphi$ and, for any $d \in \mathbb{D}$, $g^d = f \circ \varphi^d$. □

## 3. SPLINE FITTING

We keep the notations and assumptions introduced before. For the sake of simplicity, we suppose that $\Omega$ is a polygonal open set. On the other hand, we suppose that we are given

- an integer $m \geq 2$,
- a bounded subset $\mathbb{H}$ of $(0, +\infty)$ such that $0 \in \overline{\mathbb{H}}$,
- for any $h \in \mathbb{H}$, a triangulation $\mathcal{T}_h$ of $\overline{\Omega}$ by means of elements $K$ with diameters $h_K \leq h$ and a finite element space $V_h$, constructed on $\mathcal{T}_h$, such that

$$V_h \text{ is a finite-dimensional subspace of } H^m(\Omega) \cap C^k(\overline{\Omega}), \text{ with } k = 1 \text{ or } 2. \tag{3.1}$$

Concerning the choice of the generic finite element of $V_h$, the comments in Section VIII–3 remain valid.

For any $h \in \mathbb{H}$, let $X_h = (V_h)^3$. By (3.1), it is clear that

$$X_h \subset H^m(\Omega, \mathbb{R}^3) \cap C^k(\overline{\Omega}, \mathbb{R}^3).$$

Taking (2.3) into account, we can consider that, for any $d \in \mathbb{D}$, $A^d$ contains a $P_{m-1}$-unisolvent subset. Furthermore, we write $N = \operatorname{card} A^d$, we number $a_1, \ldots, a_N$ the points of $A^d$ and we introduce the linear continuous operator $\rho^d : C^0(\overline{\Omega}, \mathbb{R}^3) \to \mathbb{R}^{N,3}$, defined by $\rho^d v = \big(v(a_i)\big)_{1 \leq i \leq N}$. Notice that the rows of the matrix $\rho^d g^d$ are the coordinates of the points of $\mathcal{P}^d$.

For any $(d, h, \varepsilon) \in \mathbb{D} \times \mathbb{H} \times (0, +\infty)$ and any $v_h \in X_h$, we put

$$J^d_{\varepsilon h}(v_h) = \langle \rho^d v_h - \rho^d g^d \rangle^2 + \varepsilon |v_h|^2_{m,\Omega,\mathbb{R}^3}, \tag{3.2}$$

where $\langle \cdot \rangle$ is the Euclidean norm in $\mathbb{R}^{N,3}$. Then we consider the problem: find $\sigma^d_{\varepsilon h}$ solution of

$$\begin{cases} \sigma^d_{\varepsilon h} \in X_h, \\ \forall v_h \in X_h,\ J^d_{\varepsilon h}(\sigma^d_{\varepsilon h}) \le J^d_{\varepsilon h}(v_h), \end{cases} \tag{3.3}$$

and also the problem: find $\sigma^d_{\varepsilon h}$ such that

$$\begin{cases} \sigma^d_{\varepsilon h} \in X_h, \\ \forall v_h \in X_h,\ \langle \rho^d \sigma^d_{\varepsilon h}, \rho^d v_h \rangle + \varepsilon (\sigma^d_{\varepsilon h}, v_h)_{m,\Omega,\mathbb{R}^3} = \langle \rho^d g^d, \rho^d v_h \rangle, \end{cases} \tag{3.4}$$

where $\langle \cdot\, , \cdot \rangle$ stands for the Euclidean scalar product in $\mathbb{R}^{N,3}$.

Reasoning as in Theorem V–3.1, one sees that there exists a common unique solution $\sigma^d_{\varepsilon h}$ of (3.3) and (3.4), called *$X_h$-discrete smoothing $D^m$-spline relative to $A^d$, $\mathcal{P}^d$ and $\varepsilon$*. It will be proved further (cf. Theorem 4.1) that, under suitable hypotheses, the parametrization $\mathfrak{g}$ introduced in (2.4) is the strong limit of $\sigma^d_{\varepsilon h}$ in $H^m(\Omega, \mathbb{R}^3)$ as $d \to 0$. For any $d \in \mathbb{D}$, the surface $\mathfrak{S}^d$ parametrized by $\sigma^d_{\varepsilon h}$ is therefore an approximant of the surface $S$ in the following sense: there exists a parametrization of $S$ which is the limit, as $d \to 0$, of a parametrization of $\mathfrak{S}^d$. Thus, the initial problem of approximation of $S$ is solved.

**Remark 3.1** – If $\Omega$ is not polygonal, one proceeds as in Sections VI–1 or VIII–2 in order to obtain, for any $h \in \mathbb{H}$, the space $V_h$, and so $X_h = (V_h)^3$. We recall that one first chooses a polygonal open set $\widetilde{\Omega}$ which contains $\Omega$ and a triangulation $\widetilde{\mathcal{T}}_h$ of $\overline{\widetilde{\Omega}}$ made up of elements $K$ with diameter $h_K \le h$. Then one constructs on $\widetilde{\mathcal{T}}_h$ a finite element space $\widetilde{V}_h$ such that $\widetilde{V}_h \subset H^m(\widetilde{\Omega}) \cap C^k(\overline{\widetilde{\Omega}})$, with $k = 1$ or $2$. Finally, one considers the open set $\Omega_h$ defined by (VI–1.2). The space $V_h$ is then given by (VI–1.5). We remark that, once (3.3) is solved for suitable values of $d$, $h$ and $\varepsilon$, the approximating surface $\mathfrak{S}^d$ of $S$ is not the trace of $\sigma^d_{\varepsilon h}$, but that of $\sigma^d_{\varepsilon h}|_\Omega$. □

**Remark 3.2** – For any $h \in \mathbb{H}$, let $M$ and $\{w_1, \ldots, w_M\}$ be the dimension and a basis of $V_h$. The solution $\sigma^d_{\varepsilon h}$ of (3.4) can be expressed in the form

$$\sigma^d_{\varepsilon h} = \sum_{i=1}^{M} \alpha_j w_j, \tag{3.5}$$

with $\alpha_j \in \mathbb{R}^3$, for $1 \le j \le M$. Then, one shows that (3.4) is equivalent to the problem: find $\alpha = (\alpha_j)_{1 \le j \le M}$ such that

$$\begin{cases} \alpha \in \mathbb{R}^{M,3}, \\ (\mathcal{A}^T \mathcal{A} + \varepsilon \mathcal{R})\alpha = \mathcal{A}^T(\rho^d g^d), \end{cases} \tag{3.6}$$

where $\mathcal{A} = \big(w_j(a_i)\big)_{1 \le i \le N,\, 1 \le j \le M}$ and $\mathcal{R} = \big((w_j, w_i)_{m,\Omega}\big)_{1 \le i,j \le M}$. Computing $\sigma^d_{\varepsilon h}$ comes down to solving three linear systems having the same matrix $\mathcal{A}^T\mathcal{A} + \varepsilon\mathcal{R}$ of dimension $M$.

Let $\beta^1$, $\beta^2$ and $\beta^3$ be the three column vectors of $\rho^d g^d$. Then, it can be seen that $\sigma^d_{\varepsilon h}$ is the $X_h$-discrete smoothing $D^m$-spline relative to $A^d$, $\mathcal{P}^d$ and $\varepsilon$ if and only if, for $i = 1,2,3$, $p_i \circ \sigma^d_{\varepsilon h}$ is the $V_h$-discrete smoothing $D^m$-spline relative to $A^d$, $\beta^i$ and $\varepsilon$. Consequently, to approximate the surface $\mathcal{S}$ once the data points have been parametrized, it is sufficient to fit independently the sets formed by the first, second and third coordinates of the points of $\mathcal{P}^d$. □

**Remark 3.3** – For any $d \in \mathbb{D}$, the GCV method can be used to find the optimal value of $\varepsilon$ in order to solve problems (3.3) and (3.4). It is readily seen that the corresponding GCV function can be expressed as the sum of the GCV functions $\mathcal{V}_1$, $\mathcal{V}_2$ and $\mathcal{V}_3$ associated with the smoothing of the three column vectors $\beta^1$, $\beta^2$ and $\beta^3$ of $\rho^d g^d$ (i.e., for $i = 1,2,3$, $\mathcal{V}_i$ is given by (VI–3.4) with $\beta^i$ instead of $\beta$ and, with the notations of the preceding remark, $Q_\varepsilon = \mathcal{A}(\mathcal{A}^T\mathcal{A}+\varepsilon\mathcal{R})^{-1}\mathcal{A}^T$). In practice, every function $\mathcal{V}_i$ is approximated as detailed in Remark VI–3.3. □

**Remark 3.4** – If the surface $\mathcal{S}$ presents cuts, fissures or holes, one can model $\mathcal{S}$ with (2.1), taking $U$ as a non-simply connected open set. In some circumstances, it may be advantageous, instead, to assume that the parametrization $f$ is a "non-regular" function, i.e. that $f$ belongs to the space $H^{m'}(U \setminus \overline{\Phi}, \mathbb{R}^3)$ whereas it does not belong to the space $H^{m'}(U, \mathbb{R}^3)$, where $m'$ denotes any integer $\geq 2$ and $\Phi$ is a suitable subset of $U$ (cf. Section IX–1). In this case, the smoothing method developed before can be used, provided the following changes are made:

- the process of data parametrization has to fix a "discontinuity set" $F \subset \overline{\Omega}$ verifying the conditions of Subsection IX–2.1 and corresponding to the singularities of the surface $\mathcal{S}$,
- for any $h \in \mathbb{H}$, the triangulation $\mathcal{T}_h$ and the space $V_h$ have to verify hypotheses (IX–4.1), (IX–4.2) and (IX–4.3), with $\Omega' = \Omega \setminus \overline{F}$,
- $\Omega$ has to be replaced by $\Omega'$ in the definition (3.2) of $J^d_{\varepsilon h}$. □

**Remark 3.5** – Suppose that the surface $\mathcal{S}$ is regular enough so that the tangent plane exists at each of its points. This is ensured if, in (2.1), $m' \geq 3$ (then $f \in C^1(\overline{U}, \mathbb{R}^3)$), and if, for any $u \in \overline{U}$, the vector product of $\partial^{(1,0)} f(u)$ and $\partial^{(0,1)} f(u)$ is non-null.

We have already pointed out (cf. Sections VIII–5 and IX–1) that the *position data* (i.e. the sets $\mathcal{P}^d$ of points of $\mathcal{S}$) are often completed by *orientation data* (usually two angles named *dip* and *strike*) which determine the tangent planes to $\mathcal{S}$. When $\mathcal{S}$ admits an explicit representation, these data can be converted into first order Hermite data. Let us show how to use the orientation data in the present context.

For any $d \in \mathbb{D}$, suppose that the tangent planes to $\mathcal{S}$ are known at the points of a subset $\mathcal{P}^d_1$ of $\mathcal{P}^d$. Let $A^d_1 \subset A^d$ be the set of the corresponding parameters (so $g^d(A^d_1) = \mathcal{P}^d_1$). Therefore, for any $a \in A^d_1$, a unit vector $\nu(a)$ normal to the surface $\mathcal{S}$ at the point $g^d(a)$ is known.

For any $(d, h, \varepsilon, \tau) \in \mathbb{D} \times \mathbb{H} \times (0, +\infty) \times [0, +\infty)$ and for any $v_h \in X_h$, we set

$$J^d_{\varepsilon\tau h}(v_h) = \langle \rho^d v_h - \rho^d g^d \rangle^2 + \tau \sum_{a \in A^d_1} \sum_{|\alpha|=1} \langle \partial^\alpha v_h(a), \nu(a) \rangle^2_{\mathbb{R}^3} + \varepsilon |v_h|^2_{m,\Omega,\mathbb{R}^3},$$

where $\langle \cdot\,, \cdot \rangle_{\mathbb{R}^3}$ denotes the Euclidean scalar product in $\mathbb{R}^3$. Let us observe that, for any $a \in A^d_1$, the term $\sum_{|\alpha|=1} \langle \partial^\alpha v_h(a), \nu(a) \rangle^2_{\mathbb{R}^3}$ is null if $\mathrm{span}\langle \partial^{(1,0)} v_h(a), \partial^{(0,1)} v_h(a) \rangle$

(i.e. the tangent linear space of $v_h$ at the point $a$) is orthogonal to $\nu(a)$, and hence this term is null if the tangent plane to the surface parametrized by $v_h$ at the point $v_h(a)$ is parallel to the tangent plane to the surface $S$ at the point $g^d(a)$.

Then, we consider the problem: find $\sigma^d_{\varepsilon\tau h}$ such that

$$\begin{cases} \sigma^d_{\varepsilon\tau h} \in X_h, \\ \forall v_h \in X_h,\ J^d_{\varepsilon\tau h}(\sigma^d_{\varepsilon\tau h}) \le J^d_{\varepsilon\tau h}(v_h). \end{cases} \tag{3.7}$$

The solution $\sigma^d_{\varepsilon\tau h}$ of (3.7), called *$X_h$-discrete smoothing $D^m$-spline with tangent conditions*, exists, is unique and admits a variational equivalent characterization similar to (3.4).

Computing $\sigma^d_{\varepsilon\tau h}$ amounts to solving a linear system of dimension $3M$, where $M = \dim V_h$, taking account of the coupling that the orientation data establish between the three components of $\sigma^d_{\varepsilon\tau h}$. We remark that $\sigma^d_{\varepsilon 0 h}$ is precisely the solution of (3.3). These splines have been introduced by M. Pasadas (cf. [114, 115]). □

# 4. CONVERGENCE OF THE APPROXIMATION

We keep the notations and hypotheses of Sections 2 and 3. We suppose that the families $(A^d)_{d\in\mathbb{D}}$ and $(\mathcal{T}_h)_{h\in\mathbb{H}}$ are linked by the relation

$$\exists C > 0,\ \forall (d,h) \in \mathbb{D}\times\mathbb{H},\ \forall K \in \mathcal{T}_h,\ \frac{\operatorname{card}(A^d \cap K)}{\operatorname{meas} K} \le C d^{-2} \tag{4.1}$$

(cf. hypothesis (VI–3.7)) and that the integers $m'$ and $m$ verify the inequality

$$m' > m. \tag{4.2}$$

Likewise, we suppose that the family $(X_h)_{h\in\mathbb{H}}$ and the integer $m'$ are such that the following result holds:

> there exists a constant $C > 0$ and, for any $h \in \mathbb{H}$, a linear operator $\widetilde{\Pi}_h : L^2(\Omega,\mathbb{R}^3) \to X_h$ such that, for any $l = 0,\dots,m'$, one has
>
> $$\forall v \in H^{m'}(\Omega,\mathbb{R}^3),\ \left(\sum_{K\in\mathcal{T}_h} |v - \widetilde{\Pi}_h v|^2_{l,K,\mathbb{R}^3}\right)^{1/2} \le C h^{m'-l} |v|_{m',\Omega,\mathbb{R}^3}. \tag{4.3}$$

**Remark 4.1** – For any $h \in \mathbb{H}$, each one of the three components of the operator $\widetilde{\Pi}_h$ introduced in (4.3) is an operator of Clément's type (cf. Section VI–1). Concerning the conditions that this result implicitly supposes verified, we refer to (VI–1.6) and Remark VI–1.1. □

Finally, we suppose that $\varepsilon$ and $h$ are functions of $d$ such that

$$\varepsilon = o(d^{-2}),\ d \to 0, \tag{4.4}$$

and that

$$\frac{h^{2m'}}{d^2\varepsilon} = o(1), \ d \to 0. \tag{4.5}$$

It follows from (4.4) and (4.5) that $h \to 0$ as $d \to 0$. To simplify the notations, we shall write, hereafter, $\varepsilon$ and $h$ instead of $\varepsilon(d)$ and $h(d)$.

***Theorem* 4.1** – *Suppose that hypotheses (2.2), (2.3), (2.4), (3.1), (4.1), (4.2), (4.3), (4.4) and (4.5) are verified. For any $d \in \mathbb{D}$, let us denote by $\sigma^d_{\varepsilon h}$ the $X_h$-discrete smoothing $D^m$-spline relative to $A^d$, $\mathcal{P}^d$ and $\varepsilon$. Then,*

$$\lim_{d\to 0} \|\sigma^d_{\varepsilon h} - \mathfrak{g}\|_{m,\Omega,\mathbb{R}^3} = 0,$$

*where $\mathfrak{g}$ is the parametrization of $S$ introduced in (2.4).*

**Proof** –

1) First, let us observe that, by (2.4) (cf. Remark 2.2),

$$\exists C > 0, \ \forall d \in \mathbb{D}, \ |g^d|_{m',\Omega,\mathbb{R}^3} \le C. \tag{4.6}$$

From (2.4) and (4.2), taking (1.1) into account, we have

$$|g^d|_{m,\Omega,\mathbb{R}^3} = |\mathfrak{g}|_{m,\Omega,\mathbb{R}^3} + o(1), \ d \to 0. \tag{4.7}$$

Hence, using (4.3), we derive

$$|\widetilde{\Pi}_h g^d|_{m,\Omega,\mathbb{R}^3} = |\mathfrak{g}|_{m,\Omega,\mathbb{R}^3} + o(1), \ d \to 0. \tag{4.8}$$

On the other hand, if we adapt Lemma VI–3.1 to the present situation, we get

$$\exists C > 0, \ \forall (d,h) \in \mathbb{D} \times \mathbb{H}, \ \langle \rho^d(\widetilde{\Pi}_h g^d - g^d)\rangle^2 \le C\frac{h^{2m'}}{d^2}|g^d|^2_{m',\Omega,\mathbb{R}^3}. \tag{4.9}$$

For any $d \in \mathbb{D}$, taking $v_h = \widetilde{\Pi}_h g^d$ in (3.3), we deduce that

$$\langle \rho^d(\sigma^d_{\varepsilon h} - g^d)\rangle^2 + \varepsilon|\sigma^d_{\varepsilon h}|^2_{m,\Omega,\mathbb{R}^3} \le \langle \rho^d(\widetilde{\Pi}_h g^d - g^d)\rangle^2 + \varepsilon|\widetilde{\Pi}_h g^d|^2_{m,\Omega,\mathbb{R}^3}.$$

Then, we infer from (4.5), (4.6), (4.8) and (4.9) that the family $(\sigma^d_{\varepsilon h})_{d\in\mathbb{D}}$ verifies the relations

$$|\sigma^d_{\varepsilon h}|_{m,\Omega,\mathbb{R}^3} \le |\mathfrak{g}|_{m,\Omega,\mathbb{R}^3} + o(1), \ d \to 0. \tag{4.10}$$

and

$$\langle \rho^d(\sigma^d_{\varepsilon h} - g^d)\rangle^2 = O(\varepsilon), \ d \to 0. \tag{4.11}$$

2) Let $B_0 = \{b_{01}, \ldots, b_{0\mathfrak{M}}\}$ be any $P_{m-1}$-unisolvent subset of $\overline{\Omega}$. Let $r_0 > 0$ be the constant of Proposition V–1.2. Reasoning as in point 2) of the proof of Theorem VI–3.2 and taking (2.3), (4.4) and (4.11) into account, one shows the existence, for any $d \in \mathbb{D}$ sufficiently small, of a set $\{b^d_1, \ldots, b^d_{\mathfrak{M}}\}$ such that, for $j = 1, \ldots, \mathfrak{M}$, $b^d_j \in \overline{B}(b_{0j}, r_0) \cap A^d$ and that

$$|\sigma^d_{\varepsilon h}(b^d_j) - g^d(b^d_j)| = o(1), \ d \to 0. \tag{4.12}$$

Then, applying Proposition V–1.2 with $B = \{b_1^d, \ldots, b_{\mathfrak{M}}^d\}$, it follows from (4.7), (4.10) and (4.12) that

$$\exists C > 0,\ \exists d^* > 0,\ \forall d \in \mathbb{D} \cap (0, d^*],\ \|\sigma_{\varepsilon h}^d - g^d\|_{m,\Omega,\mathbb{R}^3} \leq C,$$

i.e. the family $(\sigma_{\varepsilon h}^d - g^d)_{d \in \mathbb{D} \cap (0,d^*]}$ is bounded in $H^m(\Omega, \mathbb{R}^3)$. Thus, by Corollary 1 in Preliminaries, there exists a sequence $(\sigma_{\varepsilon_l h_l}^{d_l} - g^{d_l})_{l \in \mathbb{N}}$ with, for any $l \in \mathbb{N}$, $d_l \in \mathbb{D}$, $\varepsilon_l = \varepsilon(d_l)$, $h_l = h(d_l)$ and

$$\lim_{l \to +\infty} d_l = \lim_{l \to +\infty} d_l^2 \varepsilon_l = \lim_{l \to +\infty} \frac{h_l^{2m'}}{d_l^2 \varepsilon_l} = 0,$$

extracted from the family $(\sigma_{\varepsilon h}^d - g^d)_{d \in \mathbb{D} \cap (0,d^*]}$, and an element $\mathfrak{g}^* \in H^m(\Omega, \mathbb{R}^3)$ such that, as $l \to +\infty$,

$$(\sigma_{\varepsilon_l h_l}^{d_l} - g^{d_l}) \to \mathfrak{g}^*, \quad \text{weakly in } H^m(\Omega, \mathbb{R}^3). \tag{4.13}$$

3) We continue as in the proof of Theorem VI–3.2. First, reasoning by contradiction, it is shown that $\mathfrak{g}^* = 0$. Next, it follows from (2.4) and (4.13) that

$$\sigma_{\varepsilon_l h_l}^{d_l} \to \mathfrak{g}, \quad \text{weakly in } H^m(\Omega, \mathbb{R}^3).$$

Hence, taking (4.10) into account, one deduces that

$$\lim_{l \to +\infty} \|\sigma_{\varepsilon_l h_l}^{d_l} - \mathfrak{g}\|_{m,\Omega,\mathbb{R}^3} = 0.$$

Then, reasoning again by contradiction, one completes the proof. □

## 5. *NUMERICAL RESULTS*

By means of three different examples, we shall illustrate the surface smoothing method presented in Section 3 as well as two of the parametrization methods cited in Remark 2.1. These examples will also serve to discuss some additional features, such as the introduction of periodicity conditions, the parameter correction technique or the use of non-polygonal parameter domains. All these questions are part of the topics that should be examined in the design of global efficient strategies for fitting parametric surfaces (cf., for instance, V. Weiss, L. Andor, G. Renner and T. Várady [150]).

In every example we shall consider a surface $\mathcal{S}$ satisfying (2.1). We shall write the corresponding parametrization $f : \overline{U} \to \mathbb{R}^3$. The smoothing method will take as input a set $\mathcal{P} \subset \mathcal{S}$ which results from evaluating $f$ on a finite subset $B$ of $\overline{U}$. In the first example, we shall think of $\mathcal{P}$ as an unorganized data set, while for the second and third examples, $\mathcal{P}$ will be, in fact, the vertex set of a supposedly known triangulation of $\mathcal{S}$.

In the first step of the fitting process, we shall fix an open set $\Omega$ (the unit square or the unit ball) and we shall get a set $A \subset \overline{\Omega}$ of parameter values. Then we shall apply the smoothing method, computing a suitable discrete $D^2$-spline $\sigma_{\varepsilon h}$. In all cases, the finite element spaces will be constructed from the Bogner-Fox-Schmit rectangle of class $C^1$ (cf. Section VIII–3).

In the preceding chapters we used the relative error defined by (VIII–6.3) to measure the accuracy of the fitting. In the present situation, we do not explicitly know the expression of any underlying parametrization $g : \overline{\Omega} \to \mathbb{R}^3$ of $\mathcal{S}$. Thus, given a fitting function $\sigma_{\varepsilon h}$, we can only compute the relative error $r(\mathcal{S})$ given by

$$r(\mathcal{S}) = \left( \frac{\sum_{a \in A} \langle \sigma_{\varepsilon h}(a) - q_a \rangle^2_{\mathbb{R}^3}}{\sum_{a \in A} \langle q_a \rangle^2_{\mathbb{R}^3}} \right)^{1/2}, \tag{5.1}$$

where $\langle \cdot \rangle_{\mathbb{R}^3}$ denotes the Euclidean norm in $\mathbb{R}^3$ and, for any $a \in A$, $q_a$ is the point of $\mathcal{P}$ associated with the parameter point $a$. Let us observe that $r(\mathcal{S})$ measures errors only at the data points.

Hereafter, we shall use, when needed, the notations of Section 3 dropping the superindex $d$.

### *Example 1*

Let $\mathcal{S}$ be the surface parametrized by $f : \overline{U} \to \mathbb{R}^3$, with $U = (0, 2\pi) \times (0, 11\pi/6)$ and

$$f(u,v) = \big(5(2 + \sin^2 v \cos 3v) \cos u,\ 3(2 + \sin^2 v \cos 3v) \sin u,\ 8v\big)$$

(cf. Figure 1). Let us observe that $\mathcal{S}$ is not a surface of revolution† and that $\mathcal{S}$ is bounded by the planes $z = 0$ and $z = 44\pi/3$.

Figure 2 shows the set $\mathcal{P}$ of data points, obtained by evaluating $f$ on a set $B$ of 800 points randomly distributed on $\overline{U}$ (in fact, $B$ contains 760 points of $U$, 20 points of $(0, 2\pi) \times \{0\}$ and 20 points of $(0, 2\pi) \times \{11\pi/6\}$).

We parametrize the data points by orthogonal projection on a reference surface $\mathcal{S}_R$. To be precise, we take $\mathcal{S}_R$ as the portion of the vertical circular cylinder of radius 1 centred along the $z$-axis which is limited by the planes $z = 0$ and $z = 44\pi/3$ (cf. Figure 2). The function $g_R : \overline{\Omega} \to \mathbb{R}^3$, with $\Omega = (0,1) \times (0,1)$, defined by $g_R(u,v) = \big(\cos 2\pi u,\ \sin 2\pi u,\ 44\pi v/3\big)$, is a parametrization of $\mathcal{S}_R$. Thus, we associate with every point $(x, y, z) \in \mathcal{P}$ the parameter point $(u, v) \in \overline{\Omega}$ given by

$$(u,v) = \Big(\frac{\theta(x,y)}{2\pi},\ \frac{3z}{44\pi}\Big),$$

where $\theta(x, y)$ is the argument belonging to $[0, 2\pi)$ of the complex number $x + iy$. We obtain, in this way, the set $A$ of parameter points in $\overline{\Omega}$ corresponding to the data set $\mathcal{P}$ (cf. Figure 3).

We remark that, for any $v \in [0,1]$, the intersection of $\mathcal{S}$ and the plane $z = 44\pi v/3$, which is a closed simple curve, would be mapped by the above parametrization method

† Any section of $\mathcal{S}$ parallel to the plane $xy$ is a ellipse with eccentricity $4/5$.

into the segment $(0,1] \times \{v\}$. Consequently, any parametrization $g : \overline{\Omega} \to \mathbb{R}^3$ of $\mathcal{S}$ coherent with such a parametrization method should satisfy that, for any $v \in [0,1]$, $g(0,v) = g(1,v)$, i.e. $g$ should be periodic, of period 1, in the $u$ variable. This fact implies that we must modify the smoothing method of Section 3 in order to use spaces $X_h$ of periodic functions. Let us examine one way to do this.

Let $\{u_0, \dots, u_{n_u}\}$ and $\{v_0, \dots, v_{n_v}\}$ be two subsets of $[0,1]$ such that

$$0 = u_0 < u_1 < \cdots < u_{n_u} = 1 \quad \text{and} \quad 0 = v_0 < v_1 < \cdots < v_{n_v} = 1.$$

Let $\mathcal{T}_h$ be the triangulation of $\overline{\Omega}$ by means of $n_u \times n_v$ rectangles whose vertices are the points $(u_i, v_j)$, with $i = 0, \dots, n_u$ and $j = 0, \dots, n_v$. Likewise, let $V_h$ be the finite element space constructed on $\mathcal{T}_h$ from the Bogner-Fox-Schmit rectangle of class $C^1$, and let $V_{0h}$ be the subspace of $V_h$ formed by the functions $w_h$ such that

$$\forall \alpha \in \mathbb{N}^2,\ |\alpha| \le 1,\ \forall v \in [0,1],\ \partial^\alpha w_h(0,v) = \partial^\alpha w_h(1,v). \tag{5.2}$$

Then we set $X_h = (V_{0h})^3$. Of course, the functions in $X_h$ are periodic in the $u$ variable, as required. Their first partial derivatives are also periodic, a condition needed to get globally smooth surfaces.

It is easy to find the dimension and a basis of $V_{0h}$, taking into account that the space $V_{0h}$ can be equivalently characterized as follows: a function $w_h \in V_h$ belongs to $V_{0h}$ if and only if

$$\forall j = 0, \dots, n_v,\ \forall \alpha \in I,\ \partial^\alpha w_h(0, v_j) = \partial^\alpha w_h(1, v_j), \tag{5.3}$$

where $I = \{(0,0), (1,0), (0,1), (1,1)\}$. Now, let $M = 4(n_u + 1)(n_v + 1)$, which is the dimension of $V_h$, let $w_1, \dots, w_M$ denote the basis functions of $V_h$ and assume that they are numbered so that, for $i = 0, \dots, n_u$ and $j = 0, \dots, n_v$, $\partial^\alpha w_{\nu+l}(u_i, v_j) = 1$, with $\nu = 4(j + (n_v + 1)i)$ and $\alpha = (0,0), (1,0), (0,1), (1,1)$ for $l = 1, 2, 3, 4$, respectively. Let $M_0 = 4n_u(n_v + 1)$. Finally, for $i = 1, \dots, 4(n_v + 1)$, let $w_{0i} = w_i + w_{i+M_0}$ and, for $i = 4(n_v + 1) + 1, \dots, M_0$, let $w_{0i} = w_i$. We deduce from (5.3) that the dimension of $V_{0h}$ is $M_0$ and that $\{w_{01}, \dots, w_{0M_0}\}$ is a basis of $V_{0h}$.

Returning to the fitting problem considered in this example, it suffices to make up $\mathcal{T}_h$ with $6 \times 6$ equal squares (i.e. $n_u = n_v = 6$) and then construct $V_{0h}$ and $X_h$ just as explained above. Since the parameter set $A$ contains a $P_1$-unisolvent subset, it is readily seen that, for any $\varepsilon > 0$, there exists the $X_h$-discrete smoothing $D^2$-spline $\sigma_{\varepsilon h}$ relative to $A$, $\mathcal{P}$ and $\varepsilon$, i.e. the common unique solution of problems (3.3) and (3.4) for $m = 2$. To find $\sigma_{\varepsilon h}$, we proceed as detailed in Remark 3.2, with $V_{0h}$, $M_0$ and $w_{01}, \dots, w_{0M_0}$ instead of $V_h$, $M$ and $w_1, \dots, w_M$. Let us observe that we have to solve a linear system of dimension $M_0 = 168$.

Using the GCV method (cf. Remark 3.3), we choose $\varepsilon = 10^{-5}$. We compute the corresponding $X_h$-discrete smoothing $D^2$-spline $\sigma_{\varepsilon h}$, whose trace, presented in Figure 4, is finally the approximating surface $\mathfrak{S}$ of $\mathcal{S}$ which we were looking for. Of course, the fitting can be improved, for example, by applying parameter correction (see the next example), increasing the dimension of the finite element space or sampling $\mathcal{S}$ at more points.

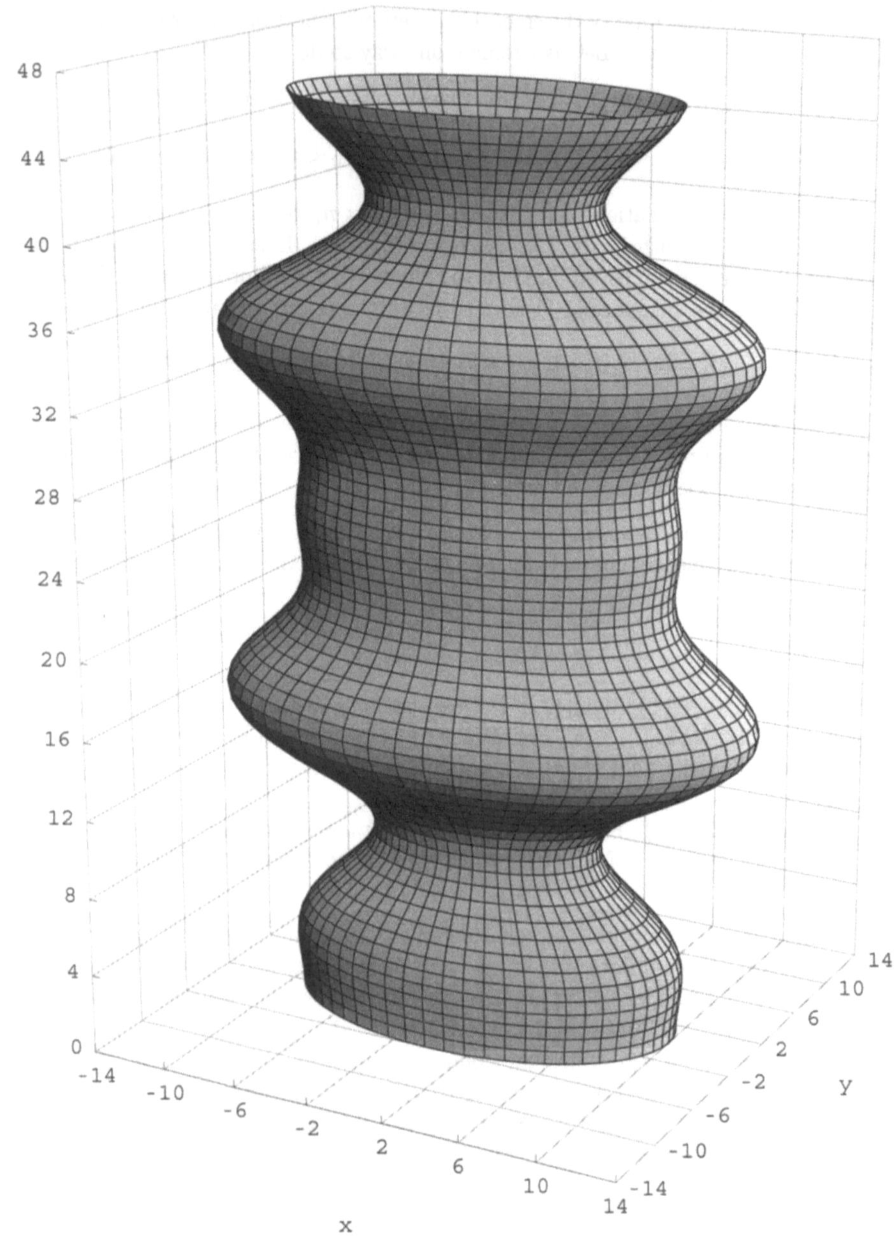


Figure 1: Example 1. Surface $\mathcal{S}$.

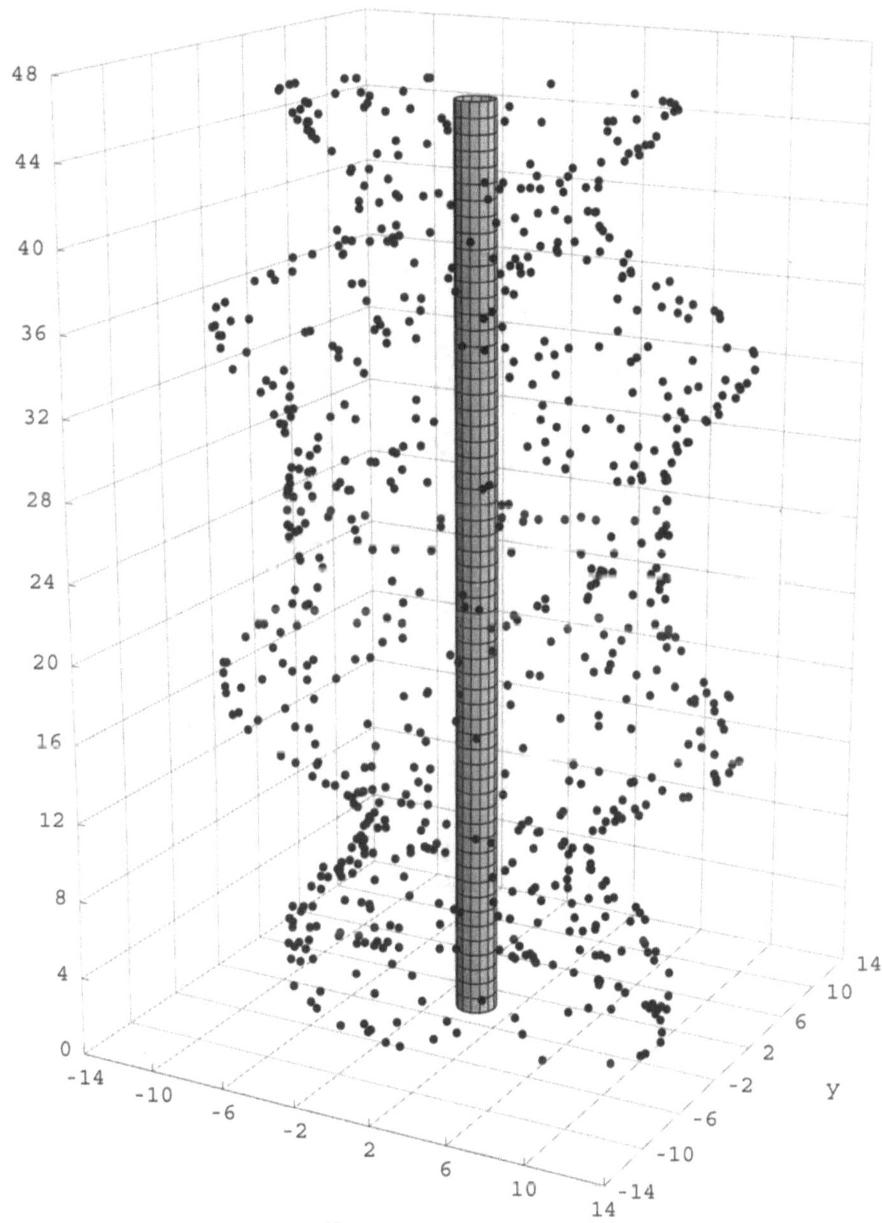


Figure 2: Example 1. Set $\mathcal{P}$ of data points and reference surface $\mathcal{S}_R$.

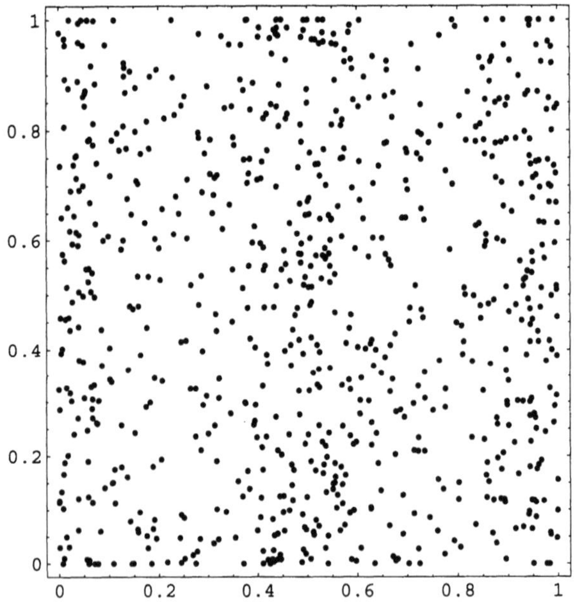


Figure 3: Example 1. Set $A$ of parameter points.

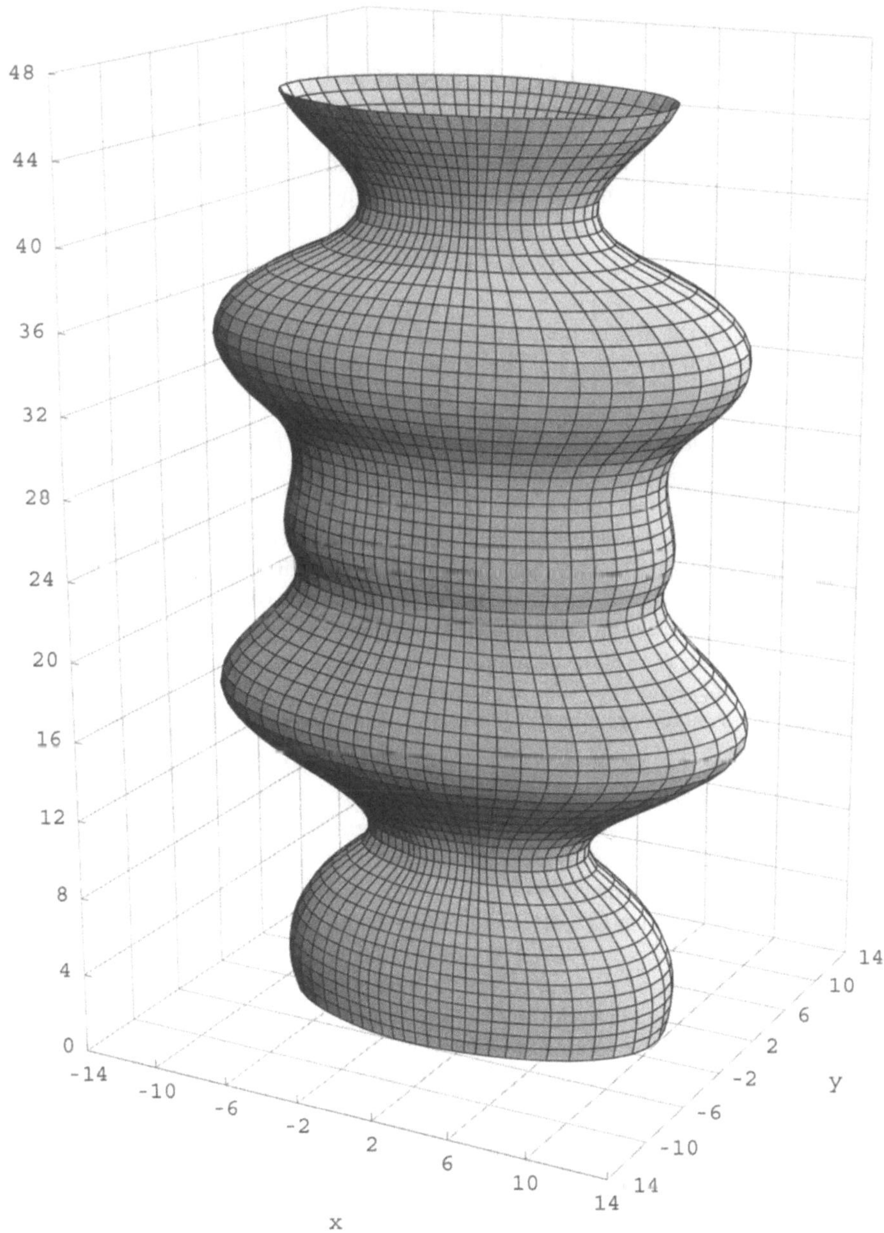


Figure 4: Example 1. Trace of the $X_h$-discrete smoothing $D^2$-spline $\sigma_{\varepsilon h}$ relative to $A$, $\mathcal{P}$ and $\varepsilon = 10^{-5}$. Relative error: $r(\mathcal{S}) = 0.00462458$.

***Example 2***

Let $\mathcal{S}$ be the Bézier surface given by

$$f(u,v) = \sum_{i=0}^{6}\sum_{j=0}^{9} b_{i,j} B_i^6(u) B_j^9(v),$$

with $(u,v) \in \overline{U} = [0,1] \times [0,1]$, where, for $\mu = 6$ and 9 and $l = 0, \ldots, \mu$, $B_l^\mu$ denotes the $l$th Bernstein polynomial of degree $\mu$, defined by

$$B_l^\mu(t) = \binom{\mu}{l} t^l (1-t)^{\mu-l}.$$

For $i = 0,1,2,3$, the Bézier points $b_{i,j}$ are given in Table 1, whereas, for $i = 4,5,6$, we let $b_{i,j} = (8,0,0) - b_{6-i,j}$. The surface $\mathcal{S}$, depicted in Figure 5, emulates an overlapping fold.

To generate the data set $\mathcal{P}$, we first choose a set $B \subset \overline{U}$ formed by the points $(0,0)$, $(1,0)$, $(1,1)$ and $(0,1)$, 18 additional random points on every side of $\overline{U}$, and 724 random points in $U$, in such a way that the Euclidean distance between any two different points of $B$ is greater than 0.01 (this constraint ensures a better distribution of the data points on $\mathcal{S}$). Then we take $\mathcal{P} = f(B)$. In this example, we treat $\mathcal{P}$ as a structured data set, so it is assumed that we also know the triangulation $\mathbb{T}$ of $\mathcal{S}$ shown in Figure 6, induced by the Delaunay triangulation of $B$. For convenience, we denote by $q_1, \ldots, q_N$, with $N = 800$, the points of $\mathcal{P}$ and we suppose that they are numbered so that $q_1, \ldots, q_\nu$, with $\nu = 724$, are the interior vertices of $\mathbb{T}$ and $q_{\nu+1}, \ldots, q_N$, the boundary vertices, are arranged in increasing order along the boundary of $\mathcal{S}$.

We next obtain the set $A = \{a_1, \ldots, a_N\}$ of parameter points using the *shape-preserving parametrization method* introduced by M. S. Floater (cf. [65]), valid only for disk-like surfaces. In two steps, this method transforms the surface triangulation $\mathbb{T}$ into a planar triangulation whose vertices make up the set $A$. In the first step, the closed polygonal line with vertices $q_{\nu+1}, \ldots, q_N$ is mapped onto the boundary of a bounded convex open set $\Omega \subset \mathbb{R}^2$. This yields the parameter points $a_{\nu+1}, \ldots, a_N$. In the second step, for $i = 1, \ldots, \nu$, every data point $q_i$ is associated with a parameter point $a_i \in \Omega$ so that, if $q_{j_1}, \ldots, q_{j_i}$ are the neighbours of $q_i$ in the triangulation $\mathbb{T}$, then $a_i$ is a strict linear convex combination of $a_{j_1}, \ldots, a_{j_i}$. The points $a_1, \ldots, a_\nu$ are obtained, in practice, by solving two sparse linear systems of dimension $\nu$ that share the same system matrix. The reader is referred to [65] for a full description of the method.

In our example, the first step of the shape-preserving parametrization method is accomplished as follows: first we take $\Omega = (0,1) \times (0,1)$, then we assign the four "corner" points of $\mathcal{S}$ to the four corners of $\overline{\Omega}$ and, finally, we parametrize by chord length the remaining boundary points along the sides of $\overline{\Omega}$. Figure 7 shows the output of the parametrization method. The parameter points are more clearly visible in Figure 8.

To continue with the smoothing process, we consider the triangulation $\mathcal{T}_h$ of $\overline{\Omega}$ by means of rectangles whose vertices are the points in the set $\{0, 0.2, 0.4, 0.6, 0.8, 1\} \times$

| | $i=0$ | $i=1$ | $i=2$ | $i=3$ |
|---|---|---|---|---|
| $j=0$ | $(0,0,2.5)$ | $(4/3,0,2.6)$ | $(8/3,0,2.6)$ | $(4,0,2.8)$ |
| $j=1$ | $(0,3.3,3.4)$ | $(4/3,3.7,3.8)$ | $(8/3,3.7,3.8)$ | $(4,3.7,4)$ |
| $j=2$ | $(0,6.4,4.5)$ | $(4/3,7.4,5)$ | $(8/3,7.6,5.2)$ | $(4,7.5,5)$ |
| $j=3$ | $(0,6.8,4)$ | $(4/3,7.9,4.6)$ | $(8/3,8.1,4.7)$ | $(4,7.9,4.6)$ |
| $j=4$ | $(0,5.8,2.8)$ | $(4/3,6.2,3)$ | $(8/3,6.2,3)$ | $(4,5.9,2.8)$ |
| $j=5$ | $(0,4.2,1.8)$ | $(4/3,3.4,1.7)$ | $(8/3,2.8,1.6)$ | $(4,3.2,1.6)$ |
| $j=6$ | $(0,4.1,1.2)$ | $(4/3,3,1.2)$ | $(8/3,2.4,1)$ | $(4,2.4,1.1)$ |
| $j=7$ | $(0,4.5,0.9)$ | $(4/3,3.1,0.8)$ | $(8/3,2.7,0.5)$ | $(4,2.6,0.5)$ |
| $j=8$ | $(0,6.4,0.8)$ | $(4/3,6.2,0.7)$ | $(8/3,6.3,0.7)$ | $(4,6.3,0.6)$ |
| $j=9$ | $(0,8,1)$ | $(4/3,8,1)$ | $(8/3,8,1.1)$ | $(4,8,1.2)$ |

Table 1: Example 2. Bézier points $b_{i,j}$ of $\mathcal{S}$, for $i=0,1,2,3$.

$\{0,0.2,0.4,0.5,0.6,0.7,0.8,1\}$ (cf. Figure 8). Let us observe that the size of the rectangles is smaller between $v=0.4$ and $v=0.8$, to take account of the greater density of parameter points in that zone.

We construct on $\mathcal{T}_h$ the finite element space $V_h$ from the Bogner-Fox-Schmit rectangle of class $C^1$. The dimension of $V_h$ is $M=192$. Then we take $X_h=(V_h)^3$, we set $\varepsilon=10^{-6}$ with the help of the GCV method and, finally, we compute the $X_h$-discrete smoothing $D^2$-spline $\sigma_{\varepsilon h}$ relative to $A$, $\mathcal{P}$ and $\varepsilon$ (cf. Figure 9).

The trace $\mathfrak{S}$ of $\sigma_{\varepsilon h}$ may be a satisfactory approximating surface of $\mathcal{S}^\dagger$. However, the accuracy of the fitting can be significantly increased, with no loss of smoothness, using the so-called *parameter correction* (cf. J. Hoschek and D. Lasser [82]), whose principles adapted to the present context we shall briefly explain. For any $i=1,\ldots,N$, the quantity $\langle\sigma_{\varepsilon h}(a_i)-q_i\rangle_{\mathbb{R}^3}$ is quite "small", since $\sigma_{\varepsilon h}$ is the solution of (3.3), but it does not necessarily measure the Euclidean distance from the data point $q_i$ to $\mathfrak{S}$, as can be checked. We are seeking, in fact, a smooth surface as close as possible to $\mathcal{S}$ at the data points. Therefore, it would be preferable to assign $q_i$ to the parameter point $a_i^*$ for which $\sigma_{\varepsilon h}(a_i^*)$ is the point of $\mathfrak{S}$ closest to $q_i$. If $q_i$ is an interior vertex (i.e., if $1\le i\le n$), $\sigma_{\varepsilon h}(a_i^*)$ is just the orthogonal projection of $q_i$ on $\mathfrak{S}$, since $q_i$ usually lies in a small neighbourhood of $\mathfrak{S}$. In this case, $a_i^*$ is the solution of the non-linear system

$$\begin{cases}\langle\partial^{(1,0)}\sigma_{\varepsilon h}(a_i^*),\sigma_{\varepsilon h}(a_i^*)-q_i\rangle_{\mathbb{R}^3}=0,\\ \langle\partial^{(0,1)}\sigma_{\varepsilon h}(a_i^*),\sigma_{\varepsilon h}(a_i^*)-q_i\rangle_{\mathbb{R}^3}=0,\end{cases}$$

†The graphics of $\mathcal{S}$ and $\mathfrak{S}$ shown in Figures 5 and 9 are obtained by drawing several $u$- and $v$-parameter curves of $f$ and $\sigma_{\varepsilon h}$ for equi-spaced values of the parameters $u$ and $v$. Due to the different parametrization of the data points, it is clear that the parameter curves of $f$ are not necessarily close to the corresponding parameter curves of $\sigma_{\varepsilon h}$. This explains the obvious differences visible at the top of the folded part of $\mathcal{S}$ and $\mathfrak{S}$. Two ray-traced images of these surfaces would be much more similar.

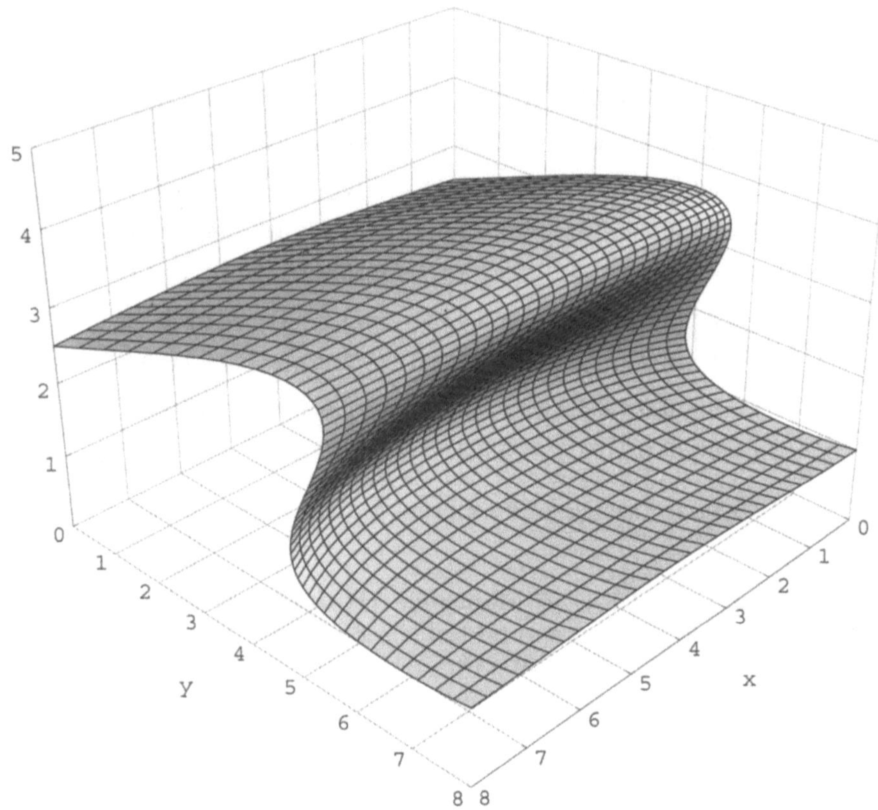


Figure 5: Example 2. Surface $\mathcal{S}$.

which can be solved by Newton iteration, choosing $a_i$ as an initial guess. If $q_i$ is a boundary vertex, we proceed analogously, taking into account that $a_i^*$ should be constrained to belong to $\partial\Omega$. In this way, we obtain a new set $A^*$ of parameter points. Then, selecting, if required, a new smoothing parameter $\varepsilon^*$, we find a new approximating surface $\mathfrak{S}^*$ by computing the $X_h$-discrete smoothing $D^2$-spline $\sigma_{\varepsilon^* h}$ relative to $A^*$, $\mathcal{P}$ and $\varepsilon^*$. The complete procedure (parameter correction and new surface fitting) can be repeated to progressively improve the exactness of the approximation. Figure 10 shows the approximating surface obtained after only one iteration. The value of $\varepsilon^*$, fixed to $10^{-6}$, is also obtained with the aid of the GCV method. In comparison with Figure 9, there are not many visual differences. We remark, however, that the relative error is almost halved.

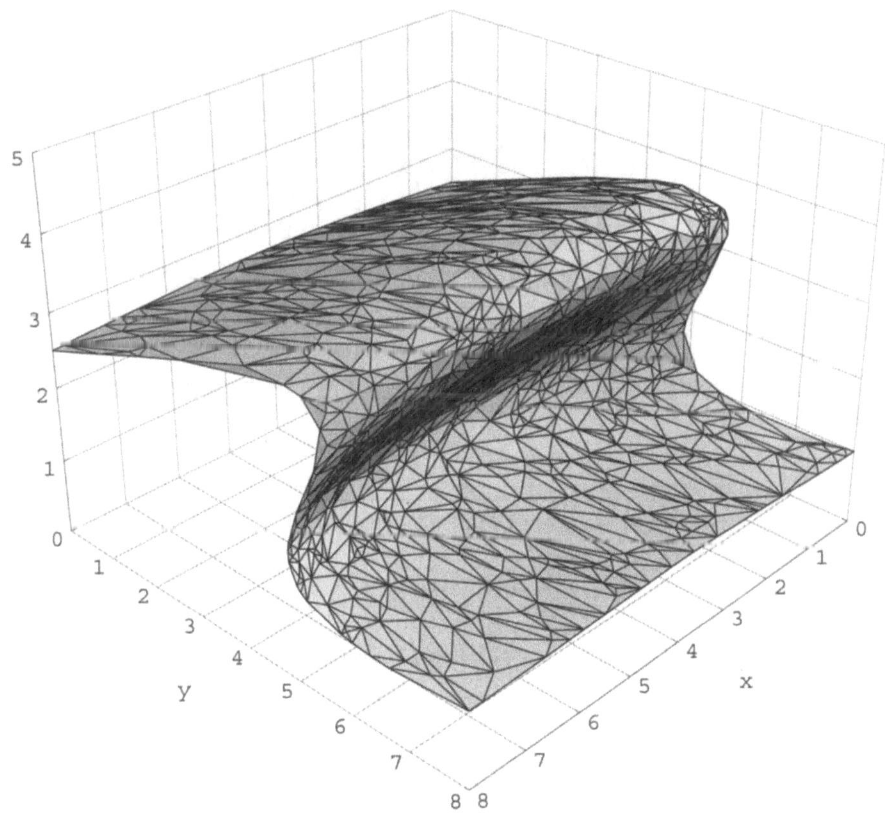


Figure 6: Example 2. Surface triangulation $\mathbb{T}$.

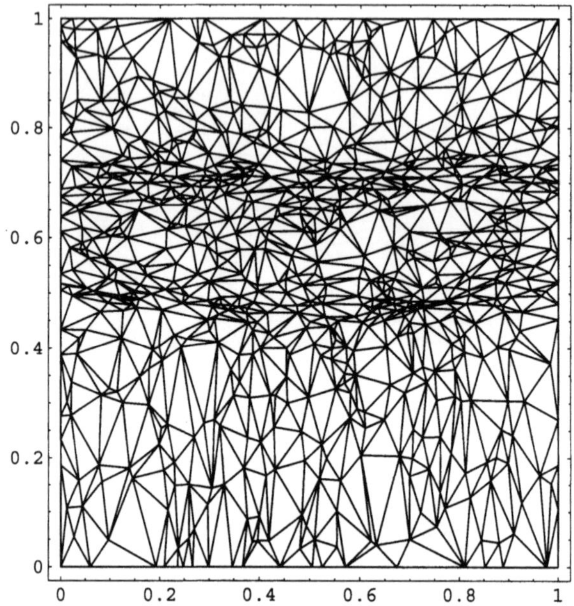


Figure 7: Example 2. Planar triangulation yielded by the shape-preserving parametrization method.

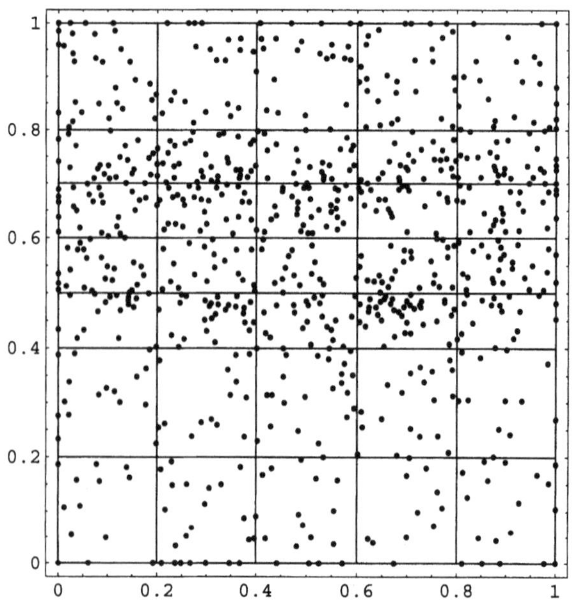


Figure 8: Example 2. Set $A$ of parameter points and triangulation $\mathcal{T}_h$.

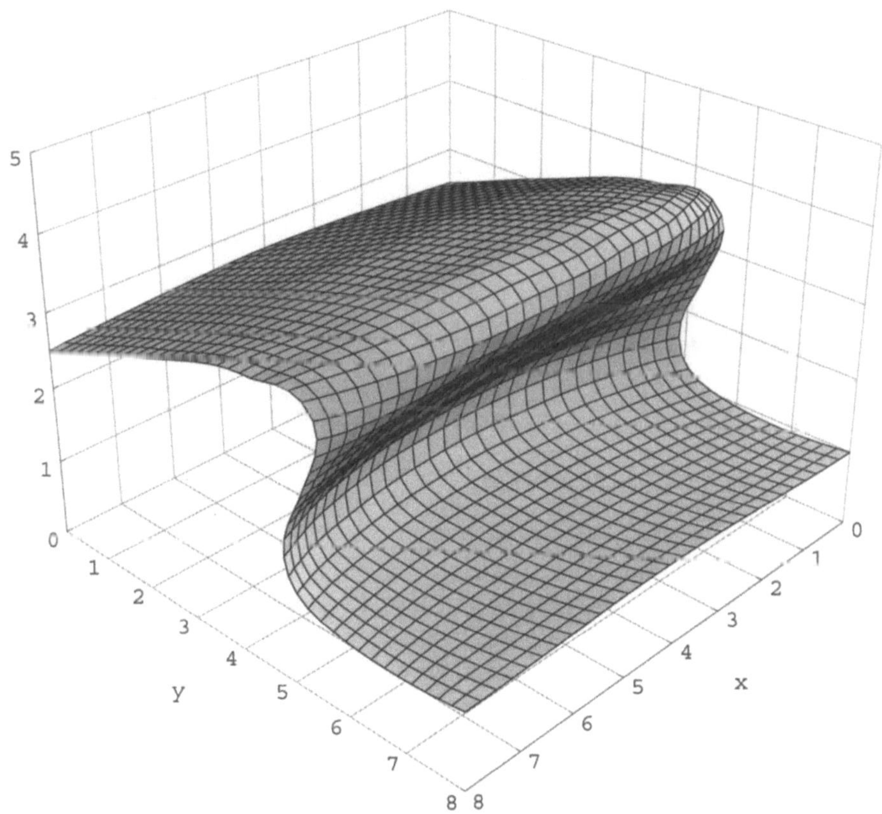


Figure 9: Example 2. Trace of the $X_h$-discrete smoothing $D^2$-spline $\sigma_{\varepsilon h}$ relative to $A$, $\mathcal{P}$ and $\varepsilon = 10^{-6}$. Relative error: $r(\mathcal{S}) = 0.00251027$.

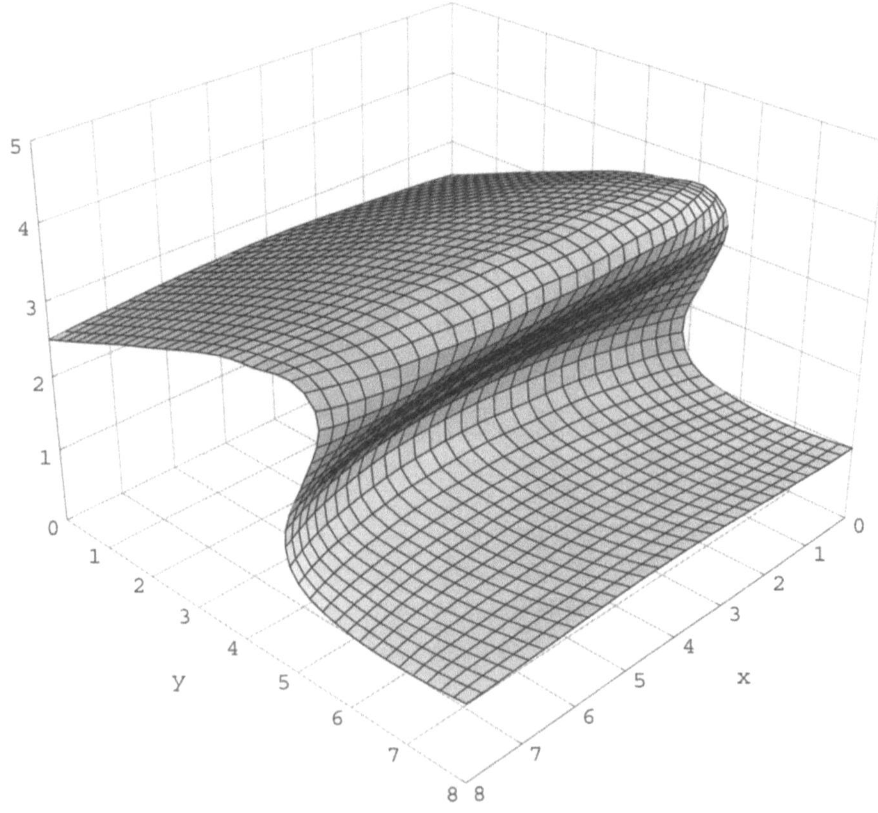


Figure 10: Example 2. Trace of the $X_h$-discrete smoothing $D^2$-spline $\sigma_{\varepsilon^* h}$ relative to $A^*$, $\mathcal{P}$ and $\varepsilon^* = 10^{-6}$. Relative error: $r(\mathcal{S}) = 0.00136998$.

***Example 3***

In this last example, we consider a surface $\mathcal{S}$ similar to a salt dome (cf. Figure 11), which is a surface frequently encountered in Geology. The parametrization $f : \overline{U} \to \mathbb{R}^3$ of $\mathcal{S}$, with $U = (0,1) \times (0,1)$, is now defined by

$$f(u,v) = \big(\varphi(v)\cos 2\pi u,\ \varphi(v)\sin 2\pi u,\ -5\phi_1(v) + 7\phi_2(v) + 7/\sqrt{2}\big),$$

where

$$\varphi(v) = 5\phi_1(v) - \phi_2(v)/4 + 21\sqrt{3}/16,$$

with

$$\phi_1(v) = \sin\big((7v+8)\pi/12\big)\cos\big((7v+8)\pi/6\big),$$
$$\phi_2(v) = \cos\big((7v+8)\pi/12\big)\sin\big((7v+8)\pi/6\big).$$

In a way similar to that in Example 2, we obtain a data set $\mathcal{P}$ and a surface triangulation $\mathbb{T}$ (cf. Figure 12). In this case, $\mathcal{P}$ contains 1000 points, of which 60 points are boundary vertices of $\mathbb{T}$ and the remaining 940 points are the interior vertices. Likewise, we use again the shape-preserving parametrization method, taking now $\Omega = B\big((0,0),1\big)$. The boundary vertices of $\mathbb{T}$ are parametrized by chord length along $\partial\Omega$. Figure 13 shows the resulting planar triangulation. The set $A$ of parameter points can also be seen in Figure 14.

Since $\Omega$ is not polygonal, we proceed as explained in Remark 3.1 to get the vector finite element space $X_h$. Specifically, we take $\widetilde{\Omega} = (-1,1) \times (-1,1)$, whereas the triangulation $\widetilde{\mathcal{T}}_h$ of $\overline{\widetilde{\Omega}}$ is made up of the rectangles whose vertices form the set $I \times I$, with $I = \{-1, -0.65, -0.35, -0.15, 0, 0.15, 0.35, 0.65, 1\}$ (cf. Figure 14). The finite element space $\widetilde{V}_h$ is constructed on $\widetilde{\mathcal{T}}_h$ from the Bogner-Fox-Schmit rectangle of class $C^1$. Since any rectangle of $\widetilde{\mathcal{T}}_h$ intersects $\Omega$, it is clear that $\Omega_h = \widetilde{\Omega}$ and so $V_h = \widetilde{V}_h$. The dimension of $V_h$ is 324.

Applying the GCV method, we set $\varepsilon = 10^{-5}$ and compute the $X_h$-discrete smoothing $D^2$-spline $\sigma_{\varepsilon h}$ relative to $A$, $\mathcal{P}$ and $\varepsilon$. The trace of $\sigma_{\varepsilon h}|_\Omega$ is shown in Figure 15. After one parameter correction, we obtain a new set $A^*$ of parameter points. Taking $\varepsilon^* = 10^{-5}$, we find the $X_h$-discrete smoothing $D^2$-spline $\sigma_{\varepsilon^* h}$ relative to $A^*$, $\mathcal{P}$ and $\varepsilon^*$, whose restriction to $\Omega$ is the parametrization of an acceptable approximating surface of $\mathcal{S}$ (cf. Figure 16).

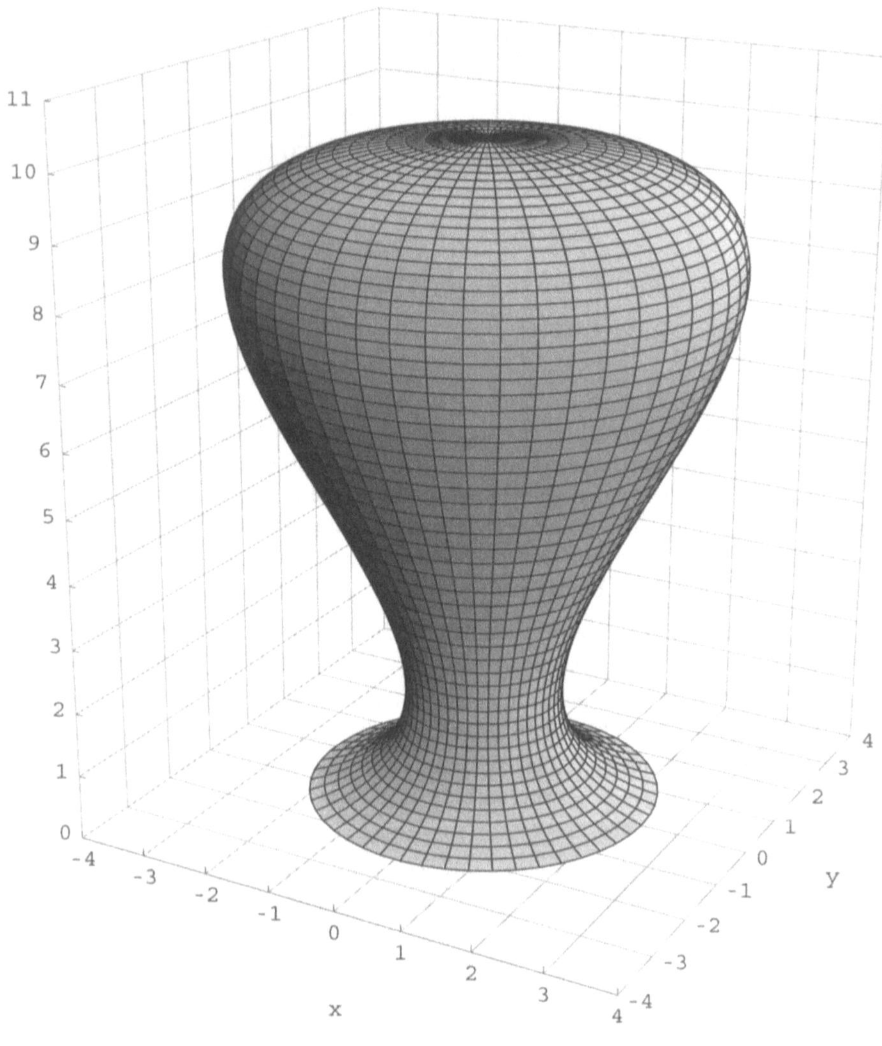


Figure 11: Example 3. Surface $\mathcal{S}$.

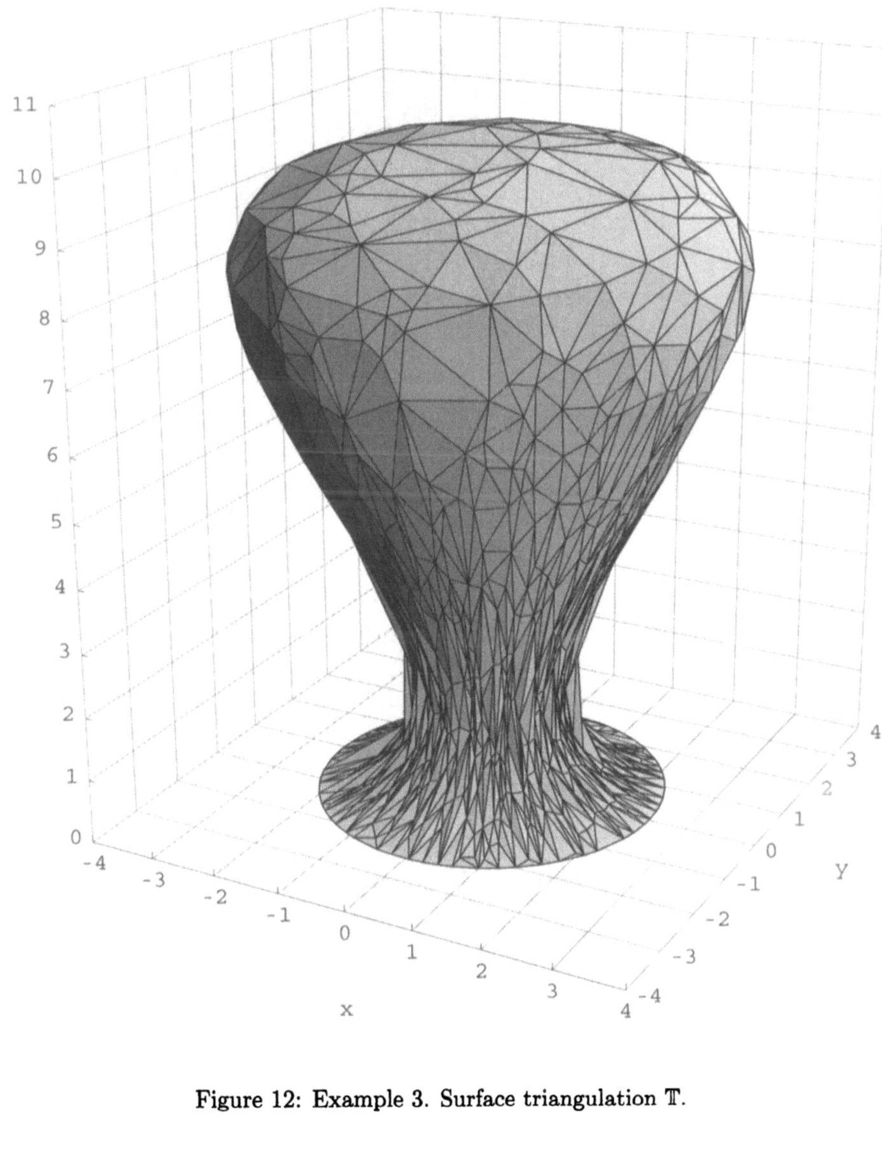


Figure 12: Example 3. Surface triangulation $\mathbb{T}$.

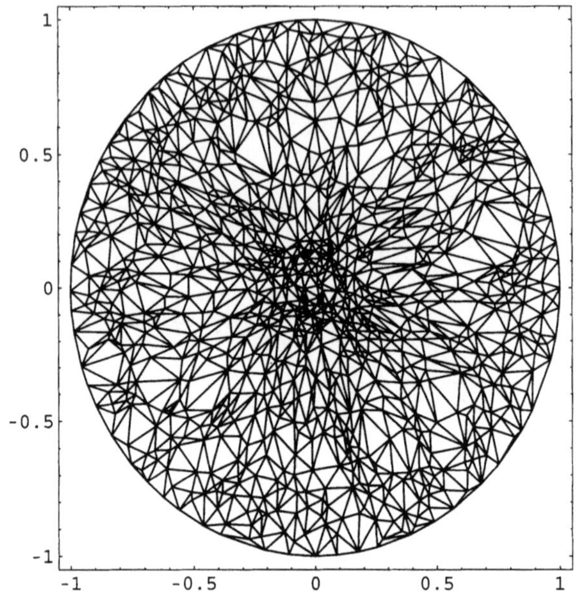


Figure 13: Example 3. Planar triangulation yielded by the shape-preserving parametrization method.

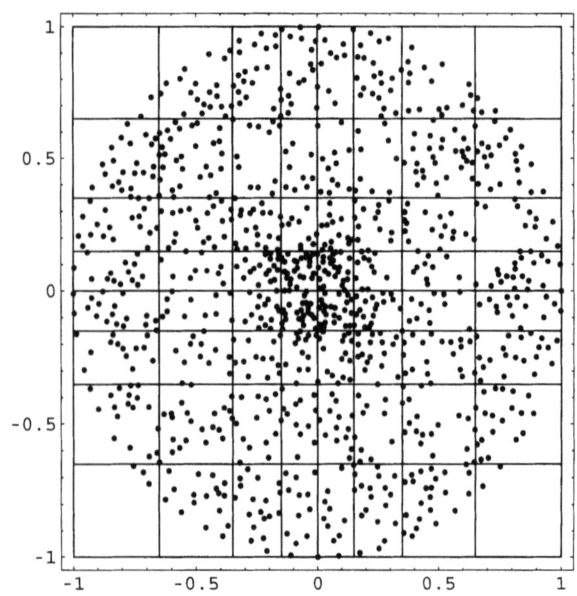


Figure 14: Example 3. Set $A$ of parameter points and triangulation $\widetilde{\mathcal{T}}_h$.

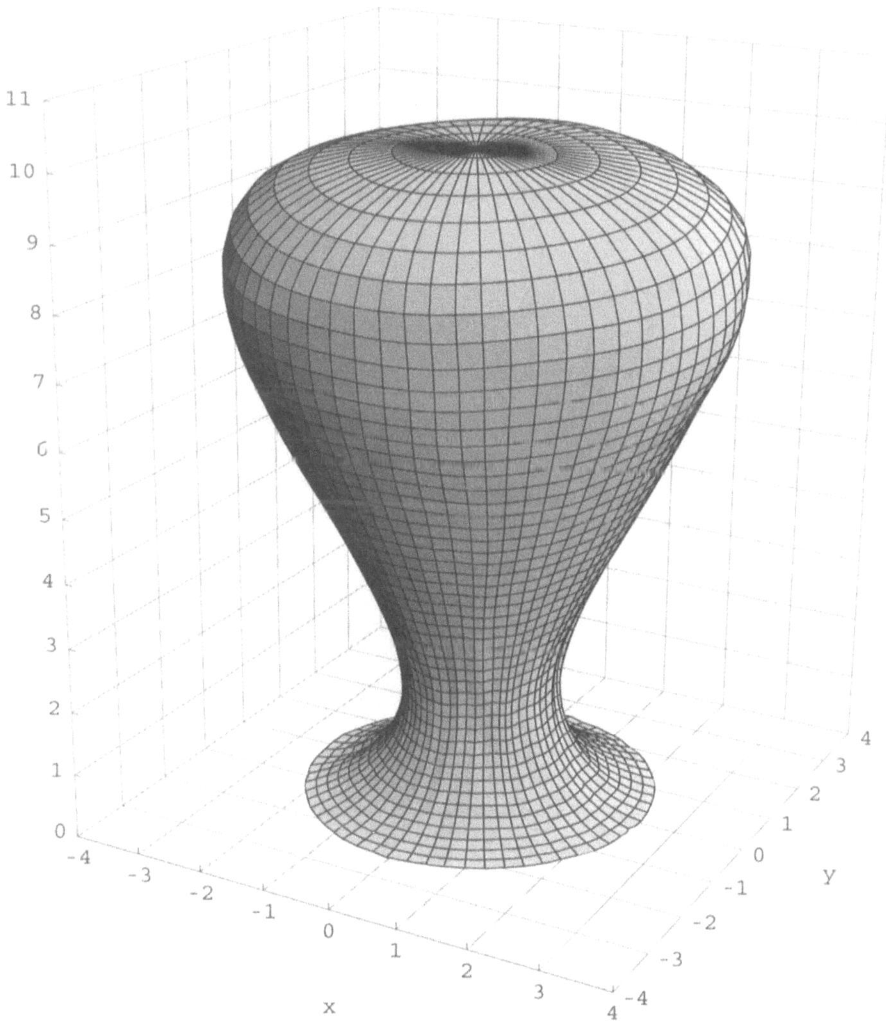


Figure 15: Example 3. Trace of the restriction to $\Omega$ of the $X_h$-discrete smoothing $D^2$-spline $\sigma_{\varepsilon h}$ relative to $A$, $\mathcal{P}$ and $\varepsilon = 10^{-5}$. Relative error: $r(\mathcal{S}) = 0.00353369$.

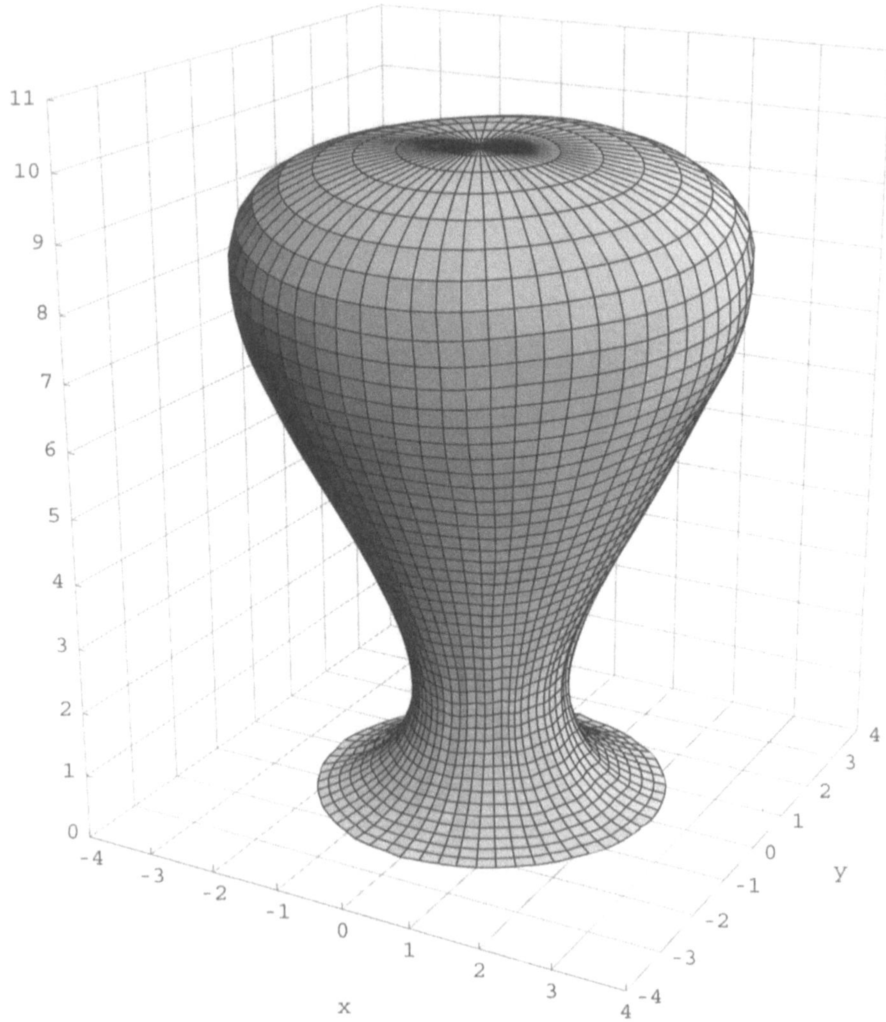


Figure 16: Example 3. Trace of the restriction to $\Omega$ of the $X_h$-discrete smoothing $D^2$-spline $\sigma_{\varepsilon^* h}$ relative to $A^*$, $\mathcal{P}$ and $\varepsilon^* = 10^{-5}$. Relative error: $r(\mathcal{S}) = 0.00229238$.

# *BIBLIOGRAPHY*


[1] R. A. Adams. *Sobolev Spaces.* Academic Press, New York, 1975.

[2] J. H. Ahlberg, E. N. Nilson and J. L. Walsh. *The Theory of Splines and their Applications.* Academic Press, New York, 1967.

[3] H. Akima. A method of bivariate interpolation and smooth surface fitting for irregularly distributed data points. *ACM Trans. Math. Soft.* **4**, 148–164, 1978.

[4] G. Allasia, R. Besenghi and A. de Rossi. A scattered data approximation scheme for the detection of fault lines. In *Mathematical Methods for Curves and Surfaces: Oslo 2000*, T. Lyche and L. L. Schumaker (eds.), pp. 25–34. Vanderbilt University Press, Nashville, 2001.

[5] D. Apprato. Étude de la convergence du produit tensoriel de fonctions spline à une variable satisfaisant à des conditions d'interpolation de Lagrange. *Ann. Fac. Sci. Toulouse* **VI**, 153–170, 1984.

[6] D. Apprato. *Approximation de Surfaces Paramétrées par Éléments Finis.* Thèse d'État, Université de Pau et des Pays de l'Adour, 1987.

[7] D. Apprato and R. Arcangéli. Approximation de surfaces paramétrées régulières par éléments finis. Partie I: Modélisation du problème. Publication URA CNRS 1204 Anal. Num. 89/6, Université de Pau et des Pays de l'Adour, 1989.

[8] D. Apprato and R. Arcangéli. Ajustement spline le long d'un ensemble de courbes. *Math. Model. Num. Anal.* **25**(2), 193–212, 1991.

[9] D. Apprato, R. Arcangéli and J. Gaches. Fonctions splines par moyennes locales sur un ouvert borné de $\mathbb{R}^n$. *Ann. Fac. Sci. Toulouse* **V**, 61–87, 1983.

[10] D. Apprato, R. Arcangéli and R. Manzanilla. Sur la construction de surfaces de classe $C^k$ à partir d'un grand nombre de données de Lagrange. *Math. Model. Num. Anal.* **21**(4), 529–555, 1987.

[11] D. Apprato and C. Gout. Spline fitting on surfaces patches. Publication URA CNRS 1204 Anal. Num. 95/2, Université de Pau et des Pays de l'Adour, 1995.

[12] D. Apprato and C. Gout. Approximation de surfaces à partir de morceaux de surfaces. *C. R. Acad. Sci. Paris, Série I,* **325**, 445–448, 1997.

[13] R. Arcangéli. $D^m$-splines sur un domaine borné de $\mathbb{R}^n$. Publication URA CNRS 1204 Anal. Num. 86/2, Université de Pau et des Pays de l'Adour, 1986.

[14] R. Arcangéli. Some applications of discrete $D^m$-splines. In *Mathematical Methods in Computer Aided Geometric Design*, T. Lyche and L. L. Schumaker (eds.), pp. 35–44. Academic Press, Boston, 1989.

[15] R. ARCANGÉLI. Splines d'ordre $(m, s)$. Publication URA CNRS 1204 Anal. Num. 90/12, Université de Pau et des Pays de l'Adour, 1990.

[16] R. ARCANGÉLI AND J. L. GOUT. Sur l'évaluation de l'erreur d'interpolation de Lagrange dans un ouvert de $\mathbb{R}^n$. *RAIRO Anal. Numer.* **10**(3), 5–27, 1976.

[17] R. ARCANGÉLI, R. MANZANILLA AND J. J. TORRENS. Approximation spline de surfaces de type explicite comportant des failles. *Math. Model. Num. Anal.* **31**(5), 643–676, 1997.

[18] R. ARCANGÉLI AND C. RABUT. Sur l'erreur d'interpolation par fonctions splines. *Math. Model. Num. Anal.* **20**(2), 191–201, 1986.

[19] R. ARCANGÉLI AND B. YCART. Almost sure convergence of smoothing $D^m$-splines for noisy data. *Numer. Math.* **66**, 281–294, 1993.

[20] E. ARGE AND M. S. FLOATER. Approximating scattered data with discontinuities. *Numer. Alg.* **8**, 149–166, 1994.

[21] N. ARONSZAJN. Theory of reproducing kernels. *Trans. AMS* **68**, 337–404, 1950.

[22] M. ATTÉIA. Fonctions "spline" définies sur un ensemble convexe. *Numer. Math.* **12**, 192–210, 1968.

[23] M. ATTÉIA. Fonctions "spline" et noyaux reproduisants d'Aronszajn-Bergman. *RIRO, R-3*, 31–43, 1970.

[24] M. ATTÉIA. *Hilbertian Kernels and Spline Functions.* Studies in Computational Mathematics. North-Holland, Amsterdam, 1992.

[25] M. ATTÉIA AND J. GACHES. *Approximation Hilbertienne: Splines-Ondelettes-Fractales.* Collection Grenoble Sciences. Presses Universitaires de Grenoble, Grenoble, 1999.

[26] M. N. BENBOURHIM. Splines elliptiques et interpolation par moyennes locales. Congrès National d'Analyse Numérique (France), 1985.

[27] M. N. BENBOURHIM. Personal communication, 1990.

[28] M. BERNADOU. *Sur l'Analyse Numérique du Modèle Linéaire de Coques Minces de W. T. Koiter. Algorithmes en Géométrie Euclidienne.* Thèse d'État, Université Paris VI, 1978.

[29] C. BERNARDI. Optimal Finite Element interpolation on curved domains. Lab. Anal. Num. Paris VI vol. 3, fasc. 3, n. 84017, Université Paris VI, 1984.

[30] R. BESENGHI AND G. ALLASIA. Scattered data near-interpolation with application to discontinuous surfaces. In *Curve and Surface Fitting: Saint-Malo 1999*, A. Cohen, C. Rabut and L. L. Schumaker (eds.), pp. 75–84. Vanderbilt University Press, Nashville, 2000.

[31] A. Y. BEZHAEV AND V. A. VASILENKO. *Variational Spline Theory.* Bulletin of the Novosibirsk Computing Center, Series on Numerical Analysis, Special Issue 3. NCC Publisher, Novosibirsk, 1993.

[32] B. D. BOJANOV, H. A. HAKOPIAN AND A. A. SAHAKIAN. *Spline Functions and Multivariate Interpolations*, vol. 248 of *Mathematics and its Applications*. Kluwer Academic Publishers, Dordrecht, 1993.

[33] C. DE BOOR. *A Practical Guide to Splines*. Springer-Verlag, Berlin-Heidelberg, 1978.

[34] C. DE BOOR AND K. HÖLLIG. B-splines from parallelepipeds. *J. Anal. Math.* **42**, 99–115, 1982–83.

[35] C. DE BOOR, K. HÖLLIG AND S. RIEMENSCHNEIDER. *Box Splines*. Springer-Verlag, New York, 1993.

[36] C. DE BOOR AND A. RON. Fourier analysis of the approximation power of principal shift-invariant spaces. *Constr. Approx.* **8**, 427–462, 1992.

[37] A. BOUHAMIDI. *Interpolation et Approximation par des Fonctions Splines Radiales à Plusieurs Variables*. Thèse, Université de Nantes, 1992.

[38] A. BOUHAMIDI AND A. LE MÉHAUTÉ. Spline curves and surfaces with tension. In *Wavelets, Images and Surface Fitting*, P.-J. Laurent, A. Le Méhauté and L. L. Schumaker (eds.), pp. 51–58. A K Peters, Wellesley, MA, 1994.

[39] A. BOUHAMIDI AND A. LE MÉHAUTÉ. Multivariate interpolating $(m, l, s)$-splines. *Adv. Comput. Math.* **11**(4), 287–314, 1999.

[40] H. BRÉZIS. *Analyse Fonctionnelle*. Masson, Paris, 1983.

[41] M. D. BUHMAN. New developments in radial basis function interpolation. In *Multivariate Approximation: From CAGD to Wavelets*, K. Jetter and F. Utreras (eds.), pp. 35–75. World Scientific, Singapour, 1993.

[42] M. D. BUHMAN, N. DYN AND D. LEVIN. On quasi-interpolation by radial basis functions with scattered centres. *Constr. Approx.* **11**(2), 239–254, 1995.

[43] M. D. BUHMAN AND A. RON. Radial basis functions: $L^p$-approximation orders with scattered centers. In *Wavelets, Images and Surface Fitting*, P.-J. Laurent, A. Le Méhauté and L. L. Schumaker (eds.), pp. 93–122. A K Peters, Wellesley, MA, 1994.

[44] C. K. CHUI. *Multivariate Splines*, vol. 54 of *CBMS-NSF Regional Conference Series in Applied Mathematics*. Society for Industrial and Applied Mathematics (SIAM), Philadelphia, 1988.

[45] P. G. CIARLET. *The Finite Element Method for Elliptic Problems*. North-Holland, Amsterdam, 1978.

[46] P. G. CIARLET AND P.-A. RAVIART. General Lagrange and Hermite interpolation in $\mathbb{R}^n$ with applications to finite element methods. *Arch. Rat. Mech. Anal.* **46**, 177–199, 1972.

[47] P. CLÉMENT. Approximation by finite element methods using local regularization. *RAIRO, R-2*, 77–84, 1975.

[48] D. COX. Multivariate smoothing spline functions. *SIAM J. Numer. Anal.* **21**(4), 789–813, 1984.

[49] P. CRAVEN AND G. WAHBA. Smoothing noisy data with spline functions. *Numer. Math.* **31**, 377–403, 1979.

[50] W. DAHMEN AND C. A. MICHELLI. Recent progress in multivariate splines. In *Approximation Theory IV*, C. K. Chui, L. L. Schumaker and J. W. Ward (eds.), pp. 27–121. Academic Press, New York, 1983.

[51] J. DENY AND J. L. LIONS. Les espaces du type de Beppo Levi. *Ann. Inst. Fourier, Grenoble,* **5**, 305–370, 1954.

[52] P. DIERCKX. *Curve and Surface Fitting with Splines.* Monographs on Numerical Analysis. Oxford University Press, New York, 1993.

[53] J. DUCHON. Fonctions-spline à énergie invariante par rotation. Rapport de recherche 27, IMAG, Université de Grenoble, 1976.

[54] J. DUCHON. Interpolation des fonctions de deux variables suivant le principe de la flexion des plaques minces. *RAIRO Anal. Numer.* **10**(12), 5–12, 1976.

[55] J. DUCHON. Splines minimizing rotation-invariant semi-norms in Sobolev spaces. *Lect. Notes in Math.* **571**, 85–100, 1977.

[56] J. DUCHON. Sur l'erreur d'interpolation des fonctions de plusieurs variables par les $D^m$-splines. *RAIRO Anal. Numer.* **12**(4), 325–334, 1978.

[57] J. DUCHON. *Fonctions-splines Homogènes à Plusieurs Variables.* Thèse, Université de Grenoble, 1980.

[58] N. DYN. Interpolation of scattered data by radial functions. In *Topics in Multivariate Approximation*, C. K. Chui, L. L. Schumaker and F. Utreras (eds.), pp. 47–61. Academic Press, New York, 1987.

[59] N. DYN. Interpolation and approximation by radial and related functions. In *Approximation Theory VI*, C. K. Chui, L. L. Schumaker and J. W. Ward (eds.), pp. 211–234. Academic Press, New York, 1989.

[60] N. DYN, D. LEVIN AND S. RIPPA. Surface interpolation and smoothing by thin plate splines. In *Approximation Theory IV*, C. K. Chui, L. L. Schumaker and J. W. Ward (eds.), pp. 445–449. Academic Press, New York, 1983.

[61] N. DYN, D. LEVIN AND S. RIPPA. Numerical procedures for surface fitting of scattered data by radial functions. *SIAM J. Sci. Stat. Comp.* **7**, 639–659, 1986.

[62] N. DYN AND A. RON. Radial basis function approximation: from gridded centres to scattered centres. *Proc. London Math. Soc. (3)* **71**(1), 76–108, 1995.

[63] W. FELLER. *An Introduction to Probability Theory and its Applications*, vol. 2. Wiley, New York, second edition, 1971.

[64] J. C. FIOROT AND P. JEANNIN. *Courbes et Surfaces Rationnelles. Applications à la CAO.* Masson, Paris, 1992. *Rational Curves and Surfaces. Applications to CAD*, english edition by John Wiley, Chichester, 1992.

[65] M. S. FLOATER. Parametrization and smooth approximation of surface triangulations. *Comput. Aided Geom. Des.* **14**, 231–250, 1997.

[66] M. S. FLOATER AND M. REIMERS. Meshless parameterization and surface reconstruction. *Comput. Aided Geom. Des.* **18**, 77–92, 2001.

[67] R. FRANKE. Scattered data interpolation: Test of some methods. *Math. Comp.* **38**(157), 181–200, 1982.

[68] R. FRANKE. Thin plate splines with tension. *Comput. Aided Geom. Des.* **2**, 87–95, 1985.

[69] R. FRANKE AND G. M. NIELSON. Surface approximation with imposed conditions. In *Surfaces in CAGD*, R. E. Barnhill and W. Boehm (eds.), pp. 135–146. North-Holland, 1983.

[70] D. GIRARD AND P.-J. LAURENT. Splines and estimation of non linear parameters. In *Mathematical Methods in Computer Aided Geometric Design*, T. Lyche and L. L. Schumaker (eds.), pp. 273–298. Academic Press, Boston, 1989.

[71] C. GOULAOUIC. *Analyse Fonctionnelle et Calcul Différentiel.* Ecole Polytechnique, Paris, 1981.

[72] C. GOUT. $C^k$ Surface approximation from surface patches. *Computers and Mathematics with Applications* **44**, 389–406, 2002.

[73] G. GREINER AND K. HORMANN. Interpolating and approximating scattered 3D data with hierarchical tensor product B-splines. In *Surface Fitting and Multiresolution Methods*, A. Le Méhauté, C. Rabut and L. L. Schumaker (eds.), pp. 163–172. Vanderbilt University Press, Nashville-London, 1997.

[74] T. N. E. GREVILLE. *Theory and Applications of Spline Functions.* Academic Press, New York, 1969.

[75] P. GRISVARD. *Elliptic Problems in Nonsmooth Domains.* Pitman, Boston, 1985.

[76] C. GU. Smoothing splines by generalized cross validation or generalized maximum likelihood. *SIAM J. Sci. Stat. Comp.* **12**, 383–398, 1991.

[77] T. GUTZMER AND A. ISKE. Detection of discontinuities in scattered data approximation. *Numerical Algorithms* **16**, 155–170, 1997.

[78] R. L. HARDY. Multiquadric equations of topography and other irregular surfaces. *J. Geophys. Res.* **76**, 1905–1915, 1971.

[79] R. L. HARDY. Theory and applications of the multiquadric biharmonic method. *Comput. Math. Applic.* **19**, 163–208, 1990.

[80] L. HÖRMANDER. *Linear Partial Differential Operators.* Springer-Verlag, Berlin, 1963.

[81] K. HORMANN AND G. GREINER. MIPS: An efficient global parametrization method. In *Curve and Surface Design: Saint-Malo 1999*, P.-J. Laurent, P. Sablonnière and L. L. Schumaker (eds.), pp. 153–162. Vanderbilt University Press, Nashville, 2000.

[82] J. HOSCHEK AND D. LASSER. *Fundamentals of Computer Aided Geometric Design.* A K Peters, Wellesley, MA, 1994.

[83] M. F. HUTCHINSON. A stochastic estimator of the trace of the influence matrix for Laplacian smoothing splines. *Comm. Statist. Simulation Comput.* **19**, 433–450, 1990.

[84] P. KLEIN. *Sur l'Approximation et la Représentation de Surfaces Explicites en Présence de Singularités.* Thèse de 3e cycle, Université Joseph Fourier-Grenoble I, 1987.

[85] M. LAGHCHIM-LAHLOU AND P. SABLONNIÈRE. Triangular finite elements of HCT type and class $C^0$. *Adv. Comput. Math.* **2**, 101–122, 1994.

[86] M. LAGHCHIM-LAHLOU AND P. SABLONNIÈRE. $C^r$-finite elements of Powell-Sabin type on the three direction mesh. *Adv. Comput. Math.* **6**, 191–206, 1996.

[87] P. LANCASTER AND K. ŠALKAUSKAS. *Curve and Surface Fitting.* Academic Press, London, 1986.

[88] P.-J. LAURENT. *Approximation et Optimisation.* Hermann, Paris, 1972.

[89] P.-J. LAURENT. Inf-convolution spline pour l'approximation de données discontinues. *Math. Model. Num. Anal.* **20**(1), 89–111, 1986.

[90] A. LE MÉHAUTÉ. *Interpolation et Approximation par des Fonctions Polynomiales par Morceaux dans* $\mathbb{R}^n$. Thèse d'État, Université de Rennes, 1984.

[91] A. LE MÉHAUTÉ AND A. BOUHAMIDI. $L^{m,l,s}$-splines in $\mathbb{R}^d$. In *Numerical Methods of Approximation Theory*, D. Braess and L. L. Schumaker (eds.), pp. 135–154. Birkhäuser Verlag, Basel, 1992.

[92] A. LE MÉHAUTÉ AND A. BOUHAMIDI. Splines in Approximation and differential operators: Interpolating $(m, l, s)$-spline. In *Spline Functions and the Theory of Wavelets (Montreal, PQ, 1996)*, pp. 77–87. Amer. Math. Soc., Providence, RI, 1999.

[93] D. LEE. Detection, classification, and measurement of discontinuities. *SIAM J. Sci. Stat. Comput.* **12**(2), 311–341, 1991.

[94] J. L. LIONS AND E. MAGENES. *Problèmes aux Limites Non Homogènes et Applications*, vol. 1. Dunod, Paris, 1968.

[95] S. K. LODHA AND R. FRANKE. Scattered data techniques for surfaces. In *Proceedings of Dagstuhl Conference on Scientific Visualization*, H. Hagen, G. M. Nielson and F. Post (eds.), pp. 182–222. IEEE Computer Society Press, 1999.

[96] M. C. LÓPEZ DE SILANES. Some results on approximation by smoothing $D^m$-splines. In *Multivariate Approximation: From CAGD to Wavelets*, K. Jetter and F. Utreras (eds.), pp. 227–237. World Scientific, Singapour, 1993.

[97] M. C. LÓPEZ DE SILANES. Convergence and error estimates for $(m, l, s)$-splines. *J. Comput. Appl. Math.* **87**, 373–384, 1997.

[98] M. C. LÓPEZ DE SILANES AND D. APPRATO. Estimations de l'erreur d'approximation sur un domaine borné de $\mathbb{R}^n$ par $D^m$-splines d'interpolation et d'ajustement discrètes. *Numer. Math.* **53**, 367–376, 1988.

[99] M. C. LÓPEZ DE SILANES, D. APPRATO AND R. ARCANGÉLI. Convergence presque sûre des $D^m$-splines d'ajustement sur un domaine borné de $\mathbb{R}^n$ pour des données entachées d'un bruit blanc. In *Actes des IIèmes Journées Saragosse-Pau de Mathématiques Appliquées*, M. C. López de Silanes and B. Ycart (eds.), pp. 223–235. Publications de l'Université de Pau et des Pays de l'Adour, 1992.

[100] M. C. LÓPEZ DE SILANES AND R. ARCANGÉLI. Estimations de l'erreur d'approximation par splines d'interpolation et d'ajustement d'ordre $(m, s)$. *Numer. Math.* **56**, 449–467, 1989.

[101] M. C. LÓPEZ DE SILANES AND R. ARCANGÉLI. Sur la convergence des $D^m$-splines d'ajustement pour des données exactes ou bruitées. *Rev. Mat. Univ. Comput. Madrid* **4**(2–3), 279–294, 1991.

[102] M. C. LÓPEZ DE SILANES, M. C. PARRA, M. PASADAS AND J. J. TORRENS. Spline approximation of discontinuous multivariate functions from scattered data. *J. Comput. Appl. Math.* **131**, 281–298, 2001.

[103] M. C. LÓPEZ DE SILANES, M. C. PARRA AND J. J. TORRENS. Vertical and oblique fault detection in explicit surfaces. *J. Comput. Appl. Math.* **140**, 559–585, 2002.

[104] W. MA AND J. P. KRUTH. Parameterization of randomly measured points for least squares fitting of B-spline curves and surfaces. *Comput.-Aided Des.* **27**, 663–675, 1995.

[105] S. MALLAT AND W. L. HWANG. Singularity detection and processing with wavelets. *IEEE Trans. Inf. Theory* **38**(2), 617–643, 1992.

[106] R. MANZANILLA. *Sur l'Approximation de Surfaces Définies par une Équation Explicite*. Thèse, Université de Pau et des Pays de l'Adour, 1986.

[107] G. MICULA AND S. MICULA. *Handbook of Splines*, vol. 462 of *Mathematics and its Applications*. Kluwer Academic Publishers, Dordrecht, 1999.

[108] C. B. MORREY. *Multiple Integrals in the Calculus of Variations*. Springer-Verlag, New York, 1966.

[109] J. NEČAS. *Les Méthodes Directes en Théorie des Équations Elliptiques.* Masson, Paris, 1967.

[110] G. M. NIELSON. A first-order blending method for triangles based upon cubic interpolation. *Int. J. Numer. Meth. Engrg.* **15**, 308–318, 1978.

[111] L. PAIHUA. *Quelques Méthodes Numériques pour le Calcul de Fonctions Spline à une et Plusieurs Variables.* Thèse de 3e cycle, Université de Grenoble, 1978.

[112] M. C. PARRA. *Sobre detección de discontinuidades y aproximación de funciones no regulares.* PhD thesis, Universidad de Zaragoza, Spain, 1999.

[113] M. C. PARRA, M. C. LÓPEZ DE SILANES AND J. J. TORRENS. Vertical fault detection from scattered data. *J. Comput. Appl. Math.* **73**, 225–239, 1996.

[114] M. PASADAS. *Aproximación de Curvas y Superficies Paramétricas con Condiciones de Tangencia.* Tesis doctoral, Universidad de Granada, 1995.

[115] M. PASADAS, M. C. LÓPEZ DE SILANES AND J. J. TORRENS. Approximation of parametric surfaces by discrete smoothing $D^m$-splines with tangent conditions. In *Mathematical Methods for Curves and Surfaces*, M. Dæhlen, T. Lyche and L. L. Schumaker (eds.), pp. 403–412. Vanderbilt University Press, Nashville-London, 1995.

[116] J. PEETRE. Espaces d'interpolation et théorème de Soboleff. *Ann. Inst. Fourier, Grenoble,* **16**, 279–317, 1966.

[117] R. PHILLIPS. *Crystals, Defects and Microstructures. Modelling across Scales.* Cambridge University Press, 2001.

[118] M. J. D. POWELL. The theory of radial basis function approximation in 1990. In *Advances in Numerical Analysis II: Wavelets, Subdivision and Radial Functions*, W. Light (ed.), pp. 105–210. Clarendon Press, Oxford, 1992.

[119] M. J. D. POWELL AND M. A. SABIN. Piecewise quadratic approximations on triangles. *ACM Trans. Math. Soft.* **3**(4), 316–325, 1977.

[120] P. M. PRENTER. *Splines and Variational Methods.* Wiley, New York, 1975.

[121] D. RAGOZIN. Error bounds for derivatives estimates based on spline smoothing of exact or noisy data. *J. Approx. Theory* **37**, 335–355, 1983.

[122] A. RON. The $L_2$-approximation orders of principal shift-invariant spaces generated by a radial basis function. In *Numerical Methods of Approximation Theory*, D. Braess and L. L. Schumaker (eds.), pp. 245–268. Birkhäuser Verlag, Basel, 1992.

[123] M. ROSSINI. Irregularity detection from noisy data in one and two dimensions. *Numerical Algorithms* **16**, 283–301, 1997.

[124] M. ROSSINI. 2-D discontinuity detection from scattered data. *Computing* **61**, 215–234, 1998.

[125] A. I. ROZHENKO AND V. A. VASILENKO. Variational approach in abstract splines: Achievements and open problems. *East J. Approx.* **1**(3), 277–308, 1995.

[126] P. SABLONNIÈRE. Composite finite elements of class $C^k$. *J. Comput. Appl. Math.* **12–13**, 541–550, 1985.

[127] A. M. SANCHEZ AND R. ARCANGÉLI. Estimations des erreurs de meilleure approximation polynomiale et d'interpolation de Lagrange dans les espaces de Sobolev d'ordre non entier. *Numer. Math.* **45**, 301–321, 1984.

[128] M. H. SCHULTZ. *Spline Analysis.* Prentice Hall, Englewood Cliffs, N. J., 1973.

[129] L. L. SCHUMAKER. *Spline Functions: Basic Theory.* Wiley, New York, 1981.

[130] L. SCHWARTZ. Sous-espaces hilbertiens d'espaces vectoriels topologiques et noyaux associés (noyaux reproduisants). *J. Anal. Math.* **13**, 115–256, 1964.

[131] L. SCHWARTZ. *Théorie des Distributions.* Hermann, Paris, 1966.

[132] D. G. SCHWEIKERT. An interpolation curve using a spline in tension. *J. Math. Phys.* **45**, 312–317, 1966.

[133] L. R. SCOTT AND S. ZHANG. Finite element interpolation of nonsmooth functions satisfying boundary conditions. *Math. Comp.* **54**(190), 483–493, 1990.

[134] C. SERRES. *Sur la Reconstruction de Surfaces de Type Explicite Présentant des Failles.* Thèse, Université de Pau et des Pays de l'Adour, 1992.

[135] J. SPRINGER. Modeling of geological surfaces using finite elements. In *Wavelets, Images and Surface Fitting*, P.-J. Laurent, A. Le Méhauté and L. L. Schumaker (eds.), pp. 467–474. A K Peters, Wellesley, MA, 1994.

[136] G. STRANG. Approximation in the Finite Element Method. *Numer. Math.* **19**, 81–98, 1972.

[137] C. TARROU. *Aspects of Smooth Surface Construction from Scattered Data.* Thèse, Institut National des Sciences Appliquées, Toulouse, 1999.

[138] A. TIKHONOV AND V. ARSENINE. *Méthodes de résolution de problèmes mal posés.* Mir, Moscow, 1976.

[139] J. J. TORRENS. *Interpolación de Superficies Paramétricas con Discontinuidades mediante Elementos Finitos. Aplicaciones.* No. 24 in Serie II, Sección 2 (Tesis doctorales). Publicaciones del Seminario Matemático García de Galdeano, Universidad de Zaragoza, 1992.

[140] J. J. TORRENS. Approximation de surfaces paramétrées présentant des discontinuités par $D^m$-splines d'ajustement discrètes. Publication URA CNRS 1204 Anal. Num. 93/9, Université de Pau et des Pays de l'Adour, 1993.

[141] J. J. TORRENS. Approximation of parametric surfaces with discontinuities by discrete smoothing $D^m$-splines. In *Wavelets, Images and Surface Fitting*, P.-J. Laurent, A. Le Méhauté and L. L. Schumaker (eds.), pp. 485–492. A K Peters, Wellesley, MA, 1994.

[142] J. J. TORRENS. Sur l'erreur d'approximation par éléments finis en utilisant la méthode des plaquettes splines. *Numer. Math.* **76**, 69–85, 1997.

[143] J. J. TORRENS. Discrete smoothing $D^m$-splines: Applications to surface fitting. In *Mathematical Methods for Curves and Surfaces II*, M. Dæhlen, T. Lyche and L. L. Schumaker (eds.), pp. 477–484. Vanderbilt University Press, Nashville-London, 1998.

[144] F. TRÈVES. *Topological Vector Spaces.* Academic Press, New York, 1967.

[145] F. UTRERAS. *Utilisation de la Méthode de Validation Croissée pour le Lissage par Fonctions Splines à une ou deux Variables.* Thèse de docteur ingénieur, Université de Grenoble, 1979.

[146] F. UTRERAS. Convergence rates for multivariate smoothing spline functions. *J. Approx. Theory* **52**, 1–27, 1988.

[147] F. UTRERAS. *Recent Results on Multivariate Smoothing Splines.* International Series in Numerical Mathematics. Birkhäuser Verlag, Basel, 1990.

[148] V. A. VASILENKO AND A. I. ROZHENKO. Discontinuity localization and spline approximation of discontinuous functions with many variables at scattered meshes. *Sov. J. Numer. Anal. Math. Modelling.* **5**(4–5), 425–434, 1990.

[149] G. WAHBA. *Spline Models for Observational Data.* Society for Industrial and Applied Mathematics (SIAM), Philadelphia, 1990.

[150] V. WEISS, L. ANDOR, G. RENNER AND T. VÁRADY. Advanced surface fitting techniques. *Comput. Aided Geom. Des.* **19**, 19–42, 2002.

[151] K. YOSIDA. *Functional Analysis.* Springer-Verlag, Berlin-Heidelberg, sixth edition, 1980.

[152] A. ŽENÍŠEK. A general theorem on triangular finite $C^{(m)}$-elements. *RAIRO, R-2*, 119–127, 1974.

# *Index*

***Front Cover Photo:***

*Composed of a plot of Richard Franke's function*
*and 3D points over a rectangular grid,*
*set against a zoom of that plot.*
*Pattern courtesy of the Authors.*

MIX
Papier aus verantwortungsvollen Quellen
Paper from responsible sources
FSC® C105338

If you have any concerns about our products,
you can contact us on
**ProductSafety@springernature.com**

In case Publisher is established outside the EU,
the EU authorized representative is:
**Springer Nature Customer Service Center GmbH**
**Europaplatz 3, 69115 Heidelberg, Germany**

Printed by Libri Plureos GmbH
in Hamburg, Germany